VOLUME SIX HUNDRED AND NINETY FOUR

METHODS IN ENZYMOLOGY

Magnetic Tweezers for the Study of Protein Structure and Function

METHODS IN ENZYMOLOGY

VOLUME SIX HUNDRED AND NINETY FOUR

METHODS IN ENZYMOLOGY

Magnetic Tweezers for the Study of Protein Structure and Function

Edited by

MIN JU SHON
Department of Physics, School of Interdisciplinary Bioscience and Bioengineering, Pohang University of Science and Technology, South Korea

TAE-YOUNG YOON
School of Biological Sciences, Institute for Molecular Biology and Genetics, Seoul National University, South Korea

ELSEVIER

ACADEMIC PRESS
An imprint of Elsevier

Academic Press is an imprint of Elsevier
50 Hampshire Street, 5th Floor, Cambridge, MA 02139, United States
525 B Street, Suite 1650, San Diego, CA 92101, United States
125 London Wall, London, EC2Y 5AS, United Kingdom

First edition 2024

ISBN: 978-0-443-22229-0
ISSN: 0076-6879

For information on all Academic Press publications
visit our website at https://www.elsevier.com/books-and-journals

Publisher: Zoe Kruze
Editorial Project Manager: Saloni Vohra
Production Project Manager: James Selvam
Cover Designer: Greg Harris

Typeset by MPS Limited, India

Contents

Contributors

Jean-François Allemand
Laboratoire de Physique de l'École normale supérieure, ENS, Université PSL, CNRS, Sorbonne Université, Université Paris-Diderot, Sorbonne Paris Cité, Paris, France

Rajendra Bhandari
Department of Physics; Vitreous State Laboratory, The Catholic University of America, Washington, DC, United States

Hu Chen
Center of Biomedical Physics, Wenzhou Institute, University of Chinese Academy of Sciences, Wenzhou; Department of Physics, Research Institute for Biomimetics and Soft Matter, Fujian Provincial Key Lab for Soft Functional Materials Research, Xiamen University, Xiamen, P.R. China

Vincent Croquette
Laboratoire de Physique de l'École normale supérieure, ENS, Université PSL, CNRS, Sorbonne Université, Université Paris-Diderot, Sorbonne Paris Cité; ESPCI Paris, Universite PSL, Paris, France

Maryne Follenfant
Institut de Biologie de l'Ecole Normale Supérieure, ENS, CNRS, INSERM, PSL Research University, Paris, France

Santosh Gaire
Department of Physics; Vitreous State Laboratory, The Catholic University of America, Washington, DC, United States

Ara Go
Engitein Research Institute, Engitein, Siheung, Republic of Korea

Charlie Gosse
Institut de Biologie de l'Ecole Normale Supérieure, ENS, CNRS, INSERM, PSL Research University, Paris, France

Zilong Guo
Center of Biomedical Physics, Wenzhou Institute, University of Chinese Academy of Sciences, Wenzhou, P.R. China

Felix Hausch
Department of Chemistry and Biochemistry; Centre for Synthetic Biology, Technical University Darmstadt, Darmstadt, Germany

Seok-Cheol Hong
Center for Molecular Spectroscopy and Dynamics, Institute for Basic Science, Department of Physics, Korea University, Seoul, Korea

Seokyun Hong
Department of Physics, Pohang University of Science and Technology (POSTECH), Pohang, Republic of Korea

Hae Jun Jung
Center for Molecular Spectroscopy and Dynamics, Institute for Basic Science, Department of Physics, Korea University, Seoul, Korea

Jaehun Jung
Department of Physics, Pohang University of Science and Technology (POSTECH), Pohang, Republic of Korea

Haesoo Kim
Engitein Research Institute, Engitein, Siheung, Republic of Korea

Seoyoon Kim
Department of Chemistry, Ulsan National Institute of Science and Technology, Ulsan, Republic of Korea

Sook Ho Kim
Center for Molecular Spectroscopy and Dynamics, Institute for Basic Science, Department of Physics, Korea University, Seoul, Korea; College of Veterinary Medicine, Chungbuk National University, Cheongju, Korea

Subin Kim
Department of Biological Sciences, Ulsan National Institute of Science and Technology (UNIST), Ulsan, Republic of Korea

Dorota Kostrz
Institut de Biologie de l'Ecole Normale Supérieure, ENS, CNRS, INSERM, PSL Research University, Paris, France

Minsuk Kwak
Center for Nanomedicine, Institute for Basic Science; Department of Nano Biomedical Engineering, Advanced Science Institute, Yonsei University, Seoul, Republic of Korea

Shimin Le
Department of Physics, Research Institute for Biomimetics and Soft Matter, Fujian Provincial Key Lab for Soft Functional Materials Research, Xiamen University, Xiamen, P.R. China

Daehyo Lee
Department of Chemistry, Ulsan National Institute of Science and Technology, Ulsan, Republic of Korea

Jayil Lee
Department of Biological Sciences, Ulsan National Institute of Science and Technology (UNIST); Institute of Basic Science Center for Genomic Integrity, Ulsan, Republic of Korea

Christian Meyners
Department of Chemistry and Biochemistry, Technical University Darmstadt, Darmstadt, Germany

Claudia Tanja Mierke
Faculty of Physics and Earth System Sciences, Peter Debye Institute for Soft Matter Physics, Biological Physics Division, Leipzig University, Leipzig, Germany

Duyoung Min
Department of Chemistry; Center for Wave Energy Materials, Ulsan National Institute of Science and Technology, Ulsan, Republic of Korea

Jessica Valle Orero
Laboratoire de Physique de l'École normale supérieure, ENS, Université PSL, CNRS, Sorbonne Université, Université Paris-Diderot, Sorbonne Paris Cité; The American University of Paris, Paris, France

Beom-Hyeon Park
Center for Molecular Spectroscopy and Dynamics, Institute for Basic Science, Department of Physics, Korea University, Seoul, Korea

Ian Pegg
Department of Physics; Vitreous State Laboratory, The Catholic University of America, Washington, DC, United States

Sebastian Pomplun
Leiden Academic Centre for Drug Research, Leiden University, Leiden, The Netherlands

Sang-Hyun Rah
Department of Physics, Pohang University of Science and Technology (POSTECH), Pohang, Republic of Korea

Martin Rieu
Department of Physics, New Biochemistry Building, University of Oxford, South Parks Road, Oxford, United Kingdom

Omar A. Saleh
Materials Department; Biomolecular Sciences and Engineering Program; Physics Department, University of California, Santa Barbara, CA, United States

Abhijit Sarkar
Department of Physics; Vitreous State Laboratory, The Catholic University of America, Washington, DC, United States

Min Ju Shon
Department of Physics; School of Interdisciplinary Bioscience and Bioengineering, Pohang University of Science and Technology (POSTECH), Pohang, Republic of Korea

François Stransky
Institut de Biologie de l'Ecole Normale Supérieure, ENS, CNRS, INSERM, PSL Research University, Paris, France

Terence Strick
Institut de Biologie de l'Ecole Normale Supérieure, ENS, CNRS, INSERM, PSL Research University, Paris, France

Hao Sun
Center of Biomedical Physics, Wenzhou Institute, University of Chinese Academy of Sciences, Wenzhou, P.R. China

Rafael Tapia-Rojo
Physics Department, King's College London, London, United Kingdom

Hoang P. Truong
Materials Department, University of California, Santa Barbara, CA, United States

Taehyun Yang
Department of Physics, Pohang University of Science and Technology (POSTECH), Pohang, Republic of Korea

Tae-Young Yoon
School of Biological Sciences and Institute for Molecular Biology and Genetics, Seoul National University; Department of Biomarker Discovery, PROTEINA Co., Ltd, Seoul, Republic of Korea

May the force be with proteins

In daily life, when we lift a heavy box or squeeze a rubber ball, we intuitively know how much force to use, hardly thinking about the energy changes involved. However, this intuitive understanding shifts dramatically when we delve into the world of biochemical studies. Here, we explore the realm of proteins at nanometer scales and the chemical energies comparable to thermal fluctuations, introducing a significant degree of stochasticity. Therefore tracking the journey of individual molecules and measuring forces at this molecular level offers us a glimpse into how energy landscapes are shaped amidst these stochastic fluctuations. By focusing on molecular forces, we open up a new, dynamic, and intuitive lens to understand these complex processes. This is where our tools, particularly magnetic tweezers, become invaluable in revealing the intricate dance of molecular interactions.

The past decades have witnessed this shift from theoretical ambition to tangible reality, with the advent of techniques capable of manipulating individual biomolecules and quantifying the minuscule forces at play, usually 12 orders of magnitude smaller than the gravitational force acting on an apple. Volume 694 in *Methods in Enzymology* brings together a comprehensive collection of methods and reviews from leading experts, exploring the various applications of magnetic tweezers technology. It covers a wide spectrum of research, from fundamental principles to advanced applications like analyzing protein–protein interactions, DNA mechanics, and protein folding. Readers will encounter cutting-edge methods for detailed manipulation and observation of single molecules, emphasizing the role of mechanical forces in structural changes.

The first two chapters are graciously contributed by pioneers in magnetic tweezers technology. Chapter 1, authored by Dr. Croquette and his team, offers a foundational overview ideal for newcomers interested in employing magnetic tweezers in biomolecular research. Chapter 2, by Dr. Gosse and colleagues, delves into a novel variant of the technique, showcasing its applicability in studying bimolecular interactions, illustrating the method's evolution and its future potential.

Chapters 3–7 explore specialized adaptations of magnetic tweezers. In Chapter 3, Dr. Tapia-Rojo discusses "magnetic tape-head tweezers," originally developed in the laboratory of Julio M. Fernandez. Although magnetic tape heads, once common in audio cassettes and VHS tapes, have faded from industrial use, their creative application in magnetic tweezers has facilitated rapid adjustments in applied force. In contrast, the high-resolution measurements of protein structural changes, described in Chapter 4, were made possible by advancements in fast industrial cameras and GPU technology. With the ongoing growth in machine vision products, a variety of new approaches, including the potential use of event cameras, are anticipated to emerge soon.

Chapters 5 and 6 highlight the integration of magnetic tweezers with fluorescence microscopy, enhancing single-molecule experiments by allowing simultaneous imaging and force spectroscopy. Specifically, Chapter 5 introduces a setup combining TIRF microscopy for single-molecule counting, visualizing protein molecules on DNA substrates. Chapter 6, led by Dr. Hong, presents the incorporation of single-molecule fluorescence resonance energy transfer, the powerful nanoscale ruler for molecular dissection. The horizontal magnetic tweezers described in Chapter 7, although less commonly used, could improve fluorescence visualization, as it allows molecules attached to tethers to be resolved when the tether is stretched laterally.

The following section, Chapters 8–11, concentrate on protein mechanics. In Chapter 8, Dr. Saleh's team presents their latest research on intrinsically disordered proteins and peptoid strands, leveraging their extensive experience in studying the mechanics of various polymers. These methods will be crucial for the ongoing exploration, as many complex biological polymers, notably glycans and their peptide or protein conjugates, along with newly engineered polymers and higher order structures, remain yet to be studied. Chapters 9–11 showcase magnetic tweezers in protein folding studies, expanding beyond traditional atomic force microscope-based pulling methods.

Chapters 12 and 13 close the volume by venturing into innovative applications of magnetic tweezers in cellular research, focusing on using magnetic particles to initiate cell signaling or investigate cell mechanics. Despite being a relatively new area compared to the extensive background in molecular studies, these applications in cellular contexts is poised for growth alongside the expanding field of mechanobiology.

We would like to sincerely thank all the authors who contributed to this volume, enabling its publication even during the holiday season. Special acknowledgement goes to Saloni Vohra, Paulo Mendoza, and Zoe Kruze for their invaluable assistance in managing the project. Finally, we are grateful to the Editors-in-Chief of *Methods in Enzymology*, Dr. David Christianson and Dr. Anna Pyle, for the opportunity to edit this volume.

Min Ju Shon
Tae-Young Yoon

CHAPTER ONE

Magnetic tweezers principles and promises

Vincent Croquette[a,b,*], Jessica Valle Orero[a,c], Martin Rieu[d], and Jean-François Allemand[a]

[a]Laboratoire de Physique de l'École normale supérieure, ENS, Université PSL, CNRS, Sorbonne Université, Université Paris-Diderot, Sorbonne Paris Cité, Paris, France
[b]ESPCI Paris, Université PSL, Paris, France
[c]The American University of Paris, Paris, France
[d]Department of Physics, New Biochemistry Building, University of Oxford, South Parks Road, Oxford, United Kingdom
*Corresponding author. e-mail address: vincent.croquette@espci.psl.eu

Contents

Abstract

Magnetic tweezers have become popular with the outbreak of single molecule micromanipulation: catching a single molecule of DNA, RNA or a single protein and applying mechanical constrains using micron-size magnetic beads and magnets turn out to be easy. Various factors have made this possible: the fact that manufacturers have been preparing these beads to catch various biological entities—the ease of use provided by magnets which apply a force or a torque at a distance thus inside a flow cell—some chance: since the forces so generated are in the right range to stretch a single molecule. This is a little less true for torque. Finally, one feature which also appears very important is the simplicity of their calibration using Brownian motion. Here we start by describing magnetic tweezers used routinely in our laboratory where we have tried to develop a

Methods in Enzymology, Volume 694
ISSN 0076-6879, https://doi.org/10.1016/bs.mie.2024.01.026

device as simple as possible so that the experimentalist can really focus on the biological aspect of the biomolecules that he/she is interested in. We discuss the implications of the various components and their important features. Next, we summarize what is easy to achieve and what is less easy. Then we refer to contributions by other groups who have brought valuable insights to improve magnetic tweezers.

1. Introduction

Magnetic tweezers are an interesting tool for single molecule studies. Understanding biological mechanisms at the molecular level has been greatly improved by the insights brought by single molecule experiments which have revealed fundamental aspects undetectable in the classical bulk experiments which average the activity of all the components of a biological system. Single molecule experiments pause, however, the simple challenge of having molecular sensitivity. Measuring the action of a single molecular motor is indeed difficult since Brownian motion dominates at the molecular scale, blurring out the trajectory of this motor. A few single molecule experiments have been successful just by visualizing the position or the angular position of a molecular motor (Noji, Yasuda, Yoshida, & Kinosita, 1997; Yildiz, Selvin, & Paul, 2005) but for most of them visualizing is not enough and one needs to reduce the blurring effect of Brownian motion. For nucleic acids like DNA, it is possible to reduce Brownian motion by just stretching the molecule which turns out to be the substrate of many enzymes. In these conditions the trajectories of these enzymes become measurable and significant.

Various tools have been proposed to stretch DNA (Moffitt, Chemla, Smith, & Bustamante, 2008; Neuman, Lionnet, & Allemand, 2007; Bustamante, Chemla, Liu, & Wang, 2021; Manosas et al., 2010), magnetic tweezers appear as one of them and bear at least two interesting properties: they are simple, and they act in parallel on many single molecules simultaneously (Dulin, 2024; Quack & Dulin, 2024). It is also possible to increase the parallelism of optical tweezers (Inman, Ye, & Hong, 2022) This later property turns out to be decisive when one wants to study, for instance, an enzyme processing a DNA molecule. The molecule template, or more precisely its attachment, is fragile and single molecule experiments nearly always result in the breaking and loss of the substrate molecule. In other words, these systems have a finite lifetime. Moreover, because of the randomness of the processes involved, catching up the activity of the enzyme on the substrate is also a matter of luck. Thus, the ability of observing a large number of single molecules in parallel is important to provide the necessary statistics to characterize the activity of the enzyme (Duderstadt et al., 2016; Visnapuu & Greene, 2009).

1.1 Principle

Stretching a molecule to reduce the effect of Brownian motion implies to be able to catch a single molecule and to apply a force to it. Although DNA may be very long, its diameter is really small 2 nm which pauses a first difficulty to the process. Once the molecule has been caught, a second difficulty is to apply a force. Thanks to molecular biology, biochemistry, self-assembly, DNA offers a variety of possibilities to attach chemical groups to a sequence-specific target and in particular to both ends of a molecule. These chemical groups can, in turn, bind to various surface chemistry. This offers the possibility to catch independently each end of the molecule. In order to apply a force and visualize the ends of the molecule, a microbead is ideal. The classical single molecule assay consists in attaching a microbead to one end of the molecule, while its second end is attached to a microscope glass slide used in the construction of a flow chamber (Figs. 1 and 2).

As mentioned, Magnetic tweezers are a simple tool that involves using magnetic beads and applying a force by positioning permanent magnets close to these beads. The next step is to visualize the bead positions by video-microscopy and track their changes over time. In essence, the experiment involves observing a flow chamber containing magnetic beads attached to DNA molecules, with magnets positioned above this chamber, using a microscope and a camera.

1.2 The various components

Indeed, although simple, the various components need to be well designed to overcome various difficulties. The most severe challenge lies in the scale of the molecule itself and the size of displacements induced by enzymes, both of which are in the nanometer range.

When our group started studying single molecule, we first advocate that measuring nanometers with an optical microscope was not possible since the resolution of an optical microscope is limited by the law of diffraction to typically a quarter of the light wavelength (Abbe, 1873). Abbe had formulated this law clearly with a very nice theoretical demonstration. The experiment did show us (as previously observed (Gelles, Schnapp, & Sheetz, 1988)) that we were able to track a micron size bead position with a few nanometers accuracy. Our interpretation of Abbe's law was incorrect: the resolution of the

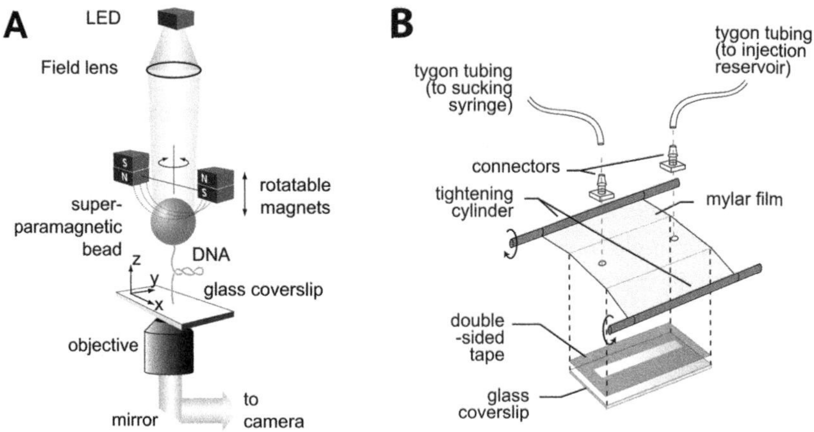

Fig. 1 Description of the magnetic tweezers apparatus (A) and of the flow-cell (B). The apparatus is an inverted microscope with the objective below the flow-cell making the image of the beads on a camera after passing through a tube lens (both not represented). The objective and the tube lens are set in an afocal system usual for a microscope. Above the flow cell containing the beads tethered to the cover slip by DNA molecules, two magnets provide a horizontal magnetic field and strong vertical field gradient centered on the optical axis. The magnets vertical position is driven by a motorized translation stage while their angle may be adjusted by a motorized rotation stage. The chip of a LED is imaged on the microscope objective by a lens so as to cover it fully. An iris(not represented at the lens position) limits the aperture of the illumination to a few degrees (A complete optical path is given in Fig. 5). The flow cell is built from a pretreated glass cover-slip, with a double-sided tape defining the chamber with a top part made of a 50 μm Mylar sheet. Holes are punched in the Mylar to provide a buffer inlet and outlet connected to Tygon tubing. The Mylar sheet is larger than the cover-slip and is used to press the flow cell on the microscope stage and to move it around in *x* and *y*.

microscope indicates that if you have beads smaller than the microscope resolution (typically $\lambda/4$) then it is impossible to distinguish whether there is a single bead or two beads close to each other. However, if you have a single bead of 1 μm in diameter, then you can measure its position with nanometer resolution, provided your image is of good quality. This is a small miracle that occurs in the single molecule world and which led, in the fluorescence field, to the Nobel prize of Betzig (Shroff, White, & Betzig, 2008) for the super-resolution discovery. In fact, the accuracy of this measurement is only limited by the signal-to-noise ratio of your imaging system. With a resolution of 1 nm one can detect, in good conditions, a step of one base pair of an enzyme which is very interesting. However, we shall see that in most cases magnetic tweezers nearly reach this resolution if extra care is taken in imaging.

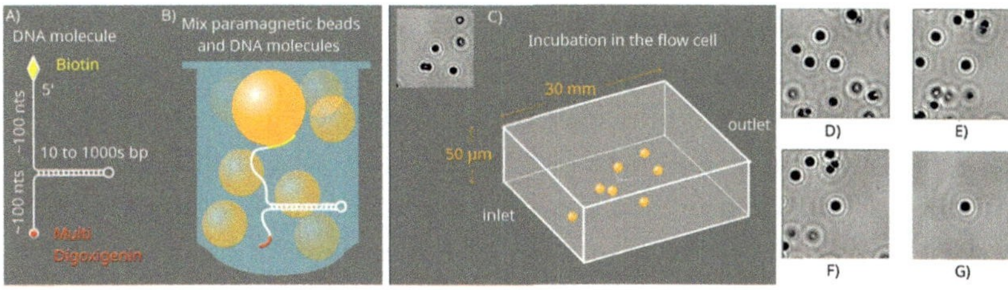

Fig. 2 Principle of the bead attachment in the flow cell. First one needs to prepare the substrate DNA or RNA molecule with specific groups at both ends allowing the bind to the bead to one end and the glass surface to the other end. Here in (A) we have sketched a DNA hairpin with biotin at the 5′end and multiple digoxigenin at the 3′ end. Next, we mix a small amount of the solution containing these molecules with rinsed paramagnetic bead in a tube (B). After typically 15 min, a few μl of the bead solution is injected in a flow chamber which has been previously treated to attach anti-digoxigenin to the bottom glass plate (C). During this phase no magnetic field is applied and the beads sediment for a few minutes, they display a visible Brownian motion. Most of the beads are not attached to a DNA molecule but some are, this can be seen since then they are not advected by a gentle flow. After incubation, a moderate flow (a few tens of microns per second) is established. All the unattached beads are rinsed out has can be seen in (D), (E), (F), but the tethered bead remain (see the central bead of these images). At the end of the rinsing phase only the tethered beads remain (G) and are ready for the experiment.

Small miracles rarely appear without limitations: the resolution achieved with the optical imaging is in the nanometer range but this means that the microscope platform, the sample holder, etc. must also be stable with this resolution and this by far not obvious. Usual microscopes have been designed to be extremely versatile with objectives mount on a revolving plate to quickly switch from one magnification to another one and sample holder that moves easily. All these mobile elements have not been designed to remain stable within a few nanometers over time (this is why active constant focus was developed on microscopes). Considering only one element of the microscope, its objective, this part is made of lenses mounted in a metallic tube. The simple thermal expansion of this tube for a 1 °c change is equal to 925 nm (1.85×10^{-5} thermal expansion of brass for a 50 mm tube)! Temperature fluctuations lead to slow drifts which are the enemy of micromanipulation experiments, they severely hamper the quality of the data. A microscope design with a minimum of moving parts is highly desirable, a temperature regulation with a millidegree stability for the sample holder and the microscope objective reduces the microscope drifts efficiently. Moreover, this temperature regulation is most welcome for biological material (Gumpp, Stahl, Strackharn, Puchner, & Gaub, 2009).

1.2.1 The molecules of interest

Single molecule micromanipulation is possible since nucleic acid polymers (DNA or RNA) and proteins may be attached specifically to a magnetic bead at one end and to the surface of the flow cell. We shall not discuss in detail the biochemical process allowing doing that now in a variety of ways. We will just say that there exist pairs of molecules like biotin and streptavidin which have a tremendous affinity for each other and that may be attached specifically to a DNA sequence or an amino acid in a protein (details may be found for instance in (Ruiz-Gutierrez et al., 2022)).

1.2.2 Magnetic beads

The magnetic beads have two purposes: their magnetization is used to generate the force applied to the molecule, so they are a kind of handle to manipulate but also a visible marker of the molecule end. Several bead properties are essential for running the experiment efficiently. First, they need to undergo the right biochemical treatment to attach the molecule of interest. For instance, they are covered with streptavidin which will bind strongly to the biotin entity link to the molecule. During the sample preparation, beads and molecules will be mixed so that the end of the molecule does bind to the bead through the streptavidin-biotin interaction. It is also essential that there is no non-specific interaction between the bead surface and the rest of the molecule, otherwise the molecule will stick to the bead surface.

Finally, the bead magnetization determines the magnitude of the force, this feature depends on the bead size or on the amount of magnetic material contained in the bead (Lipfert, Wiggin, Kerssemakers, Pedaci, & Dekker, 2011). Depending on the manufacturer, the bead magnetization displays a dispersion which becomes a problem when it is too large as it is not possible to use the parallelism of the magnetic tweezers if all beads/molecules are subject to very different forces. Some manufacturer succeeds in preparing beads with fairly homogeneous magnetization which turns out to be critical. In our hands Thermofisher Dynabeads MyOne T1 beads provide the best compromise. M270, M280 and M450 offer increased magnetization with the equivalent homogeneity but bigger size (Table 1).

1.2.3 The magnetic field gradient and the magnetic bead system

Magnets to apply a force on microbeads are very convenient but the strength of the force which is required to study enzymes imposes to use either large beads or highly optimized magnetic gradients. To be more

Table 1 A short list of magnetic beads used in magnetic tweezers and some of their properties listed with increasing radius.

	Radius r (μm)	$M_{Saturation}$ 10^{-15} A.m^2	B_s Saturation Field (mT)	Force (pN) at dB/dz = 1 T/mm	Noise generated by the bead F_L ($pN/Hz^{1/2}$)	Gravitational force $P = \rho.V.g$ (pN)
Ademtech 300 nm	0,15	1	N.a.	1 pN	0.011	0.000027
Ademtech 500 nm	0,25	4.7	N.a.	4.7 pN	0.014	0.000124
Microparticles GmbH PS-MAG-SA-S1979-1	0.268	3.63	N.a.	3.6 pN	0.015	0.000130
Magsense	0.5	79	31	79 pN	0.020	0.008
MyOne T1	0.5025	25.5	12	26 pN	0.020	0.003
M270	1.35	214	Probably 12	214 pN	0.033	0.016
M280	1.40	161	Probably 12	161 pN	0.034	0.016
M450	2.25	2 417	Probably 12	2 417 pN	0.045	0.076

The magnetizations are given from the manufacturer specifications or existing measurement. The force given in a field gradient at 1 T/mm is the maximum one that the magnets (see Fig. 4) provide. As bead size increases, the Langevin force responsible for the Brownian fluctuations in water ($\eta = 10^{-3}$ N.s.m^{-2}) increases. For small beads the weight of the bead is very small compared with the Langevin force.

Table 2 List of force values associated with a specific transition in DNA.

Characteristic force	Biological significance
0.1 pN	Force necessary to stretch a dsDNA molecule to an elongation of half its contour length
0.5 to 0.6 pN	Force at which negatively supercoiled dsDNA undergoes denaturation (depends on salt concentration)
3 to 6 pN	Force at which positively supercoiled dsDNA undergoes phase transition to scPDNA (depends on salt concentration)
5 pN	Force at which a dsDNA molecule and a ssDNA with the same number of bases are equal in extension
15 pN	Force necessary to unzip a DNA hairpin
60 pN	Force necessary to overstretch dsDNA unconstrained in rotation (Cluzel et al., 1996; Smith, Cui, & Bustamante, 1996)
120 pN	Force necessary to overstretch dsDNA constrained in rotation

specific, we need to overcome weak forces which are the rule in biology, the typical energy involves in the binding of a weak force is comparable to thermal energy k_BT, while the bond exists for distance in the range of 1 nm. These forces are thus equal to $k_BT/1\ nm = 4.1$ pN. To manipulate safely, one needs to be able to apply at least 20 pN or more, see the Table 2. The magnetic force exerted on the bead is proportional to the volume of the magnetic material contained in the bead, specifically to r^3 (r being the bead radius), but also to the strength of the magnetic field gradient. This parameter depends upon the design of the magnets used. Naturally, one would tend to use a strong, large magnet that can adhere so strongly to a piece of iron that one can barely separate them (Fig. 3).

But such a choice is just wrong! What is needed is a strong gradient, thus one needs two magnets with opposite magnetic polarity separated by a gap which will be the place where the gradient is stronger. The magnetic field produced by magnets with a given aspect ratio does not depend on their size. Therefore, to increase the field gradient, the way to go is to reduce the gap size. Good magnets (N52 material) for magnetic tweezers are thus a pair of small (≈1 mm) magnets separated by a small gap (few hundreds of μm) (HKCM *9964-71786* Magnet-Cuboid Q01.7 × 01 × 03Ep-N52). Increasing

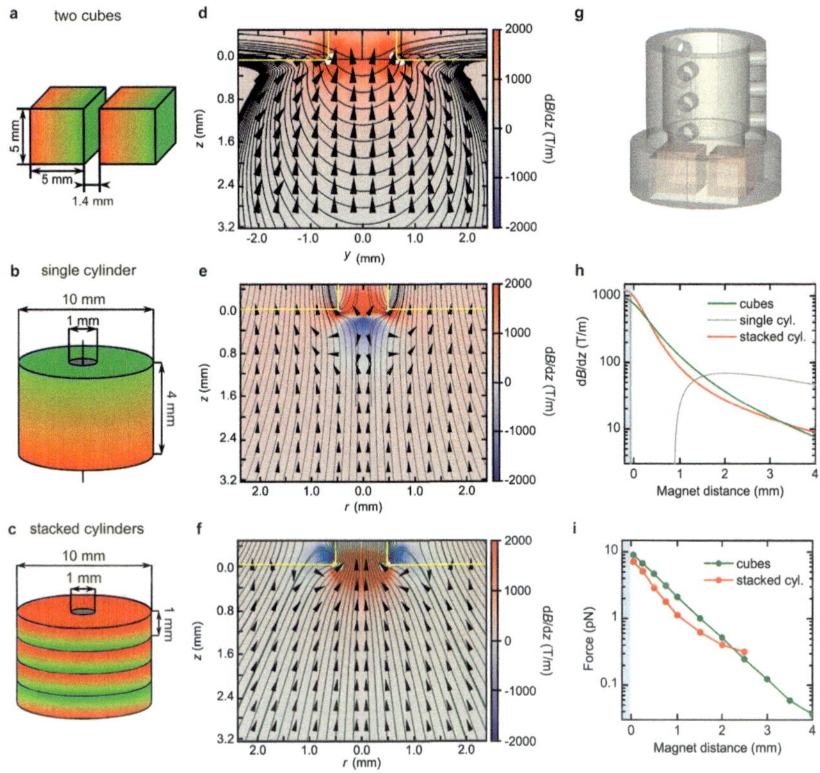

Fig. 3 Magnet configurations as discussed by Kauer (Kauert, Kurth, Liedl, and Seidel 2011) and Lipfert (Lipfert, Hao, & Dekker, 2009). In (a), (b) and (c) you find different geometries used: (a) two magnets with horizontal magnetization. In (b) a hollow cylindrical magnet with a vertical magnetization and (c) an ingenious configuration made of a stack of hollow discs with their magnetizations vertical in the same direction except the bottom disc which has its magnetization reverted. The (a) configuration leads to a very strong horizontal magnetic field inside the gap between the two cubes (hold in place by the piece described in (g). The field decays rapidly as you go downwards along z (see d) leading to a strong field gradient and thus in principle to a strong force. However, the place where the gradient culminates is very close to the surface of the magnets which cannot be reached owing to the finite thickness of the flow cell. When z is large, the gradient and the force decay roughly exponentially with z as shown in (h) and (i). When you come close to the bottom horizontal magnets surface, the force increases and at some point the force is not anymore aligned with z but bifurcates towards one the corner of the magnets becoming horizontal. Although you generally do not reach this situation, you may observe that beads are pulled on the side at high force. The cylindrical configuration in (b) with a central hole presents a peculiar behavior. As seen in e) at large distance z >> 2 mm the field is vertical and it decays roughly exponentially with z. This provides an interesting configuration to measure torque applied to a molecule. When one comes closer to the bottom surface of the cylinder, the bead starts to be pulled radially out of the center. Interestingly

the bead size is an option but not a good idea in the end because this also increases the noise of the experiment. Reducing the gap between the magnets also reduces the spatial range over which the magnetic field gradient exists. As it is a good idea to have the magnets outside the flow chamber containing the beads, there exists a compromise in the gap size: small enough to get high force but not too small so that the gradient does not vanish before reaching the beads. In many magnetic tweezers designs, the sample illumination is achieved by sending light vertically through the gap of the magnets, imposing also that the gap is not too small, which would reduce the field of view.

Experimentally we find that the magnetic field along z-axis is horizontal (along x) and its strength decreases nearly exponentially with z: $B(z) = B_0.\exp(-z/b)$ where b is a characteristic length typically equal to the gap size. Where $z = 0$ corresponds to the magnet surface. Thus, the maximum gradient that can be used is $B_0/b. \exp(-z_e/b)$, where z_e is equal to the distance between the bead to the top of the flow cell (Fig. 4).

The best magnets are those which lead to the strongest force and thus to the highest field gradient with the constrain that z cannot be smaller than z_e. This leads to choose a small gap "b" however b cannot be too small compared to z_e otherwise the exponential decrease is too strong. A good choice is to have a gap comparable to z_e, leading to b = 200 or 300 μm.

One interesting feature of magnets is their inherent simplicity: with two magnets installed properly in a holder you apply a force on a single molecule which is completely stable in time and reliable. To modulate it you need to be able to adjust the distance between these magnets and the sample that we call z_{mag}. You need a motorized stage moving the magnets vertically over a range of 10 mm at most and with the accuracy of 10 μm. Since the force decreases by a factor 2 over a distance typically of 300 μm, the force becomes negligible after 5 mm. A maximum moving speed of 5–10 mm/s is sufficient, but it is interesting to be able to achieve slow ramps where the force is increased smoothly over a few minutes. Finally,

Fig. 3—Cont'd when changing z it is possible to pull a molecule with an angle which goes smoothly from 90° (vertical) to ~0° a nearly horizontal situation with all the intermediate angle possible. However, you cannot modulate much the force in this situation. Finally, the stack cylinder configuration suppresses this inversion zone and leads to a strong vertical field gradient with a vertical field direction. As seen in f) and h) and i) this configuration provides forces nearly as strong as the horizontal magnets one, which is useful for torque measurement. Other articles discuss magnet configurations: (Aldag et al., 2023; Choi, Gyu Kim, Ju Shon, & Yoon, 2022). *This figure is reproduced from the Supplementary Materials of Kauert, Kurth, Liedl, and Seidel (2011).*

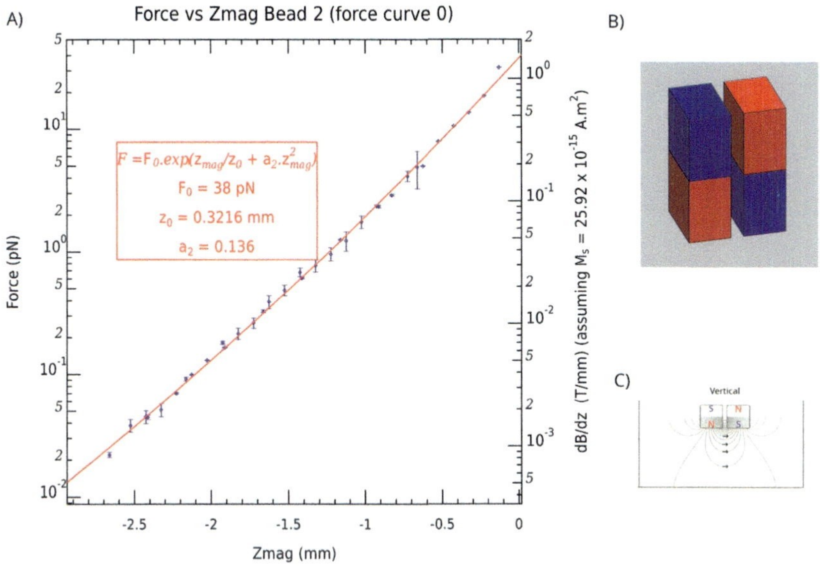

Fig. 4 (A) Measurement of the force versus the magnet distance (z_{mag}) from the sample obtained with a dsDNA molecule with a contour length $L_0 = 5.25\ \mu m$, a bead 1 μm in diameter and with a pair N52 magnets 3 × 1.7×1mm with opposite vertical magnetization and separated by a gap of ~300 μm (sketched in B). This curve is used to calibrate the magnets and provided a very good estimated force for another bead of the same type. The force decays nearly as an exponential, notice that z_{mag} must be < -0.1 mm otherwise the magnets smash the flow cell which is ~100 μm thick. At most, the force applied can reach 25 pN. Using the MyOne bead magnetization at saturation, we compute the magnetic field gradient displayed on the y-axis which reaches 1 T/mm at maximum. (C) sketch of the magnetic field lines in this situation.

we shall see that it is interesting to be able to adjust the angular position of the magnets to study the effect of the torsion constrains, the magnet holder should spin around the Z-axis with angular position that we define as n, where n is the number of turns applied. A maximum speed of rotation of a few turns per second is sufficient (Mosconi, Allemand, & Croquette, 2011). A good design should be symmetrical around the z-axis so that the pulling force remains vertical.

Using only magnets limit the field to ~1.4 T at most. Using pole pieces in ferromagnetic materials opens the possibility to reach 2.1 or 2.5 T which combined with a narrow gap and a very thin flow-cell is a solution to improve the force applied on a bead. We list below maximum field gradient reached in different groups together with some comments.

Reference	Design	Gap size	Minimum distance to the flow cell	dB/dz maximum at the level of the flow cell	Comment
Lipfert et al. (2009)	Cubic magnets with vertical magnetization	1 mm	1 mm	0.5 T/mm	Simple, large field of view
Kauert, Kurth, Liedl, and Seidel (2011)	Cubic magnets with horizontal magnetization	1.4 mm	1 mm	0.7 T/mm	Simple, large field of view
Lin and Valentine (2012)	Magnets + pole pieces	1 mm	0.1 mm	0.8 T/mm	Complex and bulky design, large field of view
Rieu et al. (2021a)	Vertical magnetization	0.3 mm	0.1 mm	1 T/mm	Small but sensitive layout. Restricted field of view.
Lansdorp (2016)	Magnets + pole pieces	0.125 mm	0.06 mm	5.5 T/mm	Complex design relatively large and require a very thin cell
Tapia-Rojo, Eckels, and Fernández (2019)	Magnetic recording head	0.025 mm	0.250 mm	0.25 T/mm	Fast switching could give stronger gradient with thinner flow cell

1.2.4 The flow cell

The principle is that the flow cell needs to be transparent typically with a glass coverslip at the bottom and transparent top surface. The chamber must be leakproof with inlet and outlet tubes allowing easy buffer exchange avoiding air bubbles. With magnet pair having a small gap, it is fundamental that the flow chamber is very thin so that the magnets can be brought as close as possible to exert strong forces when needed. As the magnets come very close to the top surface, it is a good practice to have the top part of the flow cell quite flat.

Like the beads, the flow cell glass surface must be treated to bind the end of the molecule and no other part of the molecule. Realizing surface with very low non-specific binding is fundamental and not easy. Various recipes have been proposed which all require some practice (Quack & Dulin, 2024; Ruiz-Gutierrez et al., 2022).

The buffer flows in the cell is best driven by a pump like a syringe pump sucking fluid out with the inlet connected to a small vessel where the operator deposit the various solutions.

1.2.5 The microscope visualization

The bead visualization is done with a classical optical configuration which is perfect to observe and measure the x,y components of the bead position. However, the bead interaction with the glass surface to which it is tethered needs to be minimized thus the vertical pulling direction is the simplest choice. This means that bead position of interest is along z the optical axis of the microscope. The magnification, the numerical aperture N.A. and whether the objective is made for oil, water or air immersion is important depending on the purpose of the experiment. If you want the best accuracy in measurement, an oil immersion objective with a high magnification x100 or x50 is the right choice. The N.A. does not need to be high, although immersion objectives do have already a significant N.A. there is no need to look for extreme N.A. since the illumination is parallel and the light diffused by the beads is mostly at small angles. An oil immersion objective with 100x and N.A. = 1.25 works perfectly. Switching to an immersion objective of 40x will not degrade much the accuracy of the measurement but will significantly extend the field of view which is interesting to increase the parallelism. An immersion objective improves the resolution in z (see Table [objectives) but imposes to use oil between the objective and the sample, this turns out to be a constrain in particular to avoid bubbles in the oil which can be delicate to prevent. If you aim to

Table 3 Comparison of the three types of microscope objectives.

Objective type	**Air immersion**	**Water immersion**	**Oil immersion**
Correction factor in z	$n_w \approx 1.33$	1	$n_w/n_o \approx 0.878$
Magnification	4x to 100x	40x to 100x	40x to 100x
Features	Convenient,	No correction	Best noise
Drawback	Higher noise	Evaporation issue	Bubble in oil below the sample can be a pain

An important parameter to keep in mind is that the *z measurements must be corrected by a factor for air and oil immersion objectives. This correction is best for noise in the case of oil immersion objectives*

obtain very large parallelism, the objective magnification will be decreased, and since the accuracy is not the most important feature anymore, an air immersion objective will simplify the manipulation of the flow cell on the microscope (Table 3).

While measuring the bead displacement in the horizontal plane is obvious, measuring along the z-axis is possible but involve using a specific sample illumination. With the classical Köhler illumination of a microscope, the bead image is sharp when the microscope is on focus on the bead but becomes quickly blur as the bead moves over or under the focus plane with little information on the actual z coordinate of the bead. If one uses a parallel monochromatic illumination instead of the Köhler one, the bead image is still sharp when the focus plane is on the bead but this image gains a set of diffraction rings decorating the bead image which diameter increases smoothly with the amount of defocusing. These features may be used to deduce the distance between the bead and the focus plane of the microscope. Although the ring shape is not quite similar if the bead is at a distance z above or below the focus plane, it is safer to work with the focus plane always on one side for instance above the bead to ensure that a diffraction ring size corresponds to a single bead position. Using some kind of lookup table of the ring shape at different defocusing position one can deduce the bead position from the diffraction ring image (Gosse & Croquette, 2002). With this process, we are able to access to the 3D bead position provide the bead z position is within 1–10 μm below the focus plane. All those measurements require the ability to move up and down the objective in a very accurate way. The objective of the microscope needs to

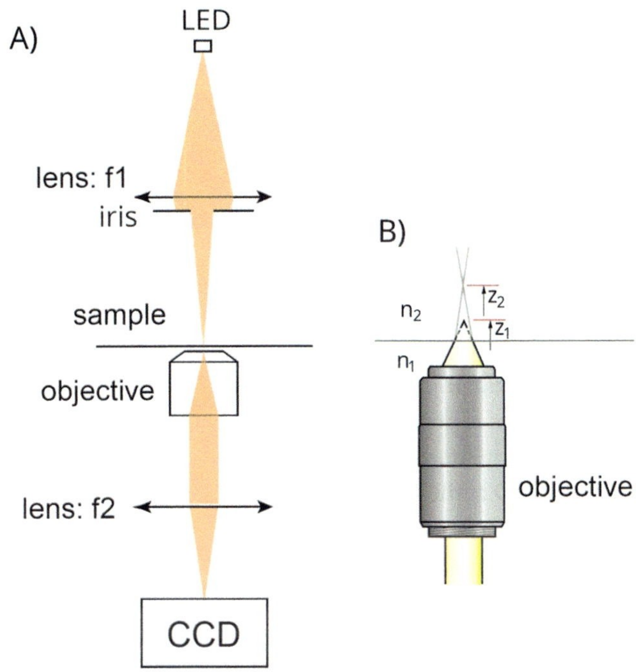

Fig. 5 (A) sketch of the light path in the magnetic tweezers microscope. Light is emitted from an LED chip which is imaged by lens f1 on the sample after passing through an iris which defines the parallelism of the illumination. The objective combined with the tube lens f2 make the image of the sample on a CCD camera. (B) *Z* extension correction due to the difference in optical index between the flow-cell and the objective media. The beads are immersed in a water buffer with a 1.33 index while the microscope objective is either an air, water or oil immersion type. If the objective is a water immersion type, n1 = n2 = 1.33 and the ray suffer no change at the interface. If you use an air immersion objective, you are in a situation corresponding to the figure with $n_2 = 1.33$ and $n_1 = 1$. The Descartes/Snell law leads to an angular change for the ray crossing the interface and the focus position $z_{n2} = (n_2/n_1)z_{n1} = 1.33\ z_{n1}$. If you use an oil immersion objective, the same phenomenon occurs leading to $z_{n2} = (n_2/n_1)\ z_{n1} = 1.33/1.52\ z_{n1} = 0.878\ z_{n1}$.

be mounted typically on an accurate piezoelectric stage with a position sensor feedback (typically a Pifoc©Physik Instrumente) ensuring the repeatability of a few nanometers of the focus plane of the objective. As the *z* measurement is made by comparison with pre-recorded images at specific focus position, the final accuracy in *z* in the measurement is directly correlated to the accuracy of the piezoelectric focus device and the interpolation process used to compute the intermediate *z* position from one reference profile to the next (Fig. 5).

The parallel illumination is also well adapted with the fact that this light must pass through the magnets gap to reach the sample. This ring pattern is indeed an interference pattern occurring between the light scattered by the bead and the parallel light beam passing close to the bead. A monochromatic LED light has just the right coherence length to observe these interferences and to average out interferences having a difference of optical path larger than a few tens of wavelengths.

Parallel illumination would be better achieved with a laser but its spatial coherence is by far too large and interference patterns appear everywhere on the image corresponding to interferences between the incoming light and any reflection occurring in the microscope. As a result, a laser cannot be used here without the addition of a decoherence filter. LEDs emit with a wide-angle beam and are not point sources, to get a quasi-parallel illumination a good choice is to place a LED at the focal point of a lens having a relatively small iris. The divergence of the light source is given by the ratio of the LED emitting chip size to the lens focal length and the iris diameter. The bead diffraction rings image quality accommodates a divergence of a few degrees. The illumination path uses a lens focusing the image of the LED chip on the sample with a focal length of 50 mm. We want to illuminate with the maximum of light so that the photon noise on the camera is minimum. One may think that the best source would be a LED with the maximum power. Again, this intuition does not provide the best choice: the LED maximum emission power actually reflects the size of the emitting surface of the LED, the light emissivity (the light emitted by a given surface of the chip) is roughly constant and there is no point in having a large LED chip since this will reduce the parallel nature of the beam.

Using micron size beads and cameras running at a few hundred Hertz, the light power of LED (Thorlabs M660D2) is just fine, it is convenient since the light power is very stable, the lifetime of LED is very long and all the system is very simple. For smaller beads and/or faster cameras, getting more power than what a LED provides is more demanding, SLED (Dulin, Barland, Hachair, & Pedaci, 2014b) offers an interesting alternative but spurious interferences are more difficult to prevent.

1.2.6 Cameras used in magnetic tweezers

All the magnetic tweezers measurements rely on the camera hook up to the microscope which turn out to be an essential element. Cameras have been improving continuously with time and this is an opportunity for the field (Fig. 6). The important features to consider for a camera are the following:

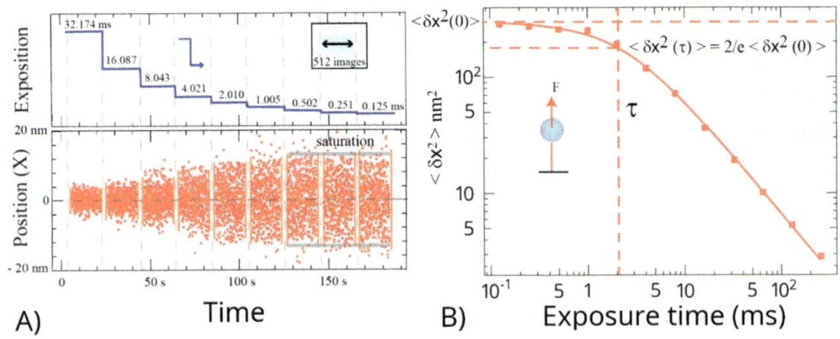

Fig. 6 Illustration of the low-pass filtering of the integration time of a camera. Here we record the fluctuations along x of a bead submitted to a strong pulling force (~15 pN) tethered to a short molecule 1.2 μm with a camera having a frame rate of 30 Hz. In (A) we show the fluctuations acquired with a variable exposure time (the maximum value of the exposure is nearly equal to the time between two frames which corresponds to the normal usage of the camera). In this latter case, the rapid fluctuations are averaged and thus filtered leading to small fluctuations. However, the level of these averaged fluctuations cannot be used to compute the force since the resulting force would be strongly overestimated. When the exposure time is reduced, the fluctuations level increases and finally saturates when the exposure time becomes smaller than the characteristic time of the fluctuations. Plotting the fluctuations strength versus the exposure time (B) allows to find both the real value of these fluctuations (which can be used to compute the applied force) and their characteristic time, here equal to 2 ms. When reducing the exposure time of the camera one should increase the illumination so as to keep the photon noise minimal (Cavatore, 2011).

- Geometry of the pixels will define the displacements measured, using square pixels make things simpler. C-MOS cameras have improved tremendously and the total number of pixels can now be huge. Of course, having a large number of pixels is interesting, but this might be at the expense of the signal-to-noise. You need to pay attention to the pixel size: in C-MOS camera aside each pixel stands some transistors and the ratio between the sensitive surface to the pixel area can be significantly smaller than one. Large pixel size generally has a better ratio, leading to a stronger well depth. As explained below, this is important in considering the signal-to-noise.
- High chip size limits the frame rate of the camera but fast cameras are now available. An interesting feature of your camera should be the ability to restrict the region of interest (ROI) and to increase the frame rate in this situation. The rate increase is generally inverted to the number of lines of your ROI and unfortunately not to the number of pixels in the ROI.

- There is no need for extremely low light level sensitivity or noise. The noise added by camera is now quite low and soon negligible compared with the photon noise. You should never increase the gain of the camera (this also increases the photon noise) but you should increase the illumination.
- Exposure time and frame rate will partially define your time resolution (if not limited by other physical parameters) and partly the averaging done by the camera. Originally camera frame rates were low restricting the time resolution. They have considerably improved and can reach kHz but there is a trade-off between parallelism related to the number of pixels of the camera and its frame rate. Considering the quality of the signal, global shutter must be used while rolling shutters should be avoided. You must keep in mind that your Brownian signal is actually time averaged by the integration time of the camera and being able to adjust this parameter is most desirable.
- Your signal-to-noise will be related to the video acquisition and to the illumination. Most of the time, the photon noise will dominate and to minimize it you need to send the maximum amount of light on the camera avoiding saturation. The interesting feature of the camera to consider is called the pixel well depth which is the maximum number of electrons that one pixel may hold before saturation. Usual cameras have a well-depth of the order of $W_d = 10^4$e, which leads to a signal-to-noise ratio of $W_d^{1/2}/W_d = 1/100$. Cameras with higher W_d will give better result provided you have enough illumination. Cameras with well-depth over 5×10^4e are not so common but can be found.
- An interesting feature of cameras is the fact that the measured signal-to-noise increases with the frame rate as far as the illumination is maintained just below saturation. A way to understand this principle is considering that if you increase the frame rate by a factor 4 while also increasing the illumination by the same factor. Your signal in the high frame rate case will rely on 4 times more photons than at the original rate improving the SNR by a factor 2. This interesting property is valid as long as the noise added by reading one frame by a camera is lower than the photon noise. As the frame rate goes-up so does the amount of data to treat and to store which soon becomes an issue.

In our laboratory we often use CMOS cameras from IDS imaging© using detectors provided by Sony but other manufacturers can be found in other companies.

1.3 Principle of the sample preparation: catching single molecules

All experiments start by preparing single molecules attached to a bead. This step is crucial and will determine the success of the experiment at the end. Having prepared the appropriate DNA or RNA molecules, a mixture of beads and these molecules will be first incubated for typically 15 min so that the molecules have a chance to bind to a bead. It is important to keep in mind that beads have a surface treatment that leads to repulsion with other beads and surfaces but also with the molecules themselves, this is important to prevent non-specific binding. As a result, the binding of the biotin entity of the molecule, to the streptavidin attached to the bead is a rare event and the binding efficiency is low. Bead manufacturers suggest using very high salt concentration in the binding buffer which reduces the repulsion but favor too much non-specific binding and is counterproductive in our hands. We prefer to use ionic conditions which are less efficient but guarantee a reduced non-specific binding. This means that in these conditions not so many beads are bound to a molecule and we must inject a high number of beads in order to find a reasonable number of molecules bound to a bead. This property is also the one insuring that most beads are attached by one molecule. After preparing the flow cell with a glass coverslip covered with the adapted molecules that can bind the second end of the molecule attached to the bead, this binding is done by injecting a high density of beads in the flow cell on the glass coverslip and by waiting for the binding to occur (see Fig. 2). The user can periodically test the advancement of this binding by starting a very slow flow which will move away unattached beads while the bound one stay at the same spot but are still animated by Brownian motion. When the density of this latter kind of bead is fine, the flow is increased but kept low enough so as to rinse unattached beads avoiding breaking the molecules attached to a bead.

1.4 The video beads tracking in 3D

Magnetic tweezers need to record the 3D bead position of the beads continuously in time. The principle is to define a piece of the image around the bead which will be large enough to collect the interference ring shape and which will permanently track the x,y bead position. This sub-image typically 128 pixels x128 pixels will allow us to measure the x and y position with accuracy in the nm range while measuring z with the accuracy of a few nanometers (Gosse & Croquette, 2002). This process can be done in real time for every frame of the camera and for hundreds or

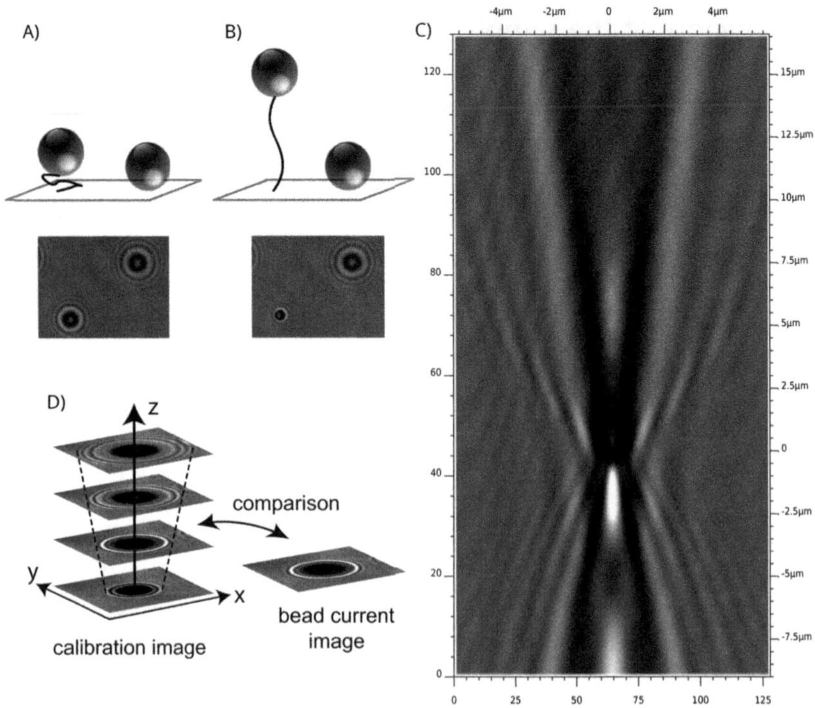

Fig. 7 Principle of video tracking in Z. In (A) and (B) we have sketched two situations with two close by beads with one attached to a molecule while the second one is stuck on the glass plate with no force applied (A) or a force applied (B) which pulls up the tethered bead. Below each sketch we show the observed image decorated by its interference rings. In (A) the two beads are at the same z position and the bead images are pretty much similar. In (B) the tethered bead z position is higher and closer to the focus plan of the objective, as a result the interference rings are smaller than the ones of the stuck bead. In (C) we have recorded the radial profile of a bead over an extended z region starting 9 µm below the microscope focus plane and extending more than 16 µm above (focus plane is at $z = 0\,\mu m$). Although the image is not really symmetrical around $z = 0$, we use only the Z > 0 side of the image and in order to keep a good accuracy we restrict z in the range [1 µm, 10 µm]. Here one profile every 0.3 µm is recorded (the image is interpolated in z) which is sufficient. In D) we sketch the principle of the z tracking: the radial profile of the current image is compared with the stack of reference profiles (forming the calibration image) to deduce the exact z position by extrapolation between the two nearest reference profile.

thousands of beads simultaneously (Dulin et al., 2015), thanks to the speed of nowadays CPU and eventually GPU (Fig. 7).

The x,y tracking starts from an estimated bead position $(x_c(t_{n-1}), y_c(t_{n-1}))$ at time t_{n-1}. First, an average profile $I(x)$ obtained by a strip of 128 pixels in x centered on this position and averaged over 16 pixels in y to determine the x position (the y position is obtained by the perpendicular profile). This profile is correlated with the inverted profile having $I_{-1}(i) = I(127_{-}i)$ using an FFT algorithm to speed up the process. The correlation function always presents a positive maximum which is offset by $2\delta x$ where δx is the shift of the bead center with the center of the strip. Doing the same process in y leads to the new position of the bead $(x_c(t_n), y_c(t_n))$. Both $2\delta x$ and $2\delta y$ are determined by a polynomial interpolation of the maximum of the correlation function. Once the bead center is found, the radial profile of the bead is computed by scanning pixels of the sub-image, computing its distance to the bead center and averaging the light intensity at the same radius. The most consuming part of this computation is the 16,384 square root calls to determine the radius associated to each pixel. The quality of the radial profile is fundamental to obtain high resolution. Key features are the contrast of the diffraction ring and the averaging on pixels along circles reducing noise (Gosse & Croquette, 2002).

The z tracking needs a more elaborate algorithm and the recording of a set of a few tens interferometric ring images recorded in advance at a well defined out of focus distance: defining a calibration image. This can be done by recording the radial profiles at a relatively high force, typically a few pN, to limit vertical fluctuations of the beads, for different positions of the objective. After finding the image bead center and the radial profile of each sub-image. This profile is compared to the radial profiles of the reference images in order to find and to extrapolate the z displacement.

1.5 The force calibration

1.5.1 Magnetic tweezers are force clamps

There are two kinds of micromanipulation device: the most natural is a system where you impose the position of the bead, for instance by a small tweezer and then you measure the force applied by recording the elastic deformation of the tweezer, this is a position clamp. Optical tweezers belong to this category, they are ideal to manipulate the bead allowing moving it to the desired place with just an optical focused beam (Bustamante et al., 2021; Wang, Yin, Landick, Gelles, & Block, 1997).

The second kind is the force clamp system where a constant force is applied and one measure the change of extension induced by this force (Strick, Allemand, Bensimon, Bensimon, & Croquette, 1996). This is precisely the case of the magnetic tweezers: the magnets impose a force to the bead and one measure the extension of the molecule under this tension.

Force clamps are less efficient to move a specific bead to a given position but they are intrinsically parallel: a large number of beads experience the same force pulling them upwards, the position of each bead will reflect the elastic property of the molecule which tether the bead to the surface. Many molecules may then be measured simultaneously.

A good force clamp should impose a well-defined force whatever the bead position and this force should be homogeneous on a large set of beads. The magnet configuration that we have described above exert a relatively homogeneous force in the x,y plane at a define z position. The homogeneity being worse when z is close to z_e where some beads are very close to the magnets. Another point to consider is that the force actually varies exponentially with z and one can argue that the force is not really constant: when the extension of the molecule increases, the bead get closer to the magnets and the force increases. This is right but the change of the extension of the molecule is typically a few microns when the force varies over a distance comparable to the gap size $b \sim 300$ microns in the configuration shown on Fig. 4. This means that the force in an experiment actually varies by less than a percent when the extension changes. One should keep in mind that in the race to high forces reducing the characteristic decay length of the field gradient by using microns size polar pieces will indeed break the constant force approximation that we shall use.

1.5.2 Absolute calibration

Single molecule experiments are quantitative only if we do know which force is applied during the experiment. The absolute calibration of the magnetic tweezers relies on the characterization of the Brownian motion of the bead molecule system. With an upwards vertical force applied, the bead pulls up the molecule and the bead molecule system behaves as an inverted, overdamped, pendulum. For a given force F, the molecule extends by l, and the bead fluctuates mostly in x and y and to a smaller extent in z. These fluctuations can be seen as the result of the Langevin random force on the bead which takes the form $f^2x,y,z = 24\ \pi\eta r\ k_BT\Delta f$. Where η is the viscosity of the buffer solution 10^{-3}N.s.m^{-2} for water, k_BT is the thermal energy at the temperature T, $k_BT = 4.1 \times 10^{-21}$J. This corresponds to a random white noise force in all three direction with a typical strength of 20 f $N\sqrt{\text{Hz}}$ for a bead *of* $0.5\ \mu m$. As a result, the bead moves out of the vertical position but the projection of the vertical force F on the tilted molecule has the tendency to bring it back towards the vertical equilibrium position. The horizontal stiffness of the pendulum is equal to $k_{x,y} = F/l$. The equipartition theorem leads to $\frac{1}{2} k_{x,y} \langle x^2 \rangle = \frac{1}{2} k_B T$. One easily sees that $F = k_B T.l / \langle x^2 \rangle$. This relation is particularly handy because to measure F, you need to determine the average of the x fluctuations $\langle x^2 \rangle$ and the molecule extension l (you know $k_B T$). To measure a force, you just need to measure distances which is quite convenient. When the force is weak, the fluctuations are large and thus easy to measure but you need some time to get an accurate measure. The reason is that the bead molecule system in water has a characteristic motion time $\tau_x = 6\pi\eta r / k_x = 6\pi\eta r l / F$ which is large at low force. This absolute calibration method is very convenient and works on a fairly large force range. Its accuracy is given by the averaging time used to measure the amplitude of the Brownian fluctuations $\langle x^2 \rangle$, *its relative accuracy is given by* $(\tau_x / t_{avg})^{1/2}$. At low forces, τ_x reaches a few seconds thus one might need to average a few hundred seconds to get a 10% accuracy. Achieving higher accuracy rapidly becomes time consuming. A calibration consists in a series of typically ten acquisitions at different magnets positions. As the magnetic force varies exponentially with their distance to the sample, choosing to record data with the magnet positions increasing by the same amount δz_m ($\sim b$) at each step gives good results. There exist two difficulties in the method:

1) measuring the absolute molecule extension l is not straightforward, more precisely it requires to determine at which z position the bead is in contact with the glass substrate which can be masked by the bead z fluctuations at low force when the molecule is dsDNA (see Fig. 8).
2) at high force τ_x becomes small, the frequency bandwidth of the fluctuations increases and may become larger than the frequency bandwidth of the camera which acts as a low-pass filter reducing the strength of the fluctuations $\langle x^2 \rangle$, leading to an overestimation of the force. A simple procedure to check that this is not occurring consists in measuring τ_x and ensuring that it remains significantly bigger than the acquisition time of the camera.

The reader can find details of corrective procedures to ensure accurate measurements in several publications (Wong & Halvorsen, 2006, 2009) See Fig. 6 describing the principle of this correction. Finally, it is a good practice to perform an absolute force calibration on a double-stranded DNA (dsDNA) molecule with a contour length L_0 significantly larger than the size of the magnetic bead. In these conditions, the Brownian motion in x and y are nearly equal in strength, allowing checking that the measurement is fine, but also

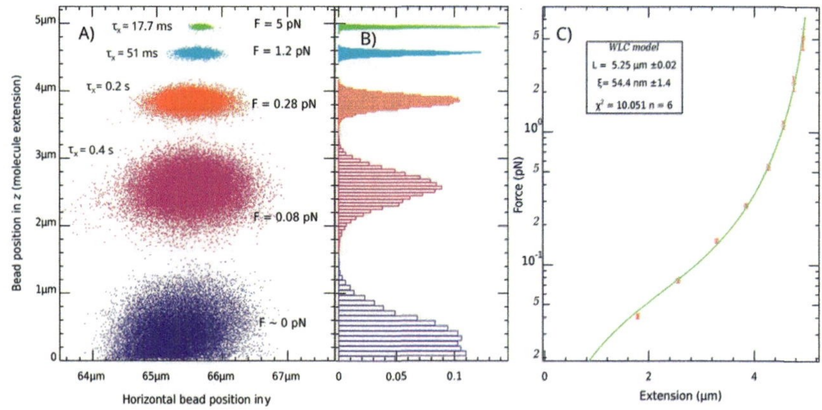

Fig. 8 Illustration of the Brownian fluctuations of the bead attached to à 5.25 µm dsDNA molecule at different forces. (A) Fluctuations in *(y,z)* for each cloud of points, we give the corresponding force on the right and the characteristic time $\tau_x(F)$ of the fluctuations on the left. The acquisition time of each cloud is chosen to be more than a hundred times larger than its characteristic time $\tau_x(F)$. (B) Histogram of fluctuations in *z*. (C) Force curve obtained for this molecule. Notice that the fluctuations are very anisotropic at high forces where extension is nearly constant. At $F = 0.08\ pN$, the molecule is at mid extension and the fluctuations in *z* are symmetrical, when force is further decreased ($F \approx 0$) the fluctuations are not anymore symmetrical because the bead excursions are limited downwards by the glass surface. In this latter situation, the mean extension of the molecule is not meaningful anymore, and this data point is not used to draw the force extension curve. (This last data point also illustrates that measuring the molecule zero of extension is not obvious for dsDNA, here if one chooses $< z(F \approx = 0) >$ we find 0.47 µm which is incorrect. The zero of extension is better described by the *z* value at which the histogram of the bead z position drops to 0.).

because the elasticity of a dsDNA molecule is very well fitted by the Worm Like Chain model which depends upon only two parameters: L_0 (which should be known if you prepare the molecule) and ξ the persistence length of DNA which is equal typically to 50 nm.

As the position of the magnets does not have to be extremely precise (compared to the typical length scale of the magnets) such a calibration does not need to be performed very often.

1.5.3 Force calibration through magnet characterization

The absolute calibration method just described is great but is time consuming: from 10 min to more than half an hour. Moreover, DNA or RNA substrates may not be convenient to follow the absolute calibration procedure. For instance, we often use a substrate made of a DNA hairpin, which has a zero-extension $l = 0$ as long as the force applied to it is smaller than a threshold value. An alternative calibration procedure is done then.

The idea is that the magnets exert a magnetic field gradient well defined versus the distance of the magnets to the bead z_{mag}. The force is equal to the product of this gradient by the bead magnetization which does not vary too much from beads to beads (typically 20% with MyOne beads). Thus if you have performed an absolute calibration with a dsDNA molecule and a specific bead, you have obtained a curve $F = F_0.f(z_{mag})$, which can be used to predict the force in future experiments assuming that the beads have the same magnetization. A good description of this behavior is given by: $F = F_0.exp(-z_{mag}/z_0 + c.\ z_{mag}^2)$, where F_0 is the maximum force exerted by the magnets on a bead if it was placed in the plane in contact with the magnet $(Z_{mag} = 0)$, z_0 is the characteristic decreasing length of the magnet (z_0 in the range of 300 μm), the $c.\ z_{mag}^2$ term is a small correction taking care of the fact that the exponential function is not a perfect description of the magnetic field gradient evolution with z_{mag} and that the bead magnetization is not anymore saturated at low fields. When the parameters F_0, z_0 and c have been determined, one can predict what we call an "estimated force" just by knowing z_{mag}. To be more accurate, one can adjust the force formula for each bead since their difference in magnetization can be taken in account by a correction factor b_i close to 1 leading to $F_i = b_i.F_0.exp(-z_{mag}/z_0 + c.\ z_{mag}^2)$. The values b_i can be determined by either performing a single point force measurement F_i using the Brownian calibration technique at a z_{mag} leading to a force of 1 or 2 pN where averaging is fast (we get $b_i = F_i/F_{estimated}$) or by using specific nucleic acid property like the opening force of a hairpin structure which is typically 15 pN.

1.6 Imposing a torsional constrain to the molecule

With the magnets configuration that we have described (Fig. 8), at a given a z_{mag} position, the magnetic field is horizontal while the force pulling the molecule is vertical. If we rotate the magnets around the z-axis, the force remains constant but the magnetic field follows the angular position. Although the magnetic beads are superparamagnetic, their angular position does follow the magnetic field rotation. So, a rotation is impinged to the molecule tethering the bead to the glass surface. Now, depending on the molecule type, this rotation will or will not have an effect. If a single molecule of single-stranded DNA (ssDNA) tethered the bead to the glass, this molecule is completely insensitive to the imposed rotation since ssDNA involves single atomic bound in the DNA polymer which acts as a swivel. If two single molecules tether the bead to the surface, then the torsion angle will entangle the two molecules around each other, involving a torsion elasticity and a change in the structure of the arrangement.

Double-stranded DNA consists of two complementary single stranded molecules, if both strands are attached to the bead and the glass substrate, the torsion angle will this time try to modify the pitch of the DNA and the elastic behavior of the molecule will be affected. Note that it is not enough that both ends of both strands are actually linked to the bead and the surface but the dsDNA molecule must have no nick (a spot where one strand is broken) to react to torsion. Twisting a dsDNA molecule can only be done if the molecule has no nick and both strands are attached to each surface (Strick et al., 1996). Nicks are common in dsDNA molecules outside the cell, the hybridization of the two strands before and after a nick maintain the dsDNA structure efficiently.

Finally, if two dsDNA molecules tether a bead to a surface, this arrangement will entangle the two molecules when a rotation is applied to the bead, as a result, the extension of the molecule will decrease with increasing rotation (Charvin, Bensimon, & Croquette, 2003). Twisting DNA turns out to be a very interesting feature of magnetic tweezers which has offered the ability to understand better the regulation of the DNA supercoiling which is fundamental in biology as we shall see (Strick, Allemand, Croquette, & Bensimon, 2000a; Charvin et al., 2003).

1.7 Being absolutely sure that you work on a single molecule

In micromanipulation it is fundamental to know that you are indeed dealing with a single molecule tethering the bead to the surface. However, this molecule is invisible and although in the way you prepare the sample you statistically favor the attachment of a single molecule, there exists exceptions and it is better to be absolutely sure that you are indeed working with only one molecule. If you are attaching a dsDNA molecule, the rotation test provides an extremely simple way to check that the bead you are considering is attached by a single molecule. You choose to impose an estimated force in the range of 1 pN and you rotate the magnets to modify the number of turns by 10% clockwise while you monitor the molecule extension. If the extension remains unchanged then you have a single molecule tethered! The test indeed tells you that either the molecule is not bound by both strands at both extremities or that it contains at least one nick.

In principle you know if you have prepared a DNA molecule with binding on both strands at both ends. If this was the case, this means you are interested in twisting dsDNA and you perform this test now in the hope to find a molecule sensitive to torsion that is which is bound properly and has no nick. Finding a bead system that is sensitive to torsion that is which extension varies with the torsion applied is interesting but does not guarantee that you are dealing with a single molecule, you are may be entangling two nicked molecules attached at both ends to the bead and the surface. You will need further testing to determine the tether nature (Charvin, Vologodskii, Bensimon, & Croquette, 2005) see Fig. 14.

If your substrate is a DNA hairpin, the molecule is inherently in ssDNA and should be insensitive to torsion. The hairpin will remain closed until the force reach 15 pN, above that force, the stem part of the hairpin will rapidly open leading to an abrupt elongation. If you have two hairpins bound to the bead, the threshold force to observe this opening transition will double, appearing at 30 pN. Alternating force cycles between 20 and 5 pN, will abruptly open and close a single hairpin leading to a very simple signal to observe and ensuring that the tether is a single molecule.

Finally, and ultimately, measuring the force extension of the molecule like in the force calibration process can be used to check the result.

1.8 What is easy and difficult to measure with magnetic tweezers

1.8.1 A limited range of z measurement

As we have explained, all the measurements done in MT are made through the x,y and z coordinates of the bead of interest which benefit from a very good resolution owing to the accuracy of video-microscopy. The range of the measurement is excellent in x and y covering the entire field of view, the range in z is more limited to typically 10 μm and requires that the user keep the focus position of the microscope in such a way as the bead diffraction image stands within the z range of the calibration image. This later restriction is not difficult for a single bead but can be more challenging when one studies a large number of beads and thus of molecules in parallel. In practice this means that the length of the molecule substrate should not exceed 5 μm so that it is possible to measure simultaneously a fully stretched molecule and a molecule with zero extension.

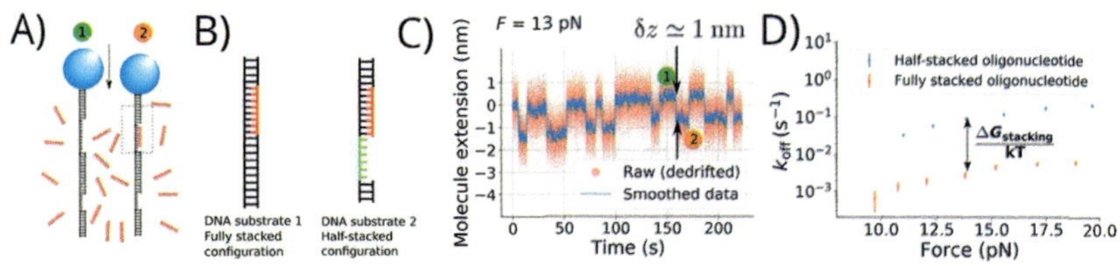

Fig. 9 Very high accuracy measurement of the hybridization of a short oligonucleotide of 8 nt and of its de-hybridization leading to changes of extension of ~1 nm steps. (A) A double-stranded molecule incorporates a small gap where an oligonucleotide in red can hybridize (B) with stacking (substrate 1) or without stacking (substrate 2, the green nucleotides remaining single-stranded). (C) The alternating hybridization followed by the departure of the oligonucleotide produce a telegraphic signal observed with a high accuracy SDI tracking technique described later. (D) Using the statistics of these events it is possible to measure the kinetic parameters of the reaction (Rieu et al., 2021a).

1.8.2 A good z accuracy to measure rapid changes

Measuring abrupt change in the molecule's extension is advantageous. For instance, detecting the transition from a closed to an open of a small stem-loop structure under force is normally an easy task to achieve given that the accuracy of z measurement equals a few nanometers. Detecting changes in extension of 10 nm is quite simple. However, it becomes more challenging if this change occurs only once, and the user does not have any means to repeat the event by decreasing and then increasing the apply force. While you may have designed your molecule to bind to the glass surface by a specific entity, non-specific binding may attach a part of the molecule, causing it to yield at some point and resulting in an abrupt extension that looks exactly like what you expect. Repetitive events involving very small changes are illustrated in Fig. 9.

1.8.3 Absolute extension measures face various issues

Measuring the absolute extension of a dsDNA molecule is also not an easy task although it is necessary to measure a force accurately. Measuring the z coordinate of the molecule when a strong force stretches it is simple but measuring the z coordinate corresponding to the situation of no extension is problematic. If we decrease the magnetic force to zero by moving magnets far from the bead, the bead is gently pulled by the elasticity of dsDNA which appears to be very weak compared with the strength of the Brownian fluctuations. As a result, the bead fluctuates above the glass surface with vertical upwards excursions that may reach one or two microns. Averaging the z coordinate of the bead does not lead to the zero-extension position which is better described by the minimum z position explore by the bead under no force (see Fig. 8). The second cause of error comes from the slow drifts of the microscope, since the full extension and the zero-extension z position must be measured one after the other, the drift occurring between the two measurements will alter the difference between the two measurements. In short, measuring a molecule absolute extension is always associated with an error which relative importance is stronger if the molecule is short. This explains why you should make your force calibration with a relatively long molecule (typically 5 μm). Another strategy used to deal with drift is to perform a differential tracking between two beads: one stuck on the glass surface with the other being the bead of interest. This reduces significantly drifts but do not cancel them completely. It is definitely worth it and requires that some care is taken to have beads really immobile on the glass, several groups have proposed recipes to get these beads (Dulin et al., 2015). Without such a preparation, we often get stuck beads in a sample which can be used but they can display some anomalous z fluctuations (because they are not completely stuck) which degrade the process. Using this strategy implies to track beads with distant positions in z which are compatible with the calibration image range meaning that the molecule contour length is quite smaller than 10 μm.

1.8.4 Anisotropy in the attachment of the molecule

Finally, locating the attachment point of the molecule on the bead and its response to the magnetic field is not straightforward. With the magnetic field horizontal and the field gradient and force vertical, the configuration turns out to provide strong force but the attachment point is usually off-centered. The reason is not obvious: the magnetic beads available are said to be superparamagnetic and one could even imagine that the magnetization of the bead $\vec{M}$ Should follow smoothly the direction of the imposed magnetic field $\vec{B}$, in this idea rotating the magnets would not rotate the bead. This is not the case, the iron oxide of the bead is paramagnetic but the material is anisotropic and the ensemble of oxide particles does present a strong anisotropy. As a result, the bead orients itself in a magnetic field and will follow the field direction with a strong stiffness. The torque required to impose a small angle θ between the magnetization and the field is too large and the bead direction is completely dictated by the magnetic field, a complete computation is given in (Mosconi et al., 2011). In Fig. 10, we illustrate the phenomenon: in A) fluctuations in x occur with the bead

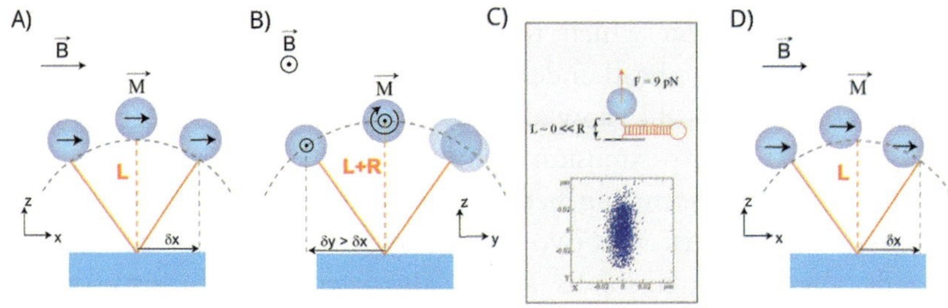

Fig. 10 With the magnetic field horizontal, the bead has a different behavior when it fluctuates along (as in A) or perpendicular (as in B) the direction of the magnetic field and this effect becomes very large for short tethers. The strength of the magnetic field imposes to the bead to follow its direction, thus when the fluctuations occur along *x*, the bead moves in a translation motion describing a circle of radius *l* (the molecule extension). But when the fluctuations occur perpendicular to the magnetic field, the bead can rotate around the magnetic field and the pendulum now has a radius of $l+r$ with *r* the bead radius. The pendulum has different stiffness in *x* and *y*: $k_x = F/l$ while $k_y = F/(l+r)$, which also means that the fluctuations are anisotropic. When calibrating the force with molecules such that *l* is greater than *r*, using k_x is interesting because it depends on *l* only (we do know the bead radius *r* but it also comes with an error). For very short molecule $l << r$, measuring force becomes an issue: k_x is very large and fluctuations are too fast for the camera. But k_y provides an interesting alternative $k_y \approx F/r$, the fluctuations are mostly in one direction (C) and k_y may be used to estimate forces. Another consequence of the anisotropy is that the anchoring point may be off centered regarding the bead bottom pole (D), this can be seen by the fact that this off centering is seen when the magnets are rotated (provide the field gradient is really vertical). As a consequence, when the bead bottom pole touch the glass surface, the molecule extension is not equals to zero! This adds one more difficulty to measuring precisely the absolute molecule extension.

doing a translation motion along a circle of radius *l*, the length of the molecule. In B) The fluctuations along *y* corresponds to a rotation along a circle of radius $l+r$, the stiffness of the bead/molecule pendulum is not the same in the two directions. An example of a very anisotropic fluctuation is given by a hairpin tethering a bead, basically $l = 0$ and fluctuations occur nearly only in y with the bead moving on a circle of radius r. In most cases, the point of attachment of the bead corresponds to the situation described in D) that is off-center compared with the bead poles, this is easily seen by observing the off-centering of the bead when the magnetic field is rotated. This further complicates the determination of the molecule extension *l*. Seidel's group have looked very carefully on the consequences of this off-centering (Klaue & Seidel, 2009).

1.8.5 Achieving an absolute force measurement with accuracy better than 10% is challenging

Tracking Brownian fluctuations allows measuring the stiffness of the bead/molecule inverted pendulum $k_x = k_BT/<x^2>$ easily. The accuracy just depends on the calibration of the x displacement and of the averaging time of the experiment. To obtain a 10% accuracy is easy but increasing this accuracy rapidly leads to very long averaging time: 5% requires to increase this averaging by a factor 4 while 1% leads to an increase by a factor 100. As a result, the absolute value of the force is typically measured with the accuracy of 10% over a wide range of force.

1.8.6 Measuring extension changes related to torsion is simple, measuring torque is hard

Measuring molecules' extension while torsion is applied is also very easy considering the measurement, the only possible trouble is that while rotating the magnets, free beads resulting from other molecule breakages who have migrated to the top of the flow chamber, may be dragged by the rotating magnetic field regrouping and obscuring the image quality. A solution to reduce this effect is to rotate magnets at low force.

When we apply a force to a molecule it responds by elongating, the equivalent in torsion is to apply a torque and observing the angular response to this constrain. However, when magnetic tweezers act as force clamp in the case of force/extension constrain, they do not act at all as torque clamp for torsion they act as rotation angle clamp. That is when you rotate the magnets, the direction of the magnetic field follows exactly this rotation and the angular position of the bead also follow exactly the impose angle. The reason behind is that the angular stiffness of the magnetic tweezers is extremely high compared with that of the molecule torsion stiffness. In other words, the torque that the molecule applies to resist the torsion constrain applied is very small and has basically no measurable effect. Torque at the molecular scale is more difficult to measure than force. The reason behind this situation is that the handle that we use, the bead, is well adapted for force measurement but not at all for torque (Gao, Hong, Ye, Inman, & Wang, 2021).

It is simple to understand if we remember that the bead by its viscous drag is the element responsible for the noise in the measurement in both cases through the magnitude of the Langevin force or torque. As we have said $f_L^2 x,y,z = 24\ \pi\eta r\ k_BT\Delta f$, thus $f_L = 20\ fN/\sqrt{Hz}$ for a bead *of* 0.5 μm, when a 1 pN is applied the Langevin noise in force is 50 times smaller than the force signal. For torque we have $\Gamma_L^2 = 32\ \pi\eta r^3\ k_B T \Delta f$, thus $\Gamma_L = 7$ pN. $nm/\sqrt{\text{Hz}}$ whereas the typical torque on a biological molecule equals 10 pN.nm. This means that the noise and the signal are just equivalent and the signal-to-noise drops to 1. Measuring torque is far more difficult than measuring force due to the size of the bead which is huge compared with the molecule diameter.

1.8.7 Detecting subnanometer extension changes is hard

In the best situation, a change of extension of one nanometer corresponds to the step of a molecular motor of one base, observing helicase or polymerase activity with such a resolution is, of course, very interesting. Optical tweezers experiments (Greenleaf & Block, 2006) and now nanopore experiments (Craig et al., 2017) have opened the way in this direction. Although the optical z accuracy on beads stuck to the glass substrate easily reaches subnanometer resolution, experiments done in real biological conditions just are less convincing. Various reasons explain the issue: first all biological experiments involve DNA or RNA substrate which as a polymer display a finite stiffness, the bead is thus attached to a spring of stiffness k_z. The fluctuations occurring in z are thus related to the Langevin fluctuating force arising from the viscous dissipation of the bead divided by k_z. To observe very small z fluctuations it is important that k_z is very large leading to a very short molecule and that the viscous dissipation is minimal that is the bead radius as small as possible. But both parameters have serious limitations: in most biological situations, the substrate contains some single-stranded part which has very limited stiffness k_z and as the molecule is very short the bead must be close to the glass surface and its dissipation tends to diverge in this situation. Both effects combine to produce larger fluctuations. One solution proposed to this issue is to use very stiff extender molecules made of DNA origami to move the bead further away from the glass substrate (Kauert et al., 2011; Nickels et al., 2016).

1.9 Single-molecule experiment without torsional constrain

1.9.1 Elasticity of DNA and polymers

One of the first experiments measuring DNA elasticity was done in Bustamante's group combining hydrodynamics and magnetic forces to deduce the elasticity of a dimer of λ-DNA molecules (Smith, Finzi, & Bustamante, 1992). This experiment provided a very nice curve of the elongation versus force which demonstrated that DNA elasticity was not described by the Free Joint Model (FJC). The force calibration used by the authors was rather complex and not easy to reproduce. Very quickly, new experiments confirmed the results of this experiment and Marko and Siggia (Bustamante, Marko, Siggia, Smith 1994) explained that the good polymer model to describe dsDNA elasticity was not the FJC model but rather the Worm Like Chain (WLC) model which indeed coincided perfectly with the experimental results with two adjustable parameters L_0 the contour length of the molecule which is known and the persistence length ξ of dsDNA which equals typically to 50 nm and varies little with other parameters within classical biological conditions (salt conditions, sequence, temperature ...) (Bouchiat et al., 1999). This finding is fortunate since dsDNA elastic curve is used as a standard allowing validating single molecule protocol. As seen on Fig. 11, the agreement between the theory and the experiment is just impressive and spans a force range of three orders of magnitude in force. This property is not true for other biological polymers where measurements exist but no model as simple and efficient explain this result. A good example is the elasticity of ssDNA which depends on many parameters and is more difficult to predict.

1.9.2 Difference in extension between double-stranded and single-stranded DNA or RNA

Aside from measuring the elastic behavior of biological polymers, magnetic tweezers are used to record in real time the action of enzymes on NA. To detect such an activity the enzyme must alter the structure of the molecule leading to a detectable change in extension. In the majority of cases, the enzyme converts either a dsDNA to ssDNA or the reverse. Knowing the elastic behavior of ssDNA versus force will be useful to determine the magnitude of the signal measured when one nucleotide is proceeded by the enzyme (Fig. 12).

If we study a DNA polymerase, the enzyme converts ssDNA to dsDNA, and thus, conducting the experiment at 5 pN it would be a bad idea, as the conversion of ssDNA to dsDNA would not produce any change in extension. The sensitivity of such an experiment is not very large at most 0.1 or 0.15 nm per base, which is well below the accuracy of the magnetic tweezers. If one study the action of a helicase opening a hairpin, the sensitivity is far better since when the dsDNA is in the form of the stem of the hairpin, its extension is null. However, when the helicase opens this base pair, it is converted in a ssDNA based on each side that is 2 times the extension of a single-stranded base, typically 0.9 nm at 12 pN, which approaches the resolution of MT.

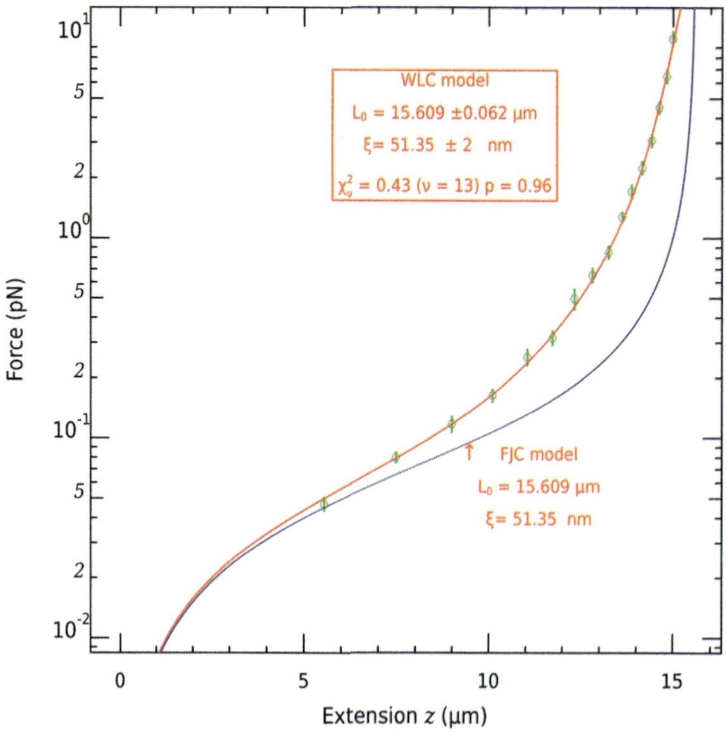

Fig. 11 Force versus extension curve of a λ-phage dsDNA measured using magnetic tweezers (diamond data points). This curve is very non-linear and the logarithmic representation is best adapted. These experimental points are fitted with a Worm Like Chain model which describes the data very well with a persistence length of 51 nm. Notice that a little less than 0.1 pN is needed to extend the molecule to half its contour length L_0. In blue the Free Jointed Chain model with the same parameter is drawn, the models differ significantly at high forces.

1.10 Single-molecule experiment involving torsional constrains

In our cells the torsion constrains of our DNA is controlled very accurately by full set of enzymes (mostly topoisomerases). In bacteria the same thing is also true, but the constrain is even stronger since the genome forms a circle avoiding the difficulty of replicating ends. One of the roles of torsion is to pack the long DNA molecule in a compact form so that it enters the small cell volume. If one applies a torsion constrain to a single DNA molecule under a constant stretching force F, first the double helix will elastically respond by increasing or decreasing its pitch depending on the fact that turns are added or removed. But this elastic response is very soon replaced

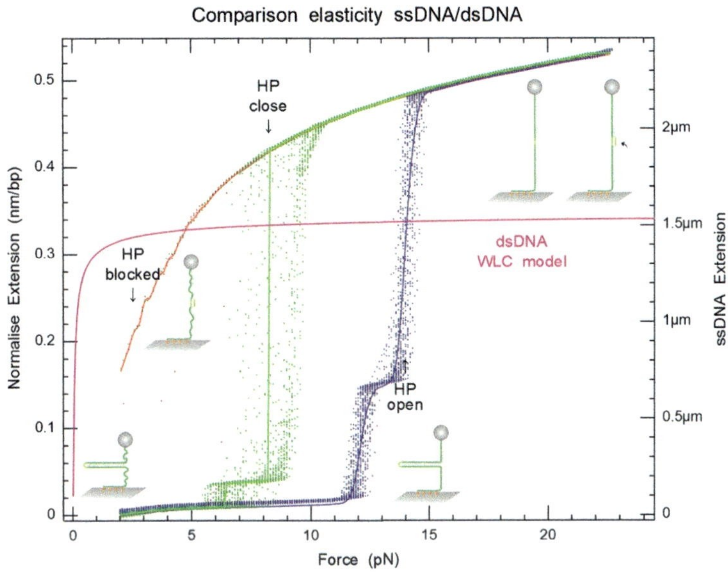

Fig. 12 we present here a typical experiment done with a hairpin DNA structure which gives access to ssDNA elasticity. In this experiment, a DNA hairpin tethers a magnetic bead and the glass substrate. When the force applied is gently increased starting at 2 pN, the molecule remains closed and its extension is very small. Upon reaching 12 and finally 15 pN, the force is sufficient to break the hydrogen bonds between the complementary bases and the hairpin opens. When the force is further increased the molecule now entirely in a ssDNA state lengthens slightly. When the force decreases, the same elastic behavior is observed until the force reaches 10 pN where the hairpin closes again with a very rapid shortening of the molecule extension. The hysteresis in force observed between the hairpin opening and closing is related to the nucleation process of the hairpin refolding. Indeed, this process starts at the apex of the hairpin and implies that a transient closing of a few bases occurs due to strong fluctuations. The force applied to the molecule hinders these fluctuations and needs to be reduced for the nucleation to happen. A way to demonstrate this is to introduce in the flow chamber a small oligonucleotide with a sequence complementary to that of the loop and of a few neighbor bases. When the hybridization occurs, the nucleation is further hindered since the nucleation initiation now requires a very strong fluctuation to bypass the hybridized region. As a result the molecule remains in ssDNA state for weaker forces, this is what happens for points of the red curve between forces 7 to 2 pN. To achieve this experiment, a series of cycles with increasing and decreasing forces are recorded since fluctuations will make variable molecule state. We have drawn in magenta the elastic behavior of dsDNA on the same graph. One can see that the extension of ssDNA is smaller than that of dsDNA at low force but the opposite occurs at high forces. ssDNA and dsDNA having the same extension around 5 pN.

by the formation of interwound structures "plectonemes" which reduces the extension of the molecule (Strick, Allemand, Bensimon, & Croquette, 1998).

As can be seen on the Fig. 13, at small forces (<0.5 pN) the change of extension is symmetrical but at higher forces the symmetry is broken and plectonemes are only seen for positive added turns. Finally, at higher force (F > 6 pN) the molecule does not seem to form anymore these interwound structures. The explanation is related to the fact that for a given stretching force, the plectonemes form only when a finite torque appears on the molecule (after sufficient turns have been applied) and they limit the value of this torque. Now the torque increases with the applied force and when too large leads to change in the DNA structure: when unwinding strongly DNA the molecule denatures and when it is highly overwound it forms a new structure (Allemand et al., 1998; Bryant et al., 2003).

Direct torque measurements are difficult but have been done (Lipfert, Kerssemakers, Jager, & Dekker, 2010; Lipfert, Klijnhout, & Dekker, 2010; Lipfert et al., 2011; Kauert et al., 2011) confirming the results of indirect torque measurements, which are easier (Mosconi, Allemand, Bensimon, & Croquette, 2009).

However, at forces below 0.5 pN, the linear, rapid with 50 nm/turn, decrease of extension of the twisted molecule provides a tremendous opportunity to study enzymes that alter the torsion constrain with extreme sensitivity (Strick, Croquette, & Bensimon, 2000b) (Fig. 14).

1.11 Improvements and adaptation of magnetic tweezers

The description made so far lists the pros and cons of classic MT. Contributors have worked either to improve various aspects or to adapt tweezers to a specific application, which we report here.

1.11.1 Other force clamps micromanipulation tools

The principle of applying a constant force for a given set of control parameters has led alternatives to magnetic tweezers. The simplest one is given by gravity, if beads are heavier than water (made of gold for instance) their weight is constant and pulls down the molecule tethering the bead with a force in the pN range for a 1-μm-diameter bead. This is a simple design with a very constant force but lacking the possibility of adjusting the force which is a very serious issue. Alternatives exist: using the pressure force exerted by ultrasonic waves is one possibility (Sitters et al., 2015; Wang et al., 2023). The second one is to use the centrifuge force which can be adjusted

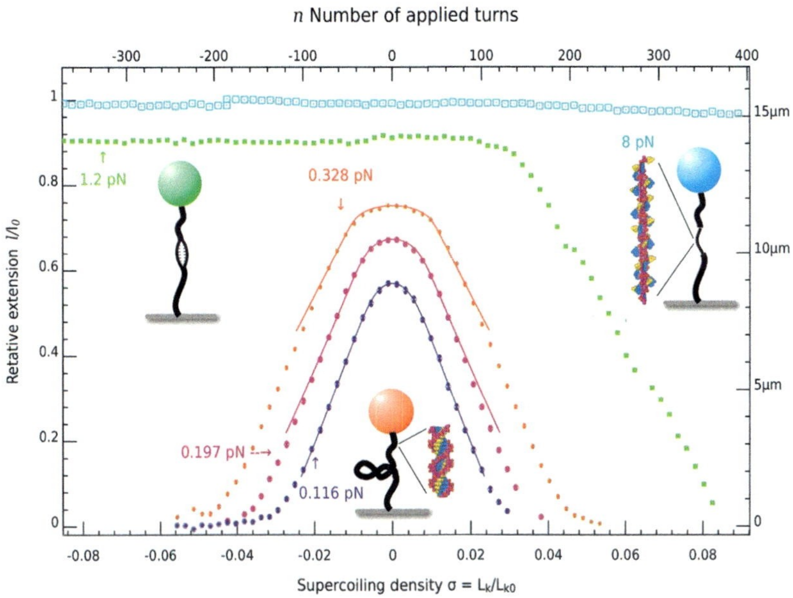

Fig. 13 Measurements of a dsDNA molecule extension versus the number of turns applied to the molecule at different stretching forces. The molecule is from a λ-phage corresponding to more than 48,000 bases and 15.6 μm which is found when no turn is applied. At low forces (F < 0.5 pN) whenever turns are added or removed from the no constrains state, the molecule first accommodates the torsion by slightly altering the pitch of DNA building up a torque in response. Rapidly this leads to a buckling instability where interwound structures (plectonemes) nucleate and shorten the extension. At low force, the phenomenon is symmetrical. At higher forces, the torque reaches a sufficient level for negative constrains first to locally separates the DNA strand building a denaturation bubble which explains why the extension does not shrink on the negative side for F > 0.5 pN. For F > 6 pN, the positive torque is sufficient to switch DNA in a new form with an overtwisted structure scP-DNA Allemand, Bensimon, Lavery, and Croquette (1998).

(Halvorsen & Wong, 2010). Third using hydrodynamic flow is also a means to apply a force to the molecule. We shall not describe these tools here but we shall just explain that these techniques are well suited for a large parallelism even larger than magnetic tweezers (which are limited by the gap size between magnets). The limitations for acoustic tweezers is that the acoustic field required to exert significant forces on beads is relatively strong and thus need a substantial level of power which has a tendency to heat up the sample. For centrifuge forces, all the experiments its camera and the buffer injection system need to be installed in a rotating stage which is challenging and reduces the accessibility of the experiment. Stretching with a hydrodynamic

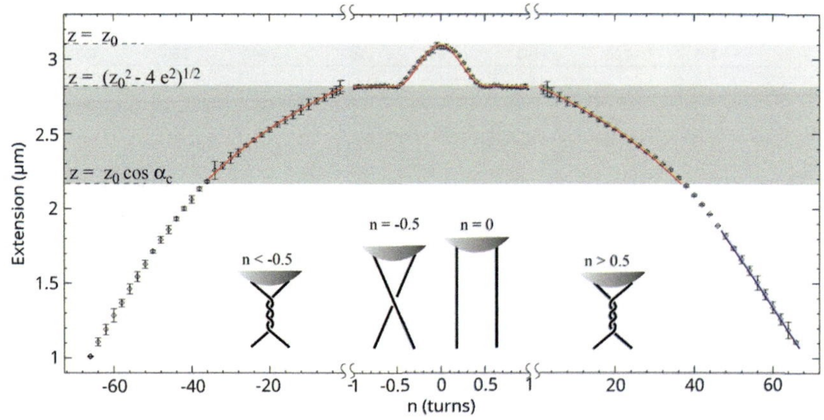

Fig. 14 Extension versus braiding for two DNA molecules; The diamond symbols correspond to experimental points, obtained at F= 2pN in 100 mM Phosphate buffer. We can distinguish two different ranges of braiding: when $|n| < 0.5$ (light gray shading), the two molecules evolve from the parallel situation without crossing to the crossing one. From the change of extension between the two situations, it is possible to deduce that the contour length of the molecule is 3.10 µm and the distance between their attachment point on the bead is $2e = 1.28$ µm. When $|n| > 0.5$ a real braiding situation occurs (heavy gray shading). A fit of the experimental points can be done with a model where the two molecules wind helically around a cylinder with a radius of 3.9 nm. In the white zone below the dark gray zone, the braiding buckles leading to plectonemes of braids exhibiting a linear decay of extension with a slope of 39 nm (Charvin et al., 2005).

flow has been widely used too, it offers a relative ease of use in a parallel system where many beads can be followed simultaneously. The difficulty here is that near the glass plate, the hydrodynamic flow decreases linearly as one approach the glass plate and the force is not constant for all beads and not very easy to calibrate (Duderstadt et al., 2016; Hill, Monachino, & Van Oijen, 2017; Graham, Marians, & Kowalczykowski, 2017; Zou et al., 2023).

In the case of Duderstadt et al. the parallelism allows reaching few 100,000's beads but the spatial resolution is more limited.

1.11.2 Magnets designs

The group of N. Dekker has done a very complete work computing and testing different magnet configurations, this work constitutes a solid base for any work on MT (Lipfert et al., 2009). Depending on the use magnets can be adapted to produce the maximum possible force (Zacchia & Valentine, 2015) or to deliver a homogeneous force on a large field of view to increase parallelism.

Electromagnets have been also used for micromanipulation. The interesting feature of electromagnets is the ability to change very quickly the driving current and thus to modulate the applied force. In the design of (Gosse & Croquette, 2002) six coils are used to actually feedback the change of the bead position compared with a reference one thus transforming the force trap into a position trap with the electromagnet current giving access to the force applied. In a completely different design, Fernandez's group takes advantage of the very narrow gap of a tape head recorder to obtain extremely strong magnetic field gradients and thus high forces that can be switched on and off very quickly (Tapia-Rojo et al., 2019).

The drawback of electromagnets is the fact that the current flowing through them provides heat which may introduce a source of drift in the system. Another point to consider is the fact that they involve magnetic material in their pole pieces which usually exhibits hysteresis with the driving current. Very low hysteresis material should be used, or a strategy to cancel them must be adopted.

1.11.3 Adapting magnetic tweezers to small molecules

We have explained that measuring the absolute extension of a molecule is not very easy in particular when the molecule is short. However, short molecules are interesting and efforts have been developed to circumvent the difficulty (Daldrop, Brutzer, Huhle, Kauert, & Seidel, 2015; Chen et al., 2011; Shon, Rah, & Yoon, 2019).

1.11.4 Horizontal magnetic tweezers

Although we have explained the feasibility of measuring a bead position in the z-direction of a microscope, measuring x and y is by far simpler. Different groups have worked at pulling the bead horizontally so that they can take advantage of this measuring facility. Very interesting results have been found with this lay out, which is very convenient for measuring but is less efficient in terms of parallelism: attaching the bead pulled horizontally requires a special attachment process which limits the number of molecules studied simultaneously (Zhanga, Chenb, Fub, Doylea, & Yana, 2012; Gaire et al., 2021; (Fabian, Tyson, Tuma, Pegg, & Sarkar, 2023); (McAndrew et al., 2016).

1.11.5 High speed acquisition

Compared with Optical Tweezers, the data acquisition of MT has long been limited by the frame rate of the video camera used to visualize beads.

This is no longer the case with at least two possibilities: using extremely fast camera dumping the frames directly in memory which allows tracking at 10 kHz as illustrated in (Otto et al., 2010; Lansdorp, Tabrizi, Dittmore, & Saleh, 2013; Huhle et al., 2015), one difficulty is that data must be post-treated and acquisition is rather limited in time. The second means is offered by C-MOS cameras which allows the user to define a small region of the full image as a Region Of Interest that is read at maximum speed. This latter solution offers some interesting features: the user can first observe the full field of view to determine which are the beads of interest, and then switch to a small ROI centered on such a bead to acquire data at 1 kHz or more with a continuous flow of data treated in real time. This method was used in (Rieu et al., 2021b) to observe rapid fluctuations between the open and the close state of a small hairpin.

1.11.6 Torque measurements with MT

We have stressed the very bad signal-to-noise issue of the torque measurement but it is not the only problem of torque measurements. We have explained that in the classical magnetic tweezers with the magnetic field horizontal and the field gradient and force vertical, the configuration turns out to be completely useless to measure torque directly. The reason being that the torsional stiffness is far too large, a detailed computation of this effect is given in (Mosconi et al., 2011).

Kinosita (Harada, Ohara, & Takatsuki, 2001) and later Dekker groups have proposed to use a different magnet configuration to circumvent this issue. Instead of having a magnetic field horizontal, they propose to use a vertical magnetic field aligned with the z-axis and thus with the molecule, the magnetic field locks the bead to have its magnetization along z but permit the rotation of the bead around the z-axis. If one has previously twisted the DNA molecule by a significant number of turns, placing the field vertically allows the bead to respond to the torque applied by the molecule leading to a slow rotation of the bead to relax the constrain (Lipfert et al., 2010, 2011; Lipfert Klijnhout et al., 2010; Kauert et al., 2011; Lipfert, Lee, Ordu, Kerssemakers, & Dekker, 2014).

Taking advantage of this possibility Dekker's group added a small horizontal magnetic field which can rotate around the z-axis. This small field induces a slight tilt in the bead magnetization which can be used to wind or unwind the molecule as could be done in the original tweezers. Keeping the horizontal magnetic field small leads to a weak angular stiffness of the rotation trap and it is then possible to detect the angle between the bead and the horizontal field component leading to a measure of the torque (Janssen et al., 2012; Van Oene et al., 2018).

1.11.7 Improved resolution of bead tracking

As we have explained, the accuracy of the measurement of the bead position is in the nanometer range (Brutzer, Schwarz, & Seidel, 2012; Dulin et al., 2015) which although impressive is close to the displacement of an enzyme stepping on one base but still larger. Improving this resolution is a most demanded challenge where the researcher imagination is at work. In the classical rings of interference used to measure z, the interference occurs between the parallel illumination beam and the light diffused by the bead with an angle which is in the range of a few tens of degrees. As a result, the interference pattern has a gentle modulation with z. In Saleh's group, the idea was proposed to build an interference between light reflected by the glass coverslip and the surface of the bead covered by a metallic coating (Kim & Saleh, 2009). This indeed made the interference pattern far more sensitive to any z modulation which improves the sensitivity. However, the burden of adding a metallic coating to the bead and the complexity of the optics make the solution difficult to use routinely. Recently we propose to filter the light diffused by beads in the Fourier plane so as to keep light diffused with a large angle and to structure the bead image by a rapid interference (Rieu et al., 2021a). This method indeed enhances the accuracy of the z measurement with minimal constrains: it requires a strong illumination and leads to a bead optical pattern less obvious for the user (Fig. 15).

1.11.8 Spectral analysis of Brownian fluctuations

Measuring the stiffness of the bead/molecule just requires to measure the bead fluctuations $<x^2>$, but when the force increases the characteristic time τ_x of these fluctuations decreases and becomes comparable to the camera acquisition time. This corresponds to a filtering of the data which will lead to an overestimation of this stiffness and of the force if no special care is taken. The right way to proceed is to compute this signal in Fourier space which offers many interesting features: this allows considering the intrinsic low-pass filtering of the camera, to compute the typical error on the measurement. More details may be found in (Te Velthuis, Kerssemakers, Lipfert, & Dekker, 2010; Van Loenhout, Kerssemakers, De Vlaminck, & Dekker, 2012; Dulin et al., 2014a; Daldrop et al., 2015; Morgan & Saleh, 2021; Ostrofet, Papini, & Dulin, 2018).

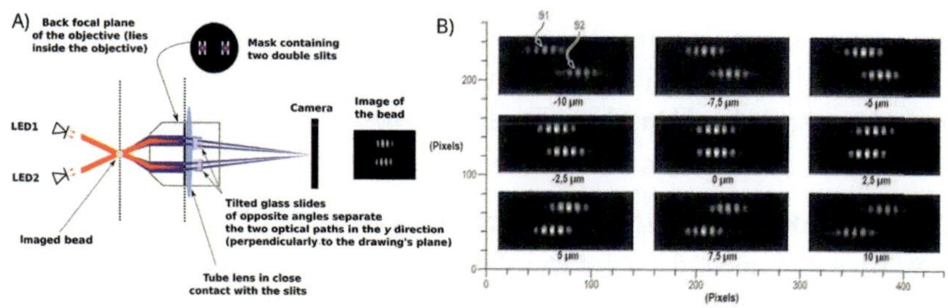

Fig. 15 Improving resolution of bead tracking by using Stereo Dark-field Interferometry (SDI). In (A) we give a sketch of the setup where the bead is illuminated by two parallel light beams tilted by θ and −θ out of the *z* axis (with $\theta \approx 30°$). An objective with a tube lens makes the image of the bead on a camera with a filtering done in the back focal plane of the objective. This filter selects the angle of the light which will form the image of the bead. The filter is made with the mask containing two pairs of slits which enabling diffused light on both sides of the illumination beams to pass through. The black part of the mask in between each pair of slits blocks both non-diffused illuminations beams leading to a dark field image. The light diffused by the bead passing through one pair of slits produces a pattern of interference fringes. Just after the mask, two small glass slides tilted with opposite angles offset the images of the fringe patterns so that they do not overlap on the camera. As a result, each bead image appears as a pair of interference fringes. When the bead moves in *z*, the two fringe patterns move in a shear manner as shown in (B). The phase displacement measured between the two patterns offers a very accurate *z* measurement because of the structured illumination pattern which spans both focus planes above and below the focus plane of the microscope. Another interesting feature is that this measurement technique does not require a calibration image procedure. The figure in (A) is a simplification of the real setup described in (Rieu, Valle-Orero, Ducos, Allemand, & Croquette, 2021b), the back focal plane of the objective is usually inside the objective and not accessible to place the mask. In the real setup, the back focal plane is imaged by a relay optical 4 f system involving two lenses which we have not represented here.

1.11.9 Coupling magnetic tweezers with fluorescence imaging

Magnetic tweezers only use light to track bead position, the intensity and the wavelength of the illumination may be chosen not to interfere with the illumination necessary to observe the fluorescence of single molecules, thus building an apparatus combining micromanipulation and single molecule fluorescence is relatively easy as done for instance in ((Madariaga-Marcos, Aldag, Kauert, & Seidel, 2024); Kemmerich, Swoboda, et al., 2016; Kemmerich, Kasaciunaite, & Seidel, 2016; Graves et al., 2015; Zou et al., 2023). Note that Aldag et al. (2023) used an interesting configuration for the coupling. One difficulty is that some magnetic beads are also fluorescent.

1.11.10 High-throughput MT

Experiments pulling on a single molecule are relatively simple and have a high success rate with just a few systems of molecules and bead. Characterizing DNA substrates and studying enzymes that produce repetitive signals, such as helicases, is already more complex. However, success can still be achieved with a few tens of the bead/molecule systems. Studying enzymes that produce irreversible changes like polymerases will limit further success rates and increasing further the parallelism of the single molecule approach is most desirable (De Vlaminck et al., 2011; Berghuis, Köber, Van Laar, & Dekker, 2016; Kuijpers, Van Laar, Janissen, & Dekker, 2022) More complex studies involving multicomponent complexes with different molecular motors and cofactors such as replisomes have a small chance to work on the molecule that you are considering. Even worse, to understand a molecular mechanism, we need sufficient statistics of events ranging from tens to hundreds, which means that you had better observe thousands of systems to see interesting events. Increasing the parallelism can be done with magnetic tweezers but this means reducing the magnification so as to observe a large field of view at the expense of the pointing accuracy. It also means that magnets cannot anymore have a small gap and the magnetic field gradient is far from optimum. Increasing the bead size compensates partially this deficiency (the force rises as r^3), it can also be associated with a hydrodynamic flow (Duderstadt et al., 2016; Hill et al., 2017).

2. Conclusion

Here we have described the classical magnetic tweezers, their limitations and give some insights on improvements which have been made by various groups. Magnetics tweezers appear well adapted for micromanipulation with high throughput owing mostly to their simplicity and stability: the permanent magnets are simple, they provide a very stable force while video-microscopy offers accurate measurements on many beads in parallel. To some extent it is not obvious that a micron size bead is the right tool to manipulate biological objects at the nanometer scale, this difference in size being quite large. As a matter of fact, the extra size of the bead is responsible for the noise in the measurement through the Langevin fluctuations related viscous dissipation occurring in translation or in rotation. While the bead size is not a real problem for forces, it becomes a real issue as soon as torque is involved. Reducing the size of the bead is definitely interesting to reduce noise but this leads to a drastic reduction of the force applied since it is proportional to the volume of the bead. Magnetic tweezers will probably improve even further to provide better signal-to-noise or higher throughput or versatility, we can give some possible suggestions of where these improvements may occur. Beads with stronger magnetization for a similar radius can be made but this is mostly the contribution of manufacturers. Simple and more efficient magnets are possible (Lansdorp, 2016), in his PhD shows an improvement by a factor of 5, which is very interesting. However, the device is not yet easy to use. Improving tracking resolution has been demonstrated by Stereo Dark-field Interferometry in (Rieu et al., 2021a)/ Again, the setup, in its present form, is less easy to implement than the classical setup, but this should be done in the near future. It is clear that cameras and faster CPUs,particularly GPUs, will increase further bandwidth and throughput of the measurement. The good news is that they will appear because these improvements are clearly awaited, not only for magnetic tweezers! Altogether, we can expect better magnetic tweezers in the future.

References

Abbe, E. (1873). Beitrage zur Theorie des Mikroskops und der mikroskopischen Wahrnehmung. *Archiv für Mikroskopische Anatomie, 9*, 413–420.

Aldag, P., Aldag, M., Rutkauskas, J., Madariaga-Marcos, I., Songailiene, T., Sinkunas, F., ... Seidel, R. (2023). Dynamic interplay between target search and recognition for a Type I CRISPR-Cas system. *Nature Communications, 14*(1), 2654.

Allemand, J. F., Bensimon, D., Lavery, R., & Croquette, V. (1998). Stretched and overwound DNA forms a Pauling-like structure with exposed bases. *Proceedings of the National Academy of Sciences of the United States of America, 24;95*(24), 14152–14157. https://doi.org/10.1073/pnas.95.24.14152. PMID: 9826669.

Berghuis, B. A., Köber, M., Van Laar, T., & Dekker, N. H. (2016). High-throughput, high-force probing of DNA-protein interactions with magnetic tweezers. *Methods (San Diego, Calif.), 105*, 90–98. https://doi.org/10.1016/j.ymeth.2016.03.025. PMID: 27038745.

Bouchiat, C., Wang, M. D., Allemand, J., Strick, T., Block, S. M., & Croquette, V. (1999). Estimating the persistence length of a worm-like chain molecule from force-extension measurements. *Biophysical Journal, 76*(1 Pt 1), 409–413. https://doi.org/10.1016/s0006-3495(99)77207-3. PMID: 9876152.

Brutzer, H., Schwarz, F. W., & Seidel, R. (2012). Scanning evanescent fields using a pointlike light source and a nanomechanical DNA gear. *Nano Letters, 12*(1), 473–478. https://doi.org/10.1021/nl203876w. PMID: 22148854.

Bryant, Z., Stone, M. D., Gore, J., Smith, S. B., Cozzarelli, N. R., Bustamante, C., et al. (2003). Structural transitions and elasticity from torque measurements on DNA. *Nature, 424*(6946), 338–341.

Bustamante, C., Marko, J. F., Siggia, E. D., & Smith, S. (1994). Entropic elasticity of lambda-phage DNA. *Science (New York, N. Y.), 265*, 1599–1601.

Bustamante, C. J., Chemla, Y. R., Liu, S., & Wang, M. D. (2021). Optical tweezers in single-molecule biophysics. *Nature Reviews Methods Primers, 1*(1), 1–29. https://doi.org/10.1038/s43586-021-00021-6.

Cavatore, E. (2011). *Cavatore, Microscopie optique appliquée à la micro-manipulation par pinces magnétiques à haute résolution et à la visualisation de nano-objets métalliques individuels* (Thèse). Sorbonne université, ⟨https://www.theses.fr/157360873⟩.

Charvin, G., Bensimon, D., & Croquette, V. (2003). Single-molecule study of DNA unlinking by eukaryotic and prokaryotic type-II topoisomerases. *PNAS, 100-17*, 9820–9825. https://doi.org/10.1073/pnas.1631550100.

Charvin, G., Vologodskii, A., Bensimon, D., & Croquette, V. (2005). Braiding DNA: Experiments, simulations, and models. *Biophysical Journal, 88*, 4124–4136.

Chen, H., Hongxia, F., Xiaoying, Z., Peiwen, C., Fumihiko, N., & Jie, Y. (2011). Improved high-force magnetic tweezers for stretching and refolding of proteins and short DNA. *Biophysical Journal, 100*, 517–523.

Choi, H.-K., Gyu Kim, H., Ju Shon, M., & Yoon, T.-Y (2022). High-resolution single-molecule magnetic tweezers. *Annual Review of Biochemistry, 91*, 33–59.

Cluzel, P., Lebrun, A., Heller, C., Lavery, R., Viovy, J.-L., Chatenay, D., & Caron, F. (1996). DNA: An extensible molecule. *Science (New York, N. Y.), 271*, 792–794.

Craig, J. M., Laszlo, A. H., Brinkerhoff, H., Derrington, I. M., Noakes, M. T., Nova, I., & Gundlach, C. J. H. (2017). Revealing dynamics of helicase translocation on single-stranded DNA using high-resolution nanopore tweezers. *PNAS, 114*(45), 11932–11937. https://doi.org/10.1073/pnas.1711282114.

Daldrop, P., Brutzer, H., Huhle, A., Kauert, D. J., & Seidel, R. (2015). Extending the range for force calibration in magnetic tweezers. *Biophysical Journal, 108*(10), 2550–2561. https://doi.org/10.1016/j.bpj.2015.04.011.

De Vlaminck, I., Henighan, T., Van Loenhout, M. T. J., Pfeiffer, I., Huijts, J., Kerssemakers, J. W. J., ... Dekker, C. (2011). Highly parallel magnetic tweezers by targeted DNA tethering. *Nano Letters, 11*(12), 5489–5493. https://doi.org/10.1021/nl203299e.

Duderstadt, K. E., Geertsema, H. J., Stratmann, S. A., Punter, C. M., Kulczyk, A. W., Richardson, C. C., & Van Oijen, A. M. (2016). Simultaneous real-time imaging of leading and lagging strand synthesis reveals the coordination dynamics of single replisomes. *Molecular Cell, 64*, 1035–1047. https://doi.org/10.1016/j.molcel.2016.10.028.

Dulin, D., Barland, S., Hachair, X., & Pedaci, F. (2014b). Efficient illumination for microsecond tracking microscopy. *PLoS One, 9*(9) e107335.

Dulin, D., Cui, T. J., Cnossen, J., Docter, M. W., Lipfert, J., & Dekker, N. H. (2015). High spatiotemporal-resolution magnetic tweezers: Calibration and applications for DNA dynamics. *Biophysical Journal, 109*(10), 2113–2125. https://doi.org/10.1016/j.bpj.2015.10.018. PMID: 26588570.

Dulin, Y. Z., Dulin, D., Cnossen, J., Köber, M., Van Oene, M. M., Ordu, O., ... Dekker, N. H. (2014a). A force calibration standard for magnetic tweezers. *The Review of Scientific Instruments, 85*(12), 123114. https://doi.org/10.1063/1.4904148. PMID: 25554279.

Dulin, D. (2024). An introduction to magnetic tweezers. In I. Heller, D. Dulin, E. J. G. Peterman (Eds.), *Single molecule analysis: Methods and protocols. Methods in molecular biology* (Chapter 18, Vol. 2694, pp. 375–401). https://doi.org/10.1007/978-1-0716-3377-9_18.

Fabian, R., Jr, Tyson, C., Tuma, P. L., Pegg, I., & Sarkar, A. (2023). A Horizontal Magnetic Tweezers and Its Use for Studying Single DNA Molecules. *Micromachines (Basel), 9*(4), 188. https://doi.org/10.3390/mi9040188. PMID: 30424121.

Gaire, F. R. Jr, Gaire, S., Tyson, C., Adhikari, R., Pegg, I., & Sarkar, A. (2021). A horizontal magnetic tweezers for studying single DNA molecules and DNA-binding proteins. *Molecules (Basel, Switzerland), 26*(16), 4781. https://doi.org/10.3390/molecules26164781. PMID: 34443369.

Gao, X., Hong, Y., Ye, F., Inman, J. T., & Wang, M. D. (2021). Torsional stiffness of extended and plectonemic DNA. *Physical Review Letters, 127*, 028101.

Gelles, J. G., Schnapp, B. J., & Sheetz, M. P. (1988). Tracking kinesin-driven movements with nanometre-scale precision. *Nature, 331*, 450–453.

Gosse, C., & Croquette, V. (2002). Magnetic tweezers: Micromanipulation and force measurement at the molecular level. *Biophysical Journal, 82*(6), 3314–3329. https://doi.org/10.1016/S0006-3495(02)75672-5. PMID: 12023254.

Graham, J. E., Marians, K. J., & Kowalczykowski, S. C. (2017). Independent and stochastic action of DNA polymerases in the replisome. *Cell, 169*, 1201–1213. https://doi.org/10.1016/j.cell.2017.05.041.

Graves, E., Duboc, C., Fan, J., Stransky, F., Leroux-Coyau, M., & Strick, T. R. (2015). A dynamic DNA-repair complex observed by correlative single-molecule nanomanipulation and fluorescence. *Nature Structural & Molecular Biology, 22*, 452–457.

Greenleaf, W. J., & Block, S. M. (2006). Single-molecule, motion-based DNA sequencing using RNA polymerase. *Science (New York, N. Y.), 313*(11), 801.

Gumpp, H., Stahl, S., Strackharn, M., Puchner, E., & Gaub, H. (2009). Ultrastable combined atomic force and total internal fluorescence microscope. *Review of Scientific Instruments, 80*(6), 063704.

Halvorsen, K., & Wong, W. P. (2010). Massively parallel single-molecule manipulation using centrifugal force. *Biophysical Journal, 98*(11), L53–L55. https://doi.org/10.1016/j.bpj.2010.03.012.

Harada, Y., Ohara, O., Takatsuki, A., et al. (2001). Direct observation of DNA rotation during transcription by *Escherichia coli* RNA polymerase. *Nature, 409*, 113–115. ⟨https://doi.org/10.1038/35051126⟩.

Hill, F. R., Monachino, E., & Van Oijen, A. M. (2017). The more the merrier: High-throughput single-molecule techniques. *Biochemical Society Transactions, 45*(3), 759–769. https://doi.org/10.1042/BST20160137.

Huhle, A., Klaue, D., Brutzer, H., Daldrop, P., Joo, S., Otto, O., ... Seidel, R. (2015). Camera-based three-dimensional real-time particle tracking at kHz rates and Ångström accuracy. *Nature Communications, 6*, 5885. https://doi.org/10.1038/ncomms6885. PMID: 25565216.

Inman, J. T., Ye, F., Hong, Y., et al. (2022). Resonator nanophotonic standing-wave array trap for single-molecule manipulation and measurement. *Nature Communications, 13*, 77. https://doi.org/10.1038/s41467-021-27709-3.

Janssen, X. J., Lipfert, J., Jager, T., Daudey, R., Beekman, J., & Dekker, N. H. (2012). Electromagnetic torque tweezers: A versatile approach for measurement of single-molecule twist and torque. *Nano Letters, 12*(7), 3634–3639. https://doi.org/10.1021/nl301330h. PMID: 22642488.

Kauert, D. J., Kurth, T., Liedl, T., & Seidel, R. (2011). Direct mechanical measurements reveal the material properties of three-dimensional DNA origami. *Nano Letters, 11*, 5558–5563 dx.doi.org/10.1021/nl203503s.

Kemmerich, F. E., Kasaciunaite, K., & Seidel, R. (2016). Modular magnetic tweezers for single-molecule characterizations of helicases. *Methods (San Diego, Calif.), 108*, 4–13. http://dx.doi.org/10.1016/j.ymeth.2016.07.004 — to be cited—.

Kemmerich, F. E., Swoboda, M., Kauert, D. J., Grieb, M. S., Hahn, S., Schwarz, F. W., ... Schlierf, M. (2016). Simultaneous single-molecule force and fluorescence sampling of DNA nanostructure conformations using magnetic tweezers. *Nano Letters, 16*(1), 381–386. https://doi.org/10.1021/acs.nanolett.5b03956. PMID: 26632021.

Kim, K., & Saleh, O. A. (2009). A high-resolution magnetic tweezer for single-molecule measurements. *Nucleic Acids Research, 37*(20), e136. https://doi.org/10.1093/nar/gkp725. Epub 2009 Sep 3. PMID: 19729511.

Klaue, D., & Seidel, R. (2009). Torsional stiffness of single superparamagnetic microspheres in an external magnetic field. *Physical Review Letters, 102*(2), 028302. https://doi.org/10.1103/PhysRevLett.102.028302. PMID: 19257322.

Kuijpers, L., Van Laar, T., Janissen, R., & Dekker, N. H. (2022). Characterizing single-molecule dynamics of viral RNA-dependent RNA polymerases with multiplexed magnetic tweezers. *STAR Protocols, 3*(3), 101606. https://doi.org/10.1016/j.xpro.2022.101606. eCollection 2022.

Lansdorp, B. (2016). *Pushing the envelope of magnetic tweezer resolution* (PhD). Santa Barbara: University of California. ⟨https://escholarship.org/content/qt0g27w21k/qt0g27w21k_noSplash_4f249cadb48f8b425da3b6754e867a53.pdf⟩.

Lansdorp, B. M., Tabrizi, S. J., Dittmore, A., & Saleh, O. A. (2013). A high-speed magnetic tweezer beyond 10,000 frames per second. *The Review of Scientific Instruments, 84*(4), 044301. https://doi.org/10.1063/1.4802678. PMID: 23635212.

Lin, J., & Valentine, M. T. (2012). High-force NdFeB-based magnetic tweezers device optimized for microrheology experiments. 053905 *Review of Scientific Instruments, 83*(5) 053905.

Lipfert, J., Hao, X., & Dekker, N. H. (2009). Quantitative modeling and optimization of magnetic tweezers. *Biophysical Journal, 96*(12), 5040–5049. https://doi.org/10.1016/j.bpj.2009.03.055. PMID: 19527664.

Lipfert, J., Klijnhout, S., & Dekker, N. H. (2010). Torsional sensing of small-molecule binding using magnetic tweezers. *Nucleic Acids Research, 38*(20), 7122–7132. https://doi.org/10.1093/nar/gkq598. PMID: 20624816.

Lipfert, J., Kerssemakers, J. W. J., Jager, T., & Dekker, N. H. (2010). Magnetic torque tweezers: measuring torsional stiffness in DNA and RecA-DNA filaments. *Nature Methods,* 7(12), 977–980. https://doi.org/10.1038/nmeth.1520.

Lipfert, J., Wiggin, M., Kerssemakers, J. W., Pedaci, F., & Dekker, N. H. (2011). Freely orbiting magnetic tweezers to directly monitor changes in the twist of nucleic acids. *Nature Communications, 2*, 439. https://doi.org/10.1038/ncomms1450. PMID: 21863006.

Lipfert, J., Lipfert, M., Ordu, O., Kerssemakers, J. W., & Dekker, N. H. (2014). Magnetic tweezers for the measurement of twist and torque. *Journal of Visualized Experiments, 87*, 51503. https://doi.org/10.3791/51503. PMID: 24894412.

Madariaga-Marcos, J., Aldag, P., Kauert, D. J., & Seidel, R. (2024). Correlated Single-Molecule Magnetic Tweezers and Fluorescence Measurements of DNA-Enzyme Interactions. *Methods in Molecular Biology, 2694*, 421–449. https://doi.org/10.1007/978-1-0716-3377-9_20. PMID: 37824016.

Manosas, M., Meglio, A., Spiering, M. M., Ding, F., Benkovic, S. J., Barre, F.-X., ... Croquette, V. (2010). *Magnetic tweezers for the study of DNA tracking motors. Methods in enzymology, l, 475*, Elsevier297–320.

McAndrew, C. P., Tyson, C., Zischkau, J., Mehl, P., Tuma, P. L., Pegg, I. L., & Sarkar, A. (2016). Simple horizontal magnetic tweezers for micromanipulation of single DNA molecules and DNA-protein complexes. *Biotechniques, 60*(1), 21–27. https://doi.org/10.2144/000114369 eCollection 2016. PMID: 26757808.

Moffitt, J. R., Chemla, Y. R., Smith, S. B., & Bustamante, C. (2008). Recent advances in optical tweezers. *Annual Review of Biochemistry,* 77, 205–228.

Morgan, T. I. L., & Saleh, O. A. (2021). A Python package for calibrating forces in single-molecule video-tracking experiments. *PLoS One, 16*(12), e0262028. https://doi.org/10.1371/journal.pone.0262028 eCollection 2021. PMID: 34972160.

Mosconi, F., Allemand, J. F., & Croquette, V. (2011). Soft magnetic tweezers: A proof of principle. *The Review of Scientific Instruments, 82*(3), 034302. https://doi.org/10.1063/1.3531959. PMID: 21456769.

Mosconi, F., Allemand, J. F., Bensimon, D., & Croquette, V. (2009). Measurement of the torque on a single stretched and twisted DNA using magnetic tweezers. 078301 *Physical Review Letters, 20;102*(7), https://doi.org/10.1103/PhysRevLett.102.078301. PMID: 19257716.

Neuman, K. C., Lionnet, T., & Allemand, J.-F. (2007). Single-molecule micromanipulation techniques. *Annual Review of Materials Research, 37*, 33–67.

Nickels, P. C., Wünsch, B., Holzmeister, P., Bae, W., Kneer, L. M., Grohmann, D., Tinnefeld, P., & Liedl, T. (2016). Molecular force spectroscopy with a DNA origami–based nanoscopic force clamp. *Science (New York, N. Y.), 354*(6310), 305.

Noji, H., Yasuda, R., Yoshida, M., & Kinosita, K. Jr (1997). Direct observation of the rotation of F1-ATPase. *Nature, 386*(6622), 299–302. https://doi.org/10.1038/386299a0.

Ostrofet, E., Papini, F. S., & Dulin, D. (2018). Correction-free force calibration for magnetic tweezers experiments. *Scientific Reports, 8*(1), 15920. https://doi.org/10.1038/s41598-018-34360-4.

Otto, O., Czerwinski, F., Gornall, J. L., Stober, G., Oddershede, L. B., Seidel, R., & Keyser, U. F. (2010). Real-time particle tracking at 10,000 fps using optical fiber illumination. *Optics Express, 18*(22), 22722–22733. https://doi.org/10.1364/OE.18.022722. PMID: 21164611.

Quack, S., & Dulin, D. (2024). Surface functionalization, nucleic acid tether characterization, and force calibration for a magnetic tweezers assay. In I. Heller, D. Dulin, E. J. G. Peterman (Eds.), *Single molecule analysis: Methods and protocols. Methods in molecular biology* (Vol. 2694, pp. 403–420). https://doi.org/10.1007/978-1-0716-3377-9_18.

Rieu, M., Valle-Orero, J., Ducos, B., Allemand, J. F., & Croquette, V. (2021b). Single-molecule kinetic locking allows fluorescence-free quantification of protein/nucleic-acid binding. *Communications Biology, 4*(1), 1083. https://doi.org/10.1038/s42003-021-02606-z. PMID: 34526657.

Rieu, M., Vieille, T., Radou, G., Jeanneret, R., Ruiz-Gutierrez, N., Ducos, B., Allemand, J. F., & Croquette, V. (2021a). Parallel, linear, and subnanometric 3D tracking of microparticles with Stereo Darkfield Interferometry. *Science Advances,* 7(6), eabe3902. https://doi.org/10.1126/sciadv.abe3902.

Ruiz-Gutierrez, N., Rieu, M., Ouellet, J., Allemand, J. F., Croquette, V., & Le Hir, H. (2022). Novel approaches to study helicases using magnetic tweezers. *Methods in Enzymology, 673*, 359–403. https://doi.org/10.1016/bs.mie.2022.03.035. PMID: 35965012.

Shon, M. J., Rah, S.-H., & Yoon, T.-Y. (2019). Submicrometer elasticity of double-stranded DNA revealed by precision force-extension measurements with magnetic tweezers. *Science Advances, 5*.

Shroff, H. White, H., & Betzig, E. (2008). Photoactivated localization microscopy (PALM) of adhesion complexes. *Current Protocols in Cell Biology/Editorial Board, Juan S. Bonifacino. [et al.]*. https://doi.org/10.1002/0471143030.cb0421s41 Chapter 4:Unit 4.21.

Sitters, G., Kamsma, D., Thalhammer, G., Ritsch-Marte, M., Peterman, E. J. G., & Wuite, G. J. L. (2015). Acoustic force spectroscopy. *Nature Methods, 12*(1), 47–50.

Smith, S. B., Finzi, L., & Bustamante, C. (1992). Direct mechanical measurements of the elasticity of single molecules using magnetic beads. *Science (New York, N. Y.), 258*(5085), 1122–1126.

Smith, S. B., Cui, Y., & Bustamante, C. (1996). Overstretching B-DNA: The elastic response of individual double-stranded and single-stranded DNA molecules. *Science (New York, N. Y.), 271*, 795–799.

Strick, T., Allemand, J., Croquette, V., & Bensimon, D. (2000a). Twisting and stretching single DNA molecules. *Progress in Biophysics and Molecular Biology, 74*(1-2), 115–140. https://doi.org/10.1016/s0079-6107(00)00018-3. PMID: 11106809.

Strick, T. R., Croquette, V., & Bensimon, D. (2000b). Single-molecule analysis of DNA uncoiling by a type II topoisomerase. *Nature, 404*(6780), 901–904. https://doi.org/10.1038/35009144.

Strick, T. R., Allemand, J. F., Bensimon, D., & Croquette, V. (1998). Behavior of supercoiled DNA. *Biophysical Journal, 74*(4), 2016–2028. https://doi.org/10.1016/S0006-3495(98)77908-1. PMID: 9545060.

Strick, T. R., Allemand, J. F., Bensimon, D., Bensimon, A., & Croquette, V. (1996). The elasticity of a single supercoiled DNA molecule. *Science (New York, N. Y.), 271*(5257), 1835–1837. https://doi.org/10.1126/science.271.5257.1835. PMID: 8596951.

Tapia-Rojo, R., Eckels, E. C., & Fernández, J. M. (2019). Ephemeral states in protein folding under force captured with a magnetic tweezers design. *Proceedings of the National Academy of Sciences of the United States of America, 116*(16), 7873–7878. https://doi.org/10.1073/pnas.1821284116.

Te Velthuis, A. J., Kerssemakers, J. W., Lipfert, J., & Dekker, N. H. (2010). Quantitative guidelines for force calibration through spectral analysis of magnetic tweezers data. *Biophysical Journal, 99*(4), 1292–1302. https://doi.org/10.1016/j.bpj.2010.06.008. PMID: 20713015.

Van Loenhout, M. T. J., Kerssemakers, J. W. J., De Vlaminck, I., & Dekker, C. (2012). Non-bias-limited tracking of spherical particles, enabling nanometer resolution at low magnification. *Biophysical Journal, 102*, 2362–2371.

Van Oene, M. M., Ha, S., Jager, T., Lee, M., Pedaci, F., Lipfert, J., & Dekker, N. H. (2018). Quantifying the precision of single-molecule torque and twist measurements using allan variance. *Biophysical Journal, 114*(8), 1970–1979. https://doi.org/10.1016/j.bpj.2018.02.039. PMID: 29694873.

Visnapuu, M. L., & Greene, E. (2009). Single-molecule imaging of DNA curtains reveals intrinsic energy landscapes for nucleosome deposition. *Nature Structural & Molecular Biology, 16*, 1056–1062. https://doi.org/10.1038/nsmb.1655.

Wang, M. D., Yin, H., Landick, R., Gelles, J., & Block, S. M. (1997). Stretching DNA with optical tweezers. *Biophysical Journal, 72*, 1335–1346.

Wang, Y. J., Valotteau, C., Aimard, A., Villanueva, L., Kostrz, D., Follenfant, M., Strick, T., Chames, P., Rico, F., Gosse, C., & Limozin, L. (2023). Combining DNA scaffolds and acoustic force spectroscopy to characterize individual protein bonds. *Biophysical Journal, 122*, 2518–2530.

Wong, W. P., & Halvorsen, K. (2006). The effect of integration time on fluctuation measurements: calibrating an optical trap in the presence of motion blur. *Optics Express, 14*(25), 12517–12531.

Wong, W. P., & Halvorsen, K. (2009). Beyond the frame rate: Measuring high-frequency fluctuations with light-intensity modulation. *Optics Letters, 34*(3), 277–279.

Yildiz, A., & Selvin, Paul, R. (2005). Fluorescence imaging with one nanometer accuracy: Application to molecular motors. *Accounts of Chemical Research, 38*, 574–582.

Zacchia, N. A., & Valentine, M. T. (2015). Design and optimization of arrays of neodymium iron boron-based magnets for magnetic tweezers applications. *The Review of Scientific Instruments, 86*, 053704. https://doi.org/10.1063/1.4921553.

Zhanga, X., Chenb, H., Fub, H., Doylea, P. S., & Yana, J. (2012). Two distinct overstretched DNA structures revealed by single-molecule thermodynamics measurements. *PNAS, 109*(21), 8103–8108.

Zou, Z., Liang, J., Jia, Q., Bai, D., Xie, W., Wu, W., Tanand, C., & Ma, J. (2023). A versatile and high-throughput flow-cell system combined with fluorescence imaging for simultaneous single-molecule force measurement and visualization. *Nanoscale, 43*(15), 17443–17454. https://doi.org/10.1039/d3nr03214k.

CHAPTER TWO

Use of DNA forceps to measure receptor-ligand dissociation equilibrium constants in a single-molecule competition assay

François Stransky[a], Dorota Kostrz[a], Maryne Follenfant[a], Sebastian Pomplun[b], Christian Meyners[c], Terence Strick[a], Felix Hausch[c,d], and Charlie Gosse[a,*]

[a]Institut de Biologie de l'Ecole Normale Supérieure, ENS, CNRS, INSERM, PSL Research University, Paris, France
[b]Leiden Academic Centre for Drug Research, Leiden University, Leiden, The Netherlands
[c]Department of Chemistry and Biochemistry, Technical University Darmstadt, Darmstadt, Germany
[d]Centre for Synthetic Biology, Technical University Darmstadt, Darmstadt, Germany
*Corresponding author. e-mail address: gosse@ens.fr

Contents

Methods in Enzymology, Volume 694
ISSN 0076-6879, https://doi.org/10.1016/bs.mie.2024.01.010

Abstract

The ability of biophysicists to decipher the behavior of individual biomolecules has steadily improved over the past thirty years. However, it still remains unclear how an ensemble of data acquired at the single-molecule level compares with the data acquired on an ensemble of the same molecules. We here propose an assay to tackle this question in the context of dissociation equilibrium constant measurements. A sensor is built by engrafting a receptor and a ligand onto a flexible dsDNA scaffold and mounting this assembly on magnetic tweezers. This way, looking at the position of the magnetic bead enables one to determine in real-time if the two molecular partners are associated or not. Next, to quantify the affinity of the scrutinized single-receptor for a given competitor, various amounts of the latter molecule are introduced in solution and the equilibrium response of the sensor is monitored throughout the titration protocol. Proofs of concept are established for the binding of three rapamycin analogs to the FKBP12 cis-trans prolyl isomerase. For each of these drugs the mean affinity constant obtained on a ten of individual receptors agrees with the one previously determined in a bulk assay. Furthermore, experimental contingencies are sufficient to explain the dispersion observed over the single-molecule values.

1. Introduction

1.1 Forceps as tools

Using single-molecule manipulation to study complexation reactions has long been delicate because of non-specific interactions. However, for more than a decade scaffolding strategies have been developed to overcome this hurdle. As an example, Fig. 1A displays the dsDNA construct at the center of the present paper. The two molecular partners are not directly attached to the micron-size handles anymore. Instead, today one relies on nanoscopic forceps for which each association or dissociation event is detected through a specific spatial signature (Fig. 1B). As a matter of fact, such fingerprinting enables the investigator to eliminate spurious signals and to only focus on the sought-after phenomena, greatly improving the quality of the collected dataset (Gosse, Strick, & Kostrz, 2019; Halvorsen, Schaak, & Wong, 2011; Johnson & Thomas, 2018; Kim, Zhang, Zhang, & Springer, 2010).

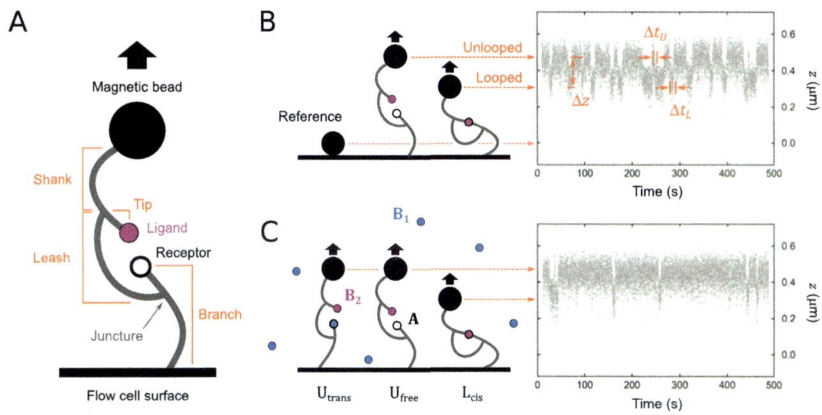

Fig. 1 Working principle of the dsDNA scaffold utilized to measure dissociation equilibrium constants on single molecules. **(A)** Structure of the so-called J-DNA (not to scale). The studied molecular partners are engrafted at the tips of the forceps-like construct which is actuated thanks to two handles fixed at the shanks, a magnetic bead and the surface of a flow cell. The leash connects the two branches and ensures the permanence of the whole assembly. The contour lengths for the various anatomical segments are 16 nm for the tips, 481 and 493 nm for the shanks, and 233 nm for the leash. **(B)** Behavior of the scaffold when a small, constant force is applied. The FKBP12 receptor and the JK313 ligand keep associating and dissociating in *cis*, the former reaction being possible because the leash maintains the two molecules in close proximity after their separation. Hence, the z-position of the bead alternates between two mean levels separated by about 80–120 nm. The amplitude of this motion is smaller than the contour length difference between the two conformations, 201 nm, because the applied force is low. This amplitude also varies from bead-to-bead because of technical uncertainties (see Section 4.1). Each event in a trajectory is characterized by a scaffold extension variation, Δz, and by a dwell time, Δt. For our analysis we further distinguish if the latter parameter corresponds to unlooped or looped J-DNA (as respectively indicated by the subscript U and L). **(C)** Behavior of the scaffold under the same force regime but with a second ligand in solution, here PPU278 at 4 nM. The presence of three molecular partners (A = FKBP12, B_1 = PPU278, and B_2 = JK313) results in three different states ($\mathrm{U}_{\mathrm{trans}}$, $\mathrm{U}_{\mathrm{free}}$, and $\mathrm{L}_{\mathrm{cis}}$) but only two conformations (unlooped and looped). Events are less numerous and dwell times in the unlooped conformation are longer because of the in-*trans* binding of the competitor. All measurements were performed at 28 °C in mTOR buffer, under an applied force of 0.1 ± 0.01 pN.

The shanks of these forceps are commonly made of dsDNA. Concerning the leash that connects the two tips where the reagents are placed, various polymer chains can be employed: polyethylene glycol (Bell & Molloy, 2023; Krasnoslobodtsev et al., 2015), polypeptides (Kim et al., 2010; Rognoni et al., 2012; Wang et al., 2019), ssDNA (Kilchherr et al., 2016; Mickolajczyk et al., 2022; Shrestha et al., 2021), as well as dsDNA, with (Halvorsen et al., 2011) or without nicks (De Bragança et al., 2023;

Kostrz et al., 2019; Ma et al., 2019; Wang et al., 2018). The material persistence length is here critical for defining the amplitude of the response observed upon complex dissociation, the force regime that can be probed, and the kinetics of association. Of course, the contour length of the chain is the other key parameter. Typically, dsDNA leashes, among the more rigid, exhibit signatures larger than 100 nm and can work at extending forces as low as 0.1 pN. On the down side, the two partners being significantly pulled apart, their rebinding via random encounter is less frequent.

So far, the scaffolding approach has been implemented on nearly all known single-molecule nanomanipulation setups: optical tweezers (Halvorsen et al., 2011; Kilchherr et al., 2016; Kim et al., 2010; Mickolajczyk et al., 2022; Rognoni et al., 2012; Shrestha et al., 2021), magnetic tweezers (Bell & Molloy, 2023; De Bragança et al., 2023; Kostrz et al., 2019; Ma et al., 2019; Shrestha et al., 2021; Wang et al., 2018; Wang et al., 2019), atomic force microscope (Krasnoslobodtsev et al., 2015), centrifuge force microscope (Yang, Ward, Halvorsen, & Wong, 2016), laminar flow chamber (Penth, Schellnhuber, Bennewitz, & Blass, 2021), and acoustic force spectroscopy apparatus (Wang et al., 2023).

In addition to the filtering out of non-specific interactions, many other interesting properties can explain the wide dissemination of such forceps. To start with, the environment around the biomolecules of interest can be finely controlled, with shanks that separate them from the surfaces of the handles (Halvorsen et al., 2011; Kim et al., 2010) and with engraftment procedures that permit precise definition of the pulling direction (Kostrz et al., 2019). Therefore, two of the artefacts reported to obscure data interpretation (Raible et al., 2006) are this way eliminated. Moreover, the leash allows the experimenter to repetitively interrogate the same pairs of interacting molecules (Halvorsen et al., 2011; Kim et al., 2010), and thus characterize them in a detailed yet statistically robust manner. Because of this robustness, any eventual alteration of the interaction metrics—for instance due to protein ageing, unfolding, or degradation—can be detected in real-time, and suspect data sorted out accordingly. Differences between molecules can also be evidenced and the homogeneity of the population verified. In brief, scaffolding potentializes the fact that single-molecule nanomanipulation is by nature an ultimate analytical tool, apt to function on individual objects.

1.2 Heterogeneity as question

The existence of diverging behaviors between isolated molecules has been questioned in a number of articles on the interactions of proteins with other proteins (Kim et al., 2010; Kim et al., 2011; Kostrz et al., 2019; Nevo et al., 2003; Wang et al., 2023), with nucleic acids (Fuhrmann et al., 2009; Raible et al., 2006), and with ligands (Rico et al., 2019). Besides dissociation reactions in bimolecular systems, similar accounts have been published for protein unfolding (Brucale et al., 2009; Kuo et al., 2010; Xia, Zuo, & Li, 2022). Nevertheless, finding the origin of heterogeneity has remained challenging. Indeed, from the efforts made to rationally tackle the problem, it appeared that various phenomena, occurring at different levels, can combine (Hinczewski, Hyeon, & Thirumalai, 2016; Raible et al., 2006; Xia et al., 2022). First, extrinsic, experimental issues may be invoked, e.g. difference in the chemical environment, poorly controlled pulling directions, instrument-related artefacts, and/or contamination of the biological material. Second, intrinsic causes may be at play. As an illustration, one can imagine molecules that are all identical although they sample different conformations; hence, if each of them is probed only once several populations will seem to emerge.

New ways to analyze data (Hinczewski et al., 2016; Raible et al., 2006) or better experimental designs (Kim et al., 2010; Wang et al., 2023; Xia et al., 2022) have been proposed. Scaffolding is one of them since, as said above, multiple single-pairs of reagents can be compared in detail thanks to long-term monitoring. With forceps researchers could evidence two different conformations for a same complex (Kim et al., 2010); resolve a mixture of species that differ by their binding properties (Yang et al., 2016) or their geometry (Shrestha et al., 2021); attribute to variation in the applied force discrepancies noticed on individual dissociation rate constants (Kostrz et al., 2019).

Whereas most of those studies are based on the measurement of either unbinding or unfolding kinetic parameters, heterogeneity for dissociation equilibrium constants (K_D) has still not received much attention.

1.3 Titrations as protocols

In single-molecule experiments one usually follows a unique object, which implies that the thermodynamic and kinetic parameters describing the biological system are often extracted from apparent monomolecular phenomena. In line with this remark, the concentration dimension that lacks when trying to evaluate the K_D of a bimolecular reaction can be introduced by a titration. In such a situation, the "single-molecule" denomination only qualifies the object we are looking at, arbitrarily referred to as the receptor. Actually, many different molecules of the reagent that has been added in solution will be allowed to bind and unbind in *trans*. By convention we will call such species the ligands.

One of the first titration procedure applied to nanomanipulation aimed at quantifying how calcium ions promote the adhesion between two calmodulins, one immobilized on the tip of an atomic force microscope (AFM) and the other on a surface (Baumgartner et al., 2000). Yet, the binding probability of the probe was measured while scanning, preventing the experimenters to study a unique pair of proteins.

Later on, two groups monitored the influence of the ligand concentration on the unfolding profile of the receptor, in a two-protein system by pulling with an AFM (Cao, Balamurali, Sharma, & Li, 2007) and in a nucleic acid/small molecule system by pulling with optical tweezers (Koirala et al., 2011). Although the fraction of complexes formed at different concentrations could be retrieved from such experiments, allowing for the determination of K_D, these protocols turn out to be quite specific.

A more general way of proceeding consists in observing, under constant force, the looping/unlooping transitions of scaffolds such as the ones described in Section 1.1. Let us first consider that the two engrafted partners, among which the receptor, are able to rapidly associate/dissociate in *cis*. Then, the addition of ligand in *trans* causes the active behavior to switch to idleness for periods that become more frequent as the concentration is ramped up (Rognoni et al., 2012). From the evolutions of the dwell times spent in the active and idle states it was possible to respectively measure the association and dissociation rate constants, and consequently the K_D, for the reaction between two domains involved in the connection of a transmembrane protein to the cytoskeleton. Relying on a variant of this protocol, some of us have been interested in a ligand, rapamycin, that triggered the activation of a sensor on which were present two proteins that do not interact naturally; hence, we could quantify the affinity of this molecular glue for one of the partners (Kostrz et al., 2019). Despite these two successes, this approach suffers from a few drawbacks: the distinction between the active and the idle periods may not be so straightforward and the fits of the dwell time distributions surely introduce some additional uncertainties.

In fact, a simpler strategy can be devised in the case where in-*trans* binding is competitive: as the titration progresses one just has to look at the linear growth of the average of the open dwell times to obtain K_D. Accordingly, two different teams could explore the self-inhibition of dsDNA looping by increasing amounts of LacI repressor, the reaction taking place in absence of mechanical tension and being followed thanks to the tethered particle motion technique (Han et al., 2009; Priest et al., 2014). Others dealt with the inhibition of DNA hairpin folding by fixation of complementary oligonucleotides to the unzipped sequence, this time a significant force being applied to a magnetic bead affixed to the nucleic acid construct (Rieu et al., 2021a). Interestingly, as predicted by theory, dissociation equilibrium constants could be measured for scaffolds the fluctuations of which occurred on a time scale either smaller or larger than competitor binding-unbinding.

Herein we will discuss a similar, albeit independently developed assay, focusing on questions related to heterogeneity. Titration is performed at constant, low force on magnetic tweezers (Choi, Kim, Shon, & Yoon, 2022; Dulin, 2024; Gosse & Croquette, 2002; Neuman & Nagy, 2008; Strick et al., 1996) and the molecular partners are attached at the tips of so-called J-DNA forceps (Kostrz et al., 2019; Wang et al., 2018; Wang et al., 2023) (Fig. 1). Our receptor, FKBP12, is a small protein that we have previously employed as a model because it is easy to handle, both experimentally (Kostrz et al., 2019; Wang et al., 2023) and numerically (Joshi, Gosse, Huang, & Lin, 2022). Furthermore, it has already been fully characterized from the structural and functional points of view (Banaszynski, Liu, & Wandless, 2005; Choi, Chen, Schreiber, & Clardy, 1996; Galat, 2013; Geiger et al., 2024; Kolos et al., 2021; Kozany, Marz, Kress, & Hausch, 2009; Purder et al., 2023). As ligand we selected three rapamycin analogs participating in a series of compounds that were synthesized to better understand the structure-activity relationship for inhibitors of FKBP proteins (Kolos, Voll, Bauder, & Hausch, 2018; Pomplun et al., 2018). They show intermediate K_D (Fig. S1) and one of them bears an alkyne group that will be mobilized to functionalize the scaffold by click chemistry (Fig. S2) (Geiger et al., 2024).

2. Theory

2.1 Chemical system

The goal of our assay is to measure the dissociation equilibrium constant K_{D1} of the reaction between receptor A and ligand B_1:

$$AB_1 \underset{k_{+1}}{\overset{k_{-1}}{\rightleftarrows}} A + B_1$$

with k_{-1} and k_{+1} the corresponding dissociation and association rate constants ($K_{D1} = k_{-1}/k_{+1}$). To do so we will utilize a DNA forceps on which receptor A has been engrafted at one of the tips and a second ligand B_2 has been attached in front of it, at the other tip (Fig. 1). Therefore, the reaction we want to characterize will compete with

$$AB_2 \underset{k_{+2}}{\overset{k_{-2}}{\rightleftarrows}} A + B_2$$

the dissociation equilibrium constant of which is $K_{D2} = k_{-2}/k_{+2}$.

In this system, one has three different states for the receptor (or equivalently for the scaffold): one in which A is not engaged in any complex, U_{free}, and two in which it is associated with a ligand, either B_1 coming from the solution in U_{trans} or B_2 located at the second tip in L_{cis} (Fig. 1C). Of these three states only L_{cis} corresponds to looped forceps, and thus to the conformation of low z-extension called L. The two other ones are characterized by unlooped forceps and belong to the conformation of high z-extension called U. Noteworthily, since we cannot experimentally detect transitions between U_{free} and U_{trans} (Figs. 1C and 2A-A"), the U conformation is referred to as a compound one (Shelley & Magleby, 2008).

From a kinetic point of view, each DNA forceps can now be described by

$$U_{trans} \underset{k_{+t}}{\overset{k_{-t}}{\rightleftarrows}} U_{free} \underset{k_{-c}}{\overset{k_{+c}}{\rightleftarrows}} L_{cis}$$

where all reactions are unimolecular. The binding side on A being far from the engraftment point, no incidence on complex formation is suspected. Writing $k_{-t} = k_{-1}$ is thus legitimate. Furthermore, in such single-molecule studies the concentration B_1 in B_1 is assumed constant and consequently we obtain $k_{+t} = k_{+1}B_1$. On the looping reaction side, in preliminary experiments we have measured by single-molecule force spectroscopy a Bell parameter around 0.4 nm (Fig. S3), which means that in our 0.1 pN conditions the dissociation rate constant value is within 1% of the one at 0 pN: as a matter of fact, $k_{-c} = k_{-2}$. Finally, concerning association, we will use some effective parameter B_2 to describe the concentration of B_2 at the tip of the forceps. Its value will primarily be related to the inverse of the volume explored by the two reagents connected by the leash. Yet, it will also account for all other factors that could influence binding in *cis*, e.g. steric and geometric constrains imposed by the engraftment, additional separation distance between reagents resulting from the applied force. Accordingly, the $k_{+c} = k_{+2}B_2$ relation will always be verified.

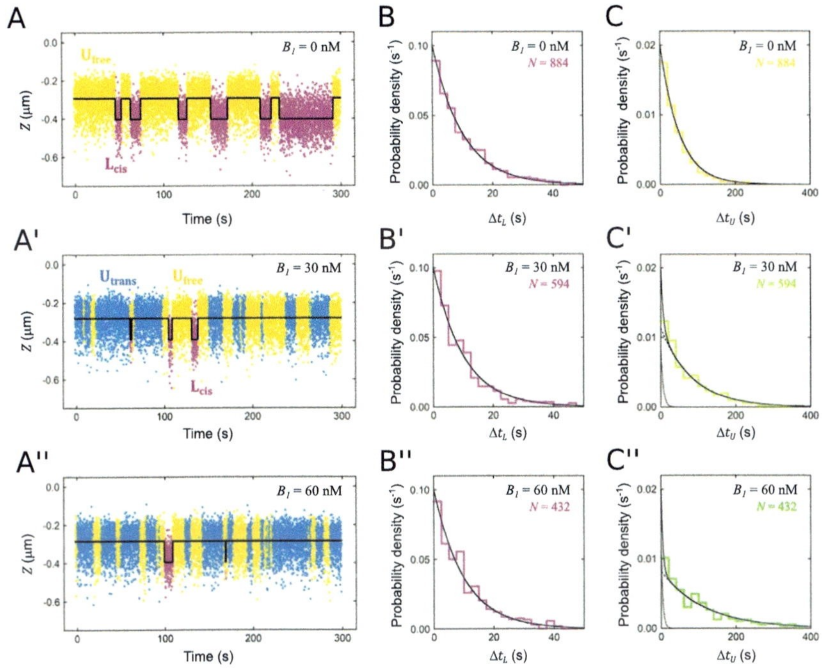

Fig. 2 Evolution, as a function of the competitor concentration, of the dwell times in the looped and unlooped conformations. More specifically, we here performed simulations for $B_1 = 0$, 30, and 60 nM. They correspond to measurements that would be performed at 28 °C in mTOR buffer, under an applied force of 0.1 ± 0.01 pN. **(A–A")** Signal trajectories with the three states, U_{trans}, U_{free}, and L_{cis}, respectively indicated in blue, gold, and lavender. The corresponding conformation trajectories are indicated with black lines. **(B–B")** Histograms for the looped dwell times, Δt_L, issued from representative state trajectories having a duration $T = 14$ h (bin size 2.5 s). The theoretical distribution of Eq. (3) is superimposed as solid black lines. **(C–C")** Histograms for the unlooped dwell times, Δt_U, plotted for the same simulations (bin size 20 s). Data for this conformation are represented with green tones when they result from the two states that were colored in gold and blue in the first panels. The theoretical distributions of Eq. (7) are superimposed as solid black lines, the + and − decays being represented with dots and dashes, respectively.

Coming to thermodynamics, the dissociation equilibrium constants in *trans* and in *cis* are simply $K_t = k_{-t}/k_{+t} = K_{D1}/B_1$ and $K_c = k_{-c}/k_{+c} = K_{D2}/B_2$.

2.2 Observables

As illustrated in Fig. 1B,C, the bead tethered by the forceps fluctuates between a high and a low z-position, respectively related to the U and L conformations. Each change is characterized by a dwell time, Δt_U if the transition occurs from an unlooped DNA molecule and Δt_L if it is from a looped one. Regarding the extension variation, Δz, it is fixed by the geometrical properties of the forceps and by the applied force.

In the present assay we shall determine, as a function of the concentration in B_1, the ratio between the arithmetic averages of the dwell times in the unlooped and looped conformations:

$$\rho = \frac{\overline{\Delta t_U}}{\overline{\Delta t_L}} \tag{1}$$

Considering a single scaffold, we conventionally call each acquisition realized at a given concentration in competitor a measurement. When the number of events for both conformation changes are equal to the same integer N, Eq. (1) identifies to $\rho = \Sigma\Delta t_U / \Sigma\Delta t_L$ with $\Sigma\Delta t_U = \Sigma\Delta t_{U_{trans}} + \Sigma\Delta t_{U_{free}}$ and $\Sigma\Delta t_L = \Sigma\Delta t_{L_{cis}}$. Independently using the mass action law expressed in terms of fractional occupancies, p_j, one can write $K_t = p_{U_{free}}/p_{U_{trans}} = \Sigma\Delta t_{U_{free}}/\Sigma\Delta t_{U_{trans}}$ and $K_c = p_{U_{free}}/p_{L_{cis}} = \Sigma\Delta t_{U_{free}}/\Sigma\Delta t_{L_{cis}}$. Hence, combining the previous expressions results in

$$\rho = K_c\left(1 + \frac{1}{K_t}\right) = \frac{K_{D2}}{B_2}\left(1 + \frac{B_1}{K_{D1}}\right) \tag{2}$$

This way, it appears possible to extract K_{D1} by plotting the linear variations of ρ with B_1. Importantly, no prior knowledge of the reaction parameters or environmental conditions at the forceps tips seems necessary. Indeed, one can easily eliminate the K_{D2}/B_2 term by dividing the intercept by the slope.

By nature, single-molecule investigations analyze discrete transitions, thereby delivering kinetic insights into the phenomenon under scrutiny. As a consequence, in addition to the above averaged values we can examine the probability density functions (PDF) for the dwell times. In the looped conformation we have a monoexponential function:

$$f_{\Delta t_L}(t) = w_L \exp(-t/\tau_L) \tag{3}$$

with time constant

$$\tau_L = \frac{1}{k_{-c}} \tag{4}$$

and pre-factor

$$w_L = k_{-c} \tag{5}$$

The mean of this PDF is given by

$$\mu_{\Delta t_L} = \int_0^\infty t\, f_{\Delta t_L}(t)\, dt = w_L \tau_L^2 = \frac{1}{k_{-c}} \tag{6}$$

For the dwell times in the unlooped conformation, which includes both indistinguishable states U_{trans} and U_{free}, the PDF is biexponential (Colquhoun & Hawkes, 1981; Rieu et al., 2021a; Shelley & Magleby, 2008):

$$f_{\Delta t_U}(t) = w_{U+}\exp[-t/\tau_{U+}] + w_{U-}\exp[-t/\tau_{U-}] \tag{7}$$

with time constants

$$\tau_{U\pm} = \frac{2}{(k_{-t} + k_{+t} + k_{+c}) \pm \sqrt{(k_{-t} + k_{+t} + k_{+c})^2 - 4k_{-t}k_{+c}}} \tag{8}$$

and pre-factors (or weights)

$$w_{U\pm} = \pm k_{+c} \frac{(1/\tau_{U\pm}) - k_{-t}}{(1/\tau_{U+}) - (1/\tau_{U-})} \tag{9}$$

Please note that the + and – signs in $U+$ and $U-$ do not correspond to the direction of any binding/unbinding reaction but just refer to some of the signs present in the above mathematical expressions. The PDF mean obtained by integration is now (Colquhoun & Hawkes, 1981; Rieu et al., 2021a)

$$\mu_{\Delta t_U} = w_{U+}\tau_{U+}^2 + w_{U-}\tau_{U-}^2 = \frac{k_{-t} + k_{+t}}{k_{-t}\, k_{+c}} \tag{10}$$

First, we can check that using $\rho = \mu_{\Delta t_U}/\mu_{\Delta t_L}$ and the expressions given in Eqs. (6) and (10) one retrieves Eq. (2). Second, we can now compute the approximate number of looping-unlooping cycles contained in a measurement of duration T as

$$N \approx \frac{T}{\mu_{\Delta t_U} + \mu_{\Delta t_L}} = \frac{T k_{-2}}{1 + \frac{K_{D2}}{B_2}\left(1 + \frac{B_1}{K_{D1}}\right)} \tag{11}$$

Both k_{-2} and K_{D2}/B_2 are included in this equation whereas neither k_{-1} nor k_{+1} are independently present, which means that this number N depends on the kinetics of the in-*cis* reaction but not on the ones of the in-*trans* reaction. In fact, the competitor modulates the overall dynamics of the scaffold only through the law of mass action: just a fraction of B_2 equal to $1/(1 + B_1/K_{D1})$ is now available for in *cis*-binding.

2.3 Uncertainties

Working with finite acquisition time and stochastic processes inevitably leads to observables that differ from their expected values. To calculate the statistical uncertainty associated with the determination of ρ we first look at the ones associated with the determination of each $\overline{\Delta t_i}$. The identity relating the standard deviation of the mean to the standard deviation of the parent distribution and to the sample size is always true provided observations are uncorrelated (Bevington & Robinson, 2002, Colquhoun, 1971). Consequently, it applies to our chemical system: $\sigma_{\overline{\Delta t_i}} = \sigma_{\Delta t_i}/\sqrt{N}$. Next, we approximate the standard deviation of the parent distributions, which are a priori unknown, by the sample standard deviations: $\sigma_{\Delta t_i} \approx s_{\Delta t_i}$. Finally, we propagate uncertainties computed for each dwell time average (Bevington & Robinson, 2002, Colquhoun, 1971) and obtain

$$\sigma_\rho = \rho\sqrt{\frac{1}{N}\left(\frac{s_{\Delta t_L}{}^2}{\overline{\Delta t_L}^2} + \frac{s_{\Delta t_U}{}^2}{\overline{\Delta t_U}^2}\right)} \tag{12}$$

3. Numerical simulations

3.1 Implementation of the Markov process

For the present paper DNA forceps were functionalized with the FKBP12 protein and the JK313 small molecule. For simulations the ligand in solution was also JK313 because it is the only compound we could characterize extensively. However, it does not infringe upon the general scope of our results.

The thermodynamic and kinetic parameters employed for these numerical investigations were selected to account realistically, yet practically, for the observed phenomena: only one significant digit was used. First, relying on literature (Pomplun et al., 2018) and on preliminary titrations at constant force we took $K_{D1,2} = 50\,\text{nM}$. Second, based on data from preparatory single-molecule force spectroscopy experiments we set $k_{-1,2} = 0.1\,\text{s}^{-1}$ (Fig. S3). Consequently, it yielded $k_{+1,2} = 0.002\,\text{nM}^{-1}\text{s}^{-1}$. Third, when ligands B_1 and B_2 are identical the expression of the ratio of the averaged dwell times reduces to $\rho = K_{D1,2}/B_2 + B_1/B_2$; hence, our early titrations also provided us an estimate for the effective concentration at the forceps tips, $B_2 = 10\,\text{nM}$. It is here important to keep in mind that the latter value is only indicative because differences in the magnetic loading of the beads (Fig. S4A, A′) (Kostrz et al., 2019), or in the magnetic field gradient (Fig. S4B), influence the applied force and may cause this B_2 parameter to vary from forceps to forceps. In conclusion, from the preceding numbers it came $k_{-t} = k_{-c} = 10^{-1}\ \mathrm{s}^{-1}$, $k_{+t} = 2 \times 10^{-3}\ B_1\ \mathrm{s}^{-1}$, and $k_{+c} = 2 \times 10^{-2}\ \mathrm{s}^{-1}$.

The chemical system was simulated using a Markov process with three-state probability vector

$$X = \begin{pmatrix} \mathrm{U_{trans}} & \mathrm{U_{free}} & \mathrm{L_{cis}} \end{pmatrix} \tag{13}$$

and transition matrix

$$A = \begin{pmatrix} 1 - k_{-t}t_s & k_{-t}t_s & 0 \\ k_{+t}t_s & 1 - (k_{+t} + k_{+c})t_s & k_{+c}t_s \\ 0 & k_{-c}t_s & 1 - k_{-c}t_s \end{pmatrix} \tag{14}$$

the $A_{i,j}$ representing the probability to find the system in state j at time $t + t_s$ given that it was in state i at time t. Please note that row probability vectors and right transition matrices are used to comply with the convention of the field and the formalism of the numerical routines we relied on. To generate model signal trajectories, it appeared practical to work with t_s equal to the inverse of the $f_s = 31.08\ \mathrm{Hz}$ sampling frequency of the camera (see Section 4.1). Since f_s is much larger than any of the present unimolecular rate constants it fully justified the adoption of the $k_{\pm t,c}t_s$ products to compute the transition probabilities (Steinfield, Francisco, & Hase, 1998).

In silico experiments were realized with the CategoricalHMM function of the Python hmmlearn package. They lasted 14 h, which corresponds to $\approx 1.5 \times 10^6$ sampling points. State trajectories were straightforwardly converted into conformation trajectories by associating the same extension Z to both $\mathrm{U_{trans}}$ and $\mathrm{U_{free}}$ and a different one to $\mathrm{L_{cis}}$ (Fig. 2A-A"). Please note that Z here corresponds to stabilized data that get located around zero after long-term drift correction (see Section 4.5.1). Relying on a representative measurement we positioned the looped conformation at −390 nm and the unlooped one at −280 nm. Finally, signal trajectories were obtained by further addition of noise, the latter one being randomly picked up from PDFs issued from experimental Z data.

3.2 Visualization of the three-state system dynamics

The simulated signal trajectories of Fig. 2A-A" display the same characteristics as the experimental ones shown in Fig. 1B,C: with increasing $\mathrm{B_1}$

concentrations the forceps spend more and more time in the unlooped conformation and the density of events diminishes. These observations are in line with Eqs. (2) and (11), respectively. Furthermore, from the coloring of the three states it is quite straightforward to qualitatively relate those two trends to a growing occupancy of the receptor by the competitor. As previously mentioned, in presence of B_1 the concentration of the receptor A able to bind in *cis* to B_2 decreases from B_2 to $B_2/(1 + B_1/K_{D1})$.

Using the *in silico* trajectories we could also plot the dwell time histograms for both L and U conformations. The first ones perfectly superimpose with the Eq. (3) distribution, that is with an exponential decay that does not vary with concentration B_1 (Fig. 2B–B"). Similarly, the second ones match Eq. (7) (Fig. 2C–C"). In the latter biexponential distributions the + decay is much faster than the − decay (Fig. 2C–C" and Fig. S5). More precisely, over the $B_1 = 0$ to 180 nM range τ_{U+} decreases from 10 to 2 s whereas τ_{U-} steadily increases from 50 to 238 s (Fig. S5A). Now looking at the number of bins that could be used to characterize each decay, $a_{U+} = w_{U+}\tau_{U+}$ remains lower than 6% of the total whereas $a_{U-} = w_{U-}\tau_{U-}$ always stays over 94% (Fig. S5C). These contrasts explain why the Δt_U histograms nearly appear monoexponential. Another implication is that we could not experimentally investigate the dynamics of our chemical system satisfactorily, the + decay remaining out of reach. However, such studies are possible, as it has been demonstrated by colleagues who worked on faster reactions with a faster detection of the bead position (Rieu et al., 2021a, 2021b).

4. Single-molecule measurements

Most experimental procedures have already been described extensively in previous articles, by us or others; consequently, they are only briefly reported here. The only exceptions are the preparation of the ligand-oligonucleotide conjugate (Section 4.2.3), the titration protocol (Section 4.4), and the signal trajectories analysis using a hidden Markov model (HMM) learning algorithm (Section 4.5.2).

4.1 Instrumentation

A practical introduction to magnetic tweezers has just been published (Dulin, 2024) and numerous reviews have been released on the topic—see for instance (Choi et al., 2022; Neuman & Nagy, 2008). Our measurements were realized on a home-built apparatus (Fig. S6) controlled with the PicoJai software suite (PicoTwist, Fleurieux-sur-l'Arbresle, France) (Gosse & Croquette, 2002; Strick et al., 1996). This setup consists in an inverted microscope built around a 100× oil-immersion objective (Plan C, 1.25 NA, Olympus, Rungis, France) mounted on a piezo scanner having a 250 μm travel range with a 0.75 nm positioning precision (P-725.2CL, Physik Instrumente, Karlsruhe, Germany). Illumination is achieved with a blue LED collimated through a simple converging lens and the image of the sample is formed on the chip of a CCD camera (CM-140 GE, JAI, Copenhagen, Denmark) through a tube lens having a focal distance of 180 mm. The 200 × 150 μm^2 field of view is analyzed at 31.08 Hz and the 1 μm diameter magnetic beads are located with ≈ 5 nm accuracy in the three spatial dimensions. The end-to-end extension of the J-DNA forceps, z, is calculated from the position of the bead of interest above the surface, using a bead stuck on the lower glass slide of the flow cell to correct for drift (see Fig. 1B and Section 4.5.1).

A pair of NdFeB magnets (10 mm long, 3 mm high, and 1 mm thick, separated by 1 mm) is inserted with opposite polar orientations on a turret located above the sample. The rotation and vertical position of the latter element are respectively controlled by a servomotor and a translation stage (M-136 and M-126.DG1, Physik Instrumente). The force applied to the J-DNA-tethered magnetic beads decreases exponentially as the distance between the magnets and the sample is increased, going from ≈ 10 pN when the turret contacts the upper glass slide of the flow cell to 0 pN when it is 5 mm away. More specifically, in our experiment the tension on the scaffolds was inferred from the magnet position, relying on a calibration force-extension curve pre-established on 11 kbp-long supercoilable dsDNA molecules (Fig. S4A). We noticed that different beads experience different forces because of differences in their magnetic content (Fig. S4A') (Kostrz et al., 2019). In fact, a same bead may even have a response that varies, depending on its position in the field of view, that is in the magnetic field gradient (Fig. S4B). As a consequence, the precision on the 0.1 pN constant force we here employed was estimated to be around ± 10%.

To control the temperature in the sample chamber the brass objective is enclosed in an aluminum block that we thermostat between 19 and 35 °C thanks to two Peltier devices connected to two other aluminum blocks used as heat-sinks or heat-sources, a temperature controller (LFI-3751, Wavelength Electronics, Bozeman, MT), and a thin film platinum RTD element (F3141, Omega, Manchester, UK). Heat transfer from the aluminum block to the objective is ensured by vacuum grease and from the objective to the solution by the immersion oil and by the lower glass slide of the flow cell. A temperature calibration in which a second probe was glued with microprocessor thermal paste to a bare coverslip placed atop the microscope objective allowed us to establish the relationship between the temperature set at the controller and the actual temperature of the solution in the immediate vicinity of the bead-DNA system. The precision of our system is around 0.5 °C. For all our titrations the temperature at the scaffolds was fixed at 28 °C.

4.2 Scaffold synthesis and functionalization

4.2.1 The bare forceps

Detailed accounts on J-DNA synthesis can be found in (Wang et al., 2018) and (Kostrz et al., 2019). For the present study we relied on our latest version of this tool, which includes the modifications of the dimensions mentioned in (Wang et al., 2023). The scaffold is made of three linear segments of dsDNA: two branches connected by a leash (Fig. 1A). Their respective lengths are 1470, 1507, and 689 bp. The junctures with the leash divide the branches in tips and shanks. The former segments are 48 bp-long and terminate with two 9-base overhangs with distinct sequences that are used for subsequent attachment of protein-oligonucleotide or ligand-oligonucleotide conjugates. The shanks are 1422 and 1459 bp-long and have their extremities ligated to dsDNA fragments multiply-labelled with either biotin or digoxigenin. These sticky ends are themselves ≈ 1000 bp-long and are employed for the respective fixation to the streptavidin-coated beads and to the anti-digoxigenin-coated surface of the flow cell.

4.2.2 The protein-oligonucleotide conjugate

Relying on unnatural amino acid incorporation (Chin et al., 2002; Lajoie et al., 2013; Young, Ahmad, Yin, & Schultz, 2010), the FKBP12 protein was labelled at its N-terminus with a 4-azido-L-phenylalanine residue. It resulted in the previously characterized $\text{FKBP12}_{\text{M0AzF}}$ mutant (Kostrz et al., 2019). Next, following the procedure described in the same paper, this receptor was conjugated by click chemistry to DbcoP1, a 9-base long oligonucleotide bearing a dibenzocyclooctyne group at its 5′ end (Fig. S2A). We here chose the strain-promoted azide-alkyne cycloaddition strategy because it does not require any catalysis, being consequently harmless for proteins.

4.2.3 The ligand-oligonucleotide conjugate

For the attachment of JK313 we once more used click chemistry, this time harnessing the copper(I)-catalyzed azide-alkyne cycloaddition strategy. More specifically, we took advantage of the alkyne group present in the ligand, at a position located far enough from the pharmacophore (Fig. S1) (Geiger et al., 2024). To demonstrate the first engraftment of a small molecule onto J-DNA, the reaction was achieved with AzP2, a 9-base long oligonucleotide bearing an azido group at its 5′ end (Fig. S2B). In a microtube we mixed 18.7 μL of water, 5 μL of DMSO, 5 μL of 1 M triethylammonium acetate buffer pH 7.0, 8.3 μL of AzP2 at 0.3 mM in water (2.5 nmol, 1 eq; Eurogentec, Seraing, Belgium), and 5 μL of JK313 at 2 mM in DMSO (4 eq; prepared according to (Pomplun et al., 2018)). Then, we introduced and gently stirred 3 μL of a freshly prepared aqueous mixture of $CuSO_4$ at 3.33 mM (4 eq; Jena Bioscience, Jena, Germany) and THPTA at 16.66 mM (20 eq; Jena Bioscience). Finally, after addition and stirring of 5 μL of sodium ascorbate at 0.1 mM in water (200 eq; Jena Bioscience), the microtube was left for 3 h 30 at 37 °C, protected from light. This general protocol, partly inspired by (Vasilyeva, Filichev, & Boutorine, 2016), has been successfully used in our team to conjugate various small molecules. It is a modified version of the one proposed by Jena Bioscience, provider of click chemistry reagents: (i) the initial phosphate buffer was replaced to avoid precipitation during the subsequent purification steps; (ii) the catalytic system was diluted 10 times to decrease the risk of biomolecule degradation and the amount of unwanted chemicals that will have to be eliminated in the following of the procedure; (iii) the reaction time was extended 3.5 times to compensate for possible slower kinetics; (iv) extra DMSO was added to favor ligand dissolution.

After incubation, the reaction was halted by addition of 500 μL (10 volumes) of a cold (−20 °C) solution of $LiClO_4$ at 3% w/v in acetone. The tube was placed at −20 °C for 30 min and the JK313-P2 conjugate was next recovered by centrifugation at 16,000 g, 4 °C for 10 min. After, removal of the supernatant 500 μL of cold acetone was added to the pellet, the tube centrifuged once more, the supernatant discarded, and the solid residue dried on the bench. The product was next dissolved in 50 μL of water and the whole precipitation/centrifugation/redissolution cycle realized a second time.

The purified conjugate was analyzed by electrophoresis on a 20% acrylamide/bisacrylamide (19:1) denaturing gel (TBE 1×, 7M urea)—see for instance (Andrus & Kuimelis, 2000). After staining with methylene blue, both conversion and recovery yields were estimated to be over 95% (a single band migrating at 0.89 times the speed of the starting reagent was detected, with roughly the same intensity). The purity of the product was further confirmed by mass spectrometry on an Exactive Orbitrap instrument (Thermo, Bremen, Germany) using the electrospray ionization HESI probe in negative ion mode, without heating and in conditions that avoid degradation. The main peak for the theoretical isotopic distribution of $(M\text{-}3\,H)^{3-}$ is at $m/z = 1158.88836$ and we experimentally found it at 1158.89146.

Please note that purification by precipitation (Gosse et al., 2004; Vasilyeva et al., 2016) is easier, faster, and results in a better overall yield than procedures relying on gel electrophoresis or on reverse-phase chromatography. The key points are to be able to reach a 100% functionalization of the initial oligonucleotide, which is usually the case with click chemistry, and to have all side reagents soluble in the organic solvent.

4.2.4 The functionalized forceps

Assembly of FKBP12-P1 and JK313-P2 conjugates onto the DNA scaffold was realized by hybridization to the complementary overhangs that had been generated by the digestion of the tips with the Nb.BbvCI restriction enzyme. Ligation with the T4 DNA ligase yielded stable and fully operational J-DNA forceps that could be stored for months at −80 °C (Kostrz et al., 2019).

4.3 Sample preparation

4.3.1 Flow cell fabrication

Microfluidic chambers functionalized with anti-digoxigenin were made along the lines given in our previous publications (Duboc, Fan, Graves, & Strick, 2017; Kostrz et al., 2019; Revyakin, Ebright, & Strick, 2005). Briefly, each of them is based on two $24 \times 60\,\text{mm}^2$ glass coverslips (Menzel-Glaeser, Braunschweig, Germany), both ≈ 180 μm thick and one having been sand-blasted to provide two 2 mm diameter access holes. After sonication once in ethanol and twice in 1 M KOH, each time followed by rinsing in water, these two windows were dried under a flow of nitrogen gas. Their surfaces were then activated in an air plasma for 90 s (PDC-002 oven, Harrick Plasma, Ithaca, NY), spin-coated with the adhesion promoter AR 300–80 New (Allresist, Strausberg, Germany), and baked for 5 min at 110 °C on a hot-plate. Finally, a 280 kDa polystyrene solution at 0.5% w/v in toluene was spin-coated and the two glass parts were allowed to dry on the bench.

The flow cell was next formed by sandwiching between the above coverslips two layers of Parafilm M (Bemis, Oshkosh, WI) into which a

$2 \times 50\ mm^2$ slit had been opened with a cutting plotter (Silhouette Portrait, Ankersmit, Bouvines, France). After softening the wax at 60 °C and applying homogeneous pressure to the top slide, this resulted in a ≈ 150 μm high microfluidic channel with a total volume of ≈ 15 μL. Reservoirs made of 5 mm-long and 5 mm-diameter sections of plastic tubing (in fact simple sections of 0.5 mL microtubes) were affixed at both inlet and outlet with epoxy glue (Sader bi-composants epoxy rapide, Bostik, Colombes, France). The maximum fill volume of each of these access ports was ≈ 100 μL.

To finalize the coating, the chamber was filled and incubated overnight at 37 °C with a solution of sheep IgG polyclonal anti-digoxigenin (11333089001, Roche Diagnostics, Mannheim, Germany) prepared at 10 μg/mL in phosphate buffer saline. Subsequently, passivation was realized by exchange for mTOR buffer (see Section 4.4.1) in which the BSA concentration has been adjusted to 50 mg/mL. After incubation for 90 min at 37 °C this solution was replaced with mTOR buffer and the flow cell could be used immediately or stored at 4 °C for a few days (provided it is kept in a container the atmosphere of which is humidified thanks to wet tissues, a tight sealing being secured with Parafilm M).

4.3.2 Bead injection and tethering through J-DNA

The scaffold was first attached to 1 μm diameter magnetic beads coated with streptavidin (MyOne Streptavidin C1, Thermo Fisher Scientific, Les Ulis, France) (Kostrz et al., 2019). More precisely, 5 μL of particle suspension at 10 mg/mL were mixed with 200 μL of mTOR buffer and the beads were agglomerated with a magnet. The supernatant was discarded and the pellet resuspended in 35 μL of mTOR buffer. Then, we added 1 μL of functionalized J-DNA scaffolds at 50 pM in the buffer used for their storage at −80 °C (see Section 4.2.4) and the mixture was incubated for 2 min at room temperature.

The beads attached to the J-DNA by the biotin sticky-ends were injected into the flow cell and left for 10 min to sediment and bind to the lower glass slide by the digoxigenin sticky-ends (Fig. 1A). Unattached beads were washed out with 2 mL of mTOR buffer under zero magnetic force.

Please note that to avoid contaminating the field of view with free beads, washing steps are always made in the direction opposite to the one of bead injection. The ad hoc reservoir is just filled with the replacement solution, which then flows through the channel by gravity. From time to time some liquid is also removed from the second reservoir, so as to guarantee a complete exchange.

4.3.3 Bead selection

In a well-chosen field of view we generally found around 60 individual beads tethered to the surface. However, from our experience not all scaffolds are able to interact, possibly because of unsuccessful ligation of both the protein- and the ligand-oligonucleotide conjugates. To decide which J-DNA were worth further analysis, force-cycling between a low (<0.01 pN) and a high (>1 pN) force was performed during 5 min, the magnets being moved every 20 s. Only the forceps displaying a two-step increment in extension upon increasing the force were retained. Indeed, it meant that we could observe, first, the extension of the scaffold, and, second, the dissociation of the FKBP12•JK313 complex—see (Kostrz et al., 2019; Wang et al., 2023) for examples. On average, only about 25% of the scaffolds were functional, that is 15 per experiment, a value in agreement with our previous statistics (Kostrz et al., 2019).

4.4 Data collection

4.4.1 General conditions

All experiments were realized in mTOR buffer, that is 20 mM HEPES, pH 7.8, 100 mM KCl, 5 mM $MgCl_2$, 0.1% w/v Tween-20, 0.5 mg/mL bovine serum albumin (BSA), and 2 mM dithiothreitol (DTT) (Kostrz et al., 2019).

The temperature was set at 28 °C, thereby enabling the ligands to dissociate with kinetics that are fast enough to collect as many events as possible yet slow enough to be easily detected.

The applied force was set around 0.1 pN (see Section 4.1). Since this parameter mainly tunes the separation distance between the tips of the unlooped forceps, the selected value appears as a compromise between the need to pull enough to distinguish the two conformations from each other and the need to have the dissociated receptor and ligand close enough to reassociate quickly (thereby once more enabling to collect as many events as possible).

For our forceps functionalized with FKBP12 and JK313 and no competitor present, such choices empirically corresponded to ratios ρ around 2 ± 1 and to densities of looping-unlooping cycles per hour around 200 ± 50. The quite large dispersion we could observe for both parameters was interpreted as a consequence of the dispersion in applied force, and thus in effective concentration B_2.

The minimal duration for acquisition at a given concentration in competitor was set at 6 h. However, depending on the schedule (day or night) longer measurements have been realized, up to $T = 18$ h. Overall, we aimed at collecting at least $N = 100$ dwell times for each condition.

A typical experiment included 5 to 9 concentration points and lasted 3 to 5 days. Keeping single-protein functioning for such a long period is challenging and only about half of the initially interacting molecules worked for the full duration of the experimental run.

4.4.2 Concentration in competitor

Initiating the titration with a ρ value of a few units at $B_1 = 0$ nM seems to leave enough space to study the linear dependence of our observable on the competitor concentration. Yet, as we have seen in Section 2.2, the number of transitions decreases hyperbolically when B_1 is added. For instance, introducing the kinetic parameters used for the simulations of Section 3.1 in Eq. (11), we find that 8 h-long measurements result in only $N \approx 120$ looping-unlooping cycles for [JK313] = 180 nM, that is 3.6 K_{D1}. As a rule of thumb, for all ligands we linearly varied competitor concentration, starting at 0 nM and ending at 3 K_{D1}^{bulk}, that is three times the dissociation equilibrium constant that had been independently measured in a competitive fluorescence polarization assay (Fig. S1). Doing so, for the last point of the titration we reached a lower N limit that did not depend on the in-*trans* reaction. Indeed, from Eq. (11) it comes $N_{\text{min}} \approx Tk_{-2}/(1 + 4K_{D2}/B_2)$. Furthermore, with increments smaller than 0.5 K_{D1}^{bulk} the B_1 concentration range covered values both below and above the sought-after single-molecule dissociation equilibrium constants, K_{D1}^{single}.

We thus worked with increasing competitor concentrations and, each time, the chamber was washed with 2 mL of the new solution, as described in Section 4.3.2. During this operation, the force was increased to ≈ 10 pN to minimize the chances for beads to get stuck to the surface by the flow. A punctual control experiment was performed to assay the completeness of the exchange: after a titration by FKBP12 up to 60 nM we reversed to pure mTOR buffer and ρ at 0 nM was measured once more. From the value obtained it appeared that FKBP12 was still present at ≈ 5 nM. With about 10% of the original solution remaining in the sample chamber it comforted us in our decision to proceed from 0 towards the largest ligand concentrations. It also allowed us to estimate a precision on B_1 ranging from 10 to 1% as the titration progress. Considering pipetting uncertainty to be around 5%, we fixed error bars at an arbitrary, but realistic, 10% for all concentrations.

Relying on improved fluidic actuation—see for instance (Gosse et al., 2020) for a state-of-the-art system—it should be possible to measure ρ with competitor concentrations set in random order. In addition, it would be advisable to double some of the concentration points to demonstrate both the efficacy of solution exchange and the absence of any temperature and/or force drifts, two parameters which can significantly affect the dwell time distributions.

4.5 Signal trajectory analysis

4.5.1 Raw data processing

Raw time traces were imported from the magnetic tweezers setup and processed on a personal computer using routines encoded in Python. First, the vertical position of a reference bead stuck onto the lower glass slide was subtracted from the one of the bead of interest, so as to correct the signal from the effects of environmental disturbances (slow mechanical relaxation for instance). It resulted in data corresponding to the scaffold extension z (Fig. 1B,C). Second, artefactual points due to errors from the tracking software or to temporary degradations of the experimental conditions were removed. More specifically, positions punctually falling outside of the expected measurement range were automatically discarded whereas longer problematic periods were identified by eye and excised manually (500 s of healthy trace being also removed on both sides of the degraded part). Third, long-term drift was once more corrected, this time by fitting a polynomial function of degree 10 to the several hours-long trace and subtracting it. We indeed noted that, even after having used the stuck bead for drift correction, slow changes in the positions of both looped and unlooped conformations remained, the associated ≈ 10 nm amplitude being problematic for subsequent automated analysis. This operation yielded stabilized extension data Z which values were now located around zero. Fourth, the signal was averaged relying on a sliding window of 50 video frames (≈ 1.6 s). Averaging significantly reduced noise, the position standard deviation associated with each conformation roughly decreasing from 70 to 20 nm. Therefore, it eased the work of the HMM learning algorithm described in the following. Incidentally, it also flattened confusing downward spikes which we attributed to non-specific interactions with the surface and which the software wrongly tagged as brief transits through the L_{cis} state.

4.5.2 Conformation trajectory determination

In our initial work (Kostrz et al., 2019) conversion of constant-force signal trajectories into conformation trajectories was achieved manually. However, visual inspection is time-consuming and somehow subjective.

Looking for an automated procedure, we first compared the thresholding and the HMM approaches—see for instance (Kinz-Thompson, Bailey, & Gonzalez, 2016) for a detailed discussion on the pros and cons. As evidenced by Fig. 1B,C, it quickly appeared that the poor separation between the L and U extensions, as well as the high signal-to-noise ratio of our acquisitions, prevented all use of the former strategy. Hence, contrary to choices made in several publications dealing with the looping-unlooping of nucleic acid scaffolds (Han et al., 2009; Rieu et al., 2021a), we turned ourselves to hidden Markov modeling.

Once more we relied on the Python hmmlearn package. The routine, which belongs to the Viterbi group of HMM learning algorithms, was trained on a limited portion of the considered signal trajectory, so as to characterize the two conformations. Typically, 2 to 20 min-long segments were employed, depending on the density of events. Furthermore, we took care to have a quite well-balanced representation of L and U (not less than 10%). Once the learning completed, the software automatically assigned a conformation, either looped or unlooped, to every time point of the whole signal trajectory. Next, we visually checked the result for major defects and, if any, corrected them by hand. Finally, all the transitions present in the obtained binary trajectory were identified and the dwell times were computed accordingly.

In line with our two conformations, we here assumed two hidden states. It may apparently be in contradiction with a Markov model including three chemical states, as displayed in Section 3.1. However, we have seen in Section 3.2 that the unlooped PDF is nearly mono-exponential; consequently, such simplification was not an issue and the program could converge satisfactorily.

5. Dissociation equilibrium constant determination

In addition to JK313, we benchmarked our technique on two competitors reported to have an affinity for the receptor respectively 5-fold larger and 5-fold smaller, PPU278 and SP334, (Pomplun et al., 2018).

All observations were realized at the single-forceps level. For each measurement, that is each concentration point, the ratio ρ was determined according to Eq. (1). Next, following Eq. (2), the collected datasets were fitted to straight lines obeying $\rho = I + S \times B_1$ (Fig. 3A–A"). Consequently, the dissociation equilibrium constant associated with each single-receptor could be computed as $K_{D1}^{\text{single}} = I/S$. As evidenced in Table 1, for all three ligands it led to mean values that agree quite well with the K_{D1}^{bulk} constants previously obtained in a bulk assay; discrepancies are within a 50% range, which is reasonable if we consider that neither the buffer nor the temperature were exactly identical. However, standard errors are larger in our case. To assess the heterogeneity of our measurements in a synoptic way, we normalized each K_{D1}^{single} by K_{D1}^{bulk} (Fig. 3B). It appears that several outliers are presents in each population, a fact which at the moment we attribute to poor fits to Eq. (2).

Considering now a pool of forceps, a second strategy to derive K_{D1} consists in taking benefit of the dispersion in effective concentration that ensues from bead-to-bead force differences (see Section 4.1 and Fig. S4). Indeed, the I parameter, which identifies with K_{D2}/B_2, can be spread over a decade (Fig. 3C). Ensemble fits to $-\log I = -\log S + pK_{D1}$ were thus achieved and it yielded equilibrium dissociation constants in fair agreement with the fluorescence polarization assay ones and with the means calculated above (Fig. 3C and Table 1). In light with the sometime wide error bars associated with I and S it is not possible at the moment to suspect any individual deviation from the expected ensemble behavior, that is any heterogeneity in the samples. In the framework of a homogeneous receptor populations, log-log plots offer the possibility of directly reading the pK_{D1} values as y-intercepts (Fig. 3C').

6. Discussion

6.1 Thermodynamic investigations on single fluctuating forceps

We introduced a DNA construct on which interacting reagents have been attached, this assembly being used as a sensor to acquire information on the binding to one of the engrafted molecules to a third partner coming from the solution. Like concepts have already been invoked in earlier works, through terms such as "detector" (Wang et al., 2019) or "probe" (Rieu et al., 2021a). As far as our titration protocol is concerned, it is very similar to the ones implemented by other groups (Han et al., 2009; Priest et al., 2014; Rieu et al., 2021a), studies we were not aware of when starting our research. These previously published methodological advances are now confirmed by our own findings.

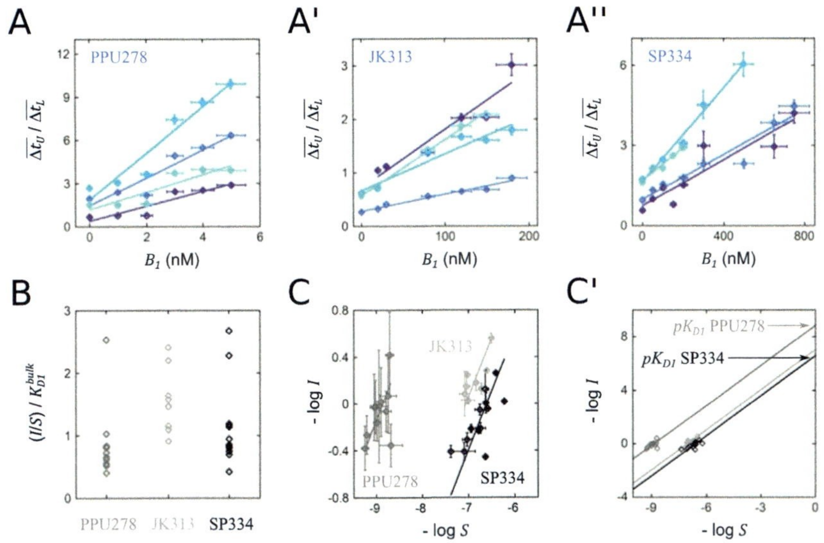

Fig. 3 Determination of the dissociation equilibrium constants for the complexes formed between the FKBP12 receptor and the PPU278, JK313, and SP334 ligands. After investigations at the single-scaffold level, additional ensemble analyses were realized to yield the quantitative results presented in Table 1. All measurements were performed at 28 °C in mTOR buffer, under an applied force of 0.1 ± 0.01 pN. **(A–A'')** Variation with the concentration in competitor of the ratio between the unlooped averaged dwell times and the looped one. For each of the three B_1 molecules, four representative datasets obtained from four single scaffolds are given. Horizontal error bars were set at 10% (see Section 4.4.2) and vertical ones were computed according to Eq. (12). A fit to a straight line yielded an intercept I and a slope S for each J-DNA. **(B)** Dispersion of the normalized dissociation equilibrium constants measured on single receptors. For each competitor the I/S ratios have been divided by the K_{D1} constant measured in bulk (Kozany et al., 2009; Pomplun et al., 2018). For PPU278 we studied $n = 11$ scaffolds in total, the sample mean m ± its standard deviation s were 0.85 ± 0.58, for JK313 we had $n = 8$ and $m \pm s = 1.56 \pm 0.40$, and for SP334 $n = 13$ and $m \pm s = 1.12 \pm 0.65$. **(C)** Variation of the intercept with the slope in log-log scale. Solid lines correspond to linear fits of slope 1 realized on the pool of single-molecule data collected for each ligand. **(C')** Large-scale view of the preceding panel, which allows one to directly read pK_{D1} on the ordinate axis (as indicated for two of the ligands).

One key feature of all these scaffold-based titrations is that the results gathered on the in-*trans* reaction do not depend on the information one can have on the in-*cis* reaction (the K_{D2}/B_2 factor of Eq. (2) can be eliminated). This observation is reminiscent of the technique developed by Schild more than 65 years ago (Arunlakshana & Schild, 1959). Indeed, he

Table 1 Dissociation equilibrium constants for the complexes formed between the FKBP12 receptor and the PPU278, JK313, and SP334 ligands. Bulk data were collected at room temperature in 20 mM HEPES, pH 8.0, 20 mM NaCl, 0.002% w/v Tween-20 (Pomplun et al., 2018). J-DNA measurements were performed at 28 °C in mTOR buffer, under an applied force of 0.1 ± 0.01 pN.

Type of determination	K_{D1} **(nM)**		
	PPU278	**JK313**	**SP334**
Bulk fluorescence polarization assay	1.83 ± 0.10	59.3 ± 07.0	249 ± 02
Mean of the single-scaffold values (±SE)	1.56 ± 0.32	92.3 ± 8.37	278 ± 44
Linear fit of the - $\log I$ vs. - $\log S$ data (±SE)	1.36 ± 0.20	87.8 ± 10.5	246 ± 33

could measure dissociation equilibrium constants for competitive antagonists using various tissues as sensors, without any a priori knowledge on the receptor physiology and functioning mechanism, or on the agonist binding reaction (Colquhoun, 2007, Colquhoun, 2006). Another similarity is that, on Fig. 3A-A", one could read K_{D1} for each forceps as the B_1 value for which ρ at zero is doubled; in Schild's method the same is true for the dose ratio $r = 1 + B_1/K_{D1}$. This remark can be explained by the fact that, when the competitor is present, a same mathematical factor accounts for the decrease of the concentration in free receptor.

Finally, it may seem odd to determine a thermodynamic parameter, K_{D1}, on a unique receptor. Actually, one expects this kind of measurement to be reliable at the sole ensemble level. However, our approach is justified if the ergodic hypothesis is verified and if acquisitions are long enough. The field of single-molecule mechanochemical sensing is new (Hu, Tahir, & Mao, 2022) and further investigations are expected to soon better define its core concepts.

6.2 Related technical hurdles

So far, we could not conclude on the presence or absence of heterogeneity in our biological system, mainly because the curves acquired on single-receptors are quite noisy (Fig. 3A–A"). One of the main difficulties we faced was to ensure stable experimental conditions over several days. For instance, both K_{Di} can be influenced by thermal drift and B_2 can be modified by force drift (itself a consequence of the mechanical drift affecting the position of the bead with respect to the one of the magnets).

In addition, B_1 may have varied during some of the measurements, or the variation of this parameter we planned to impose from measurement to measurement may have been poorly controlled, because of problematic solution exchanges or because of drug adsorption onto the flow cell surface.

Among the studies that employed an analogous titration protocol (Han et al., 2009; Priest et al., 2014; Rieu et al., 2021a) none provides a clear scaffold-to-scaffold comparison. It is therefore hard for us to draw any conclusion on future perspectives. Nevertheless, datasets in (Rieu et al., 2021a) all report on single-objects and they appear cleaner than ours. As a matter of fact, their sensor was a DNA hairpin with mean dwell times around 10 ms and their in-*trans* reactions, the hybridization of 5 to 7-base pair long duplexes, were also faster. Consequently, with a rapid tracking system (Rieu et al., 2021b) they could perform two titrations in a single day with a quality criterion of at least $N \approx 4000$ transitions per concentration point, consequently minimizing the risk of drift. The idea that, in this kind of assay, variation in the experimental conditions are the principal source of uncertainties is also backup by a series of successive measurements they realized in the same condition on a same unique scaffold: error bars obtained by bootstrapping on each single acquisition are roughly five times smaller than the discrepancies noticed between the different acquisitions.

6.3 Possible implementation on other detection platforms

As seen in Section 1, DNA forceps (Gosse et al., 2019) have already been mounted on numerous single-molecule nanomanipulation setups (Bell & Molloy, 2023; De Bragança et al., 2023; Halvorsen et al., 2011; Kilchherr et al., 2016; Kostrz et al., 2019; Ma et al., 2019; Mickolajczyk et al., 2022; Shrestha et al., 2021). In all these studies, conformation monitoring is achieved in real-time by following the position of one of the two handles, the micron-size bead. Labelling the shanks/tips with fluorophores engaged in energy transfer constitutes an alternative way towards the same goal (Bartnik et al., 2020; Nickels et al., 2016; Schickinger, Zacharias, & Dietz, 2018; Vafabakhsh & Ha, 2012). With a possibly higher grafting density at the flow cell surface, our assay could then benefit from improved multiplexing although efficient solutions are yet to be fully validated to solve the photobleaching issue—see (Schickinger et al., 2018) for instance. Furthermore, DNA constructs designed for fluorescence energy transfer (FRET) are not typically compatible with force-based strategies—with some notable exceptions like in (Nickels et al., 2016); thus, with them energy landscape mapping remains out of reach.

We have computed Eq. (2) for a single scaffold transiting back and forth between the looped and the unlopped configurations, over the entire duration of the measurement. However, the same relation could be established for an ensemble of scaffolds scrutinized at an instant in time (i.e. taking the population average rather than the time average). Hence, our assay could also be relevant for single-molecule techniques in which end-point observations are the rule, cryo-electron microscopy (Funke et al., 2016; Ke, Meyer, Shih, & Bellot, 2016) or nanopore detection (Beamish, Tabard-Cossa, & Godin, 2019; Zhu et al., 2023) to name a few. Although looking for heterogeneity would not be possible anymore, confirming the number of conformations should still be feasible.

Further extending our proposition, applying titration protocols to DNA scaffolds should even work at the ensemble level, relying on a fluorescent signal (Funke et al., 2016; Ke et al., 2016; Mak, Marszal, Matscheko, & Rant, 2023) or on the shift of a band in electrophoresis (Koussa, Halvorsen, Ward, & Wong, 2015) to determine the ratio of the unlooped population over the looped one.

Acknowledgements

We thank P. Purder and J. Kolos (TU Darmstadt) for assistance in ligand synthesis, L. Jaouen, E. Point, and M. Garrigues (IBPC, Paris) for preliminary experiments on click chemistry, C. Buré and F. Rosu (IECB, Bordeaux) for mass spectrometry measurements, and finally M. Rieu (LPENS, Paris), D. Colquhoun (University College London), D. Lacoste (ESPCI, Paris), and L. Dinis (Universidad Complutense de Madrid) for helpful discussions on data analysis. The synthesis of the ligands was supported by grant 433472263 from the DFG (to F.H.). The biophysical study was supported by grants ForcaRapa from CNRS, Gephyrip from Labex MemoLife, and jDNA from PSL valorization (to T.S. and C.G.). The Molecular Motors and Machines team at IBENS has received a "Coup d'élan" from the Fondation Bettencourt Schueller and is an "Equipe labellisée" by the Ligue Nationale Contre le Cancer. F.S. benefited from a doctoral fellowship from the Ministère de l'Enseignement Supérieur et de la Recherche.

Author contributions

C.G. conceived the project. T.S. and C.G. supervised research. S.P., C.M., and F.H. contributed ligands. D.K. and M.F. contributed protein and DNA scaffold. M.F. and C.G. prepared the oligonucleotide-ligand conjugate. F.S. performed single-molecule measurements and numerical simulations. F.S. and C.G. analyzed data. F.S. and C.G. wrote the manuscript and edited it according to the comments from all other co-authors.

Conflict of interest

PSL valorisation has submitted a patent related to the J-DNA forceps (PCT FR2018/053533) with D.K., T.S., and C.G. among the inventors.

Appendix A. Supporting information

Supplementary data associated with this article can be found in the online version at http://doi:10.1016/bs.mie.2024.01.010.

References

Andrus, A., & Kuimelis, R. G. (2000). Polyacrylamide gel electrophoresis (PAGE) of synthetic nucleic acids. *Current Protocols in Nucleic Acid Chemistry, 1*, 10.14.11 10.14.11–10.14.10.

Arunlakshana, O., & Schild, H. O. (1959). Some quantitative uses of drug antagonists. *British Journal of Pharmacology, 14*, 48–58.

Banaszynski, L. A., Liu, C. W., & Wandless, T. J. (2005). Characterization of the FKBP—Rapamycin—FRB ternary complex. *Journal of the American Chemical Society, 127*, 4715–4721.

Bartnik, K., et al. (2020). A DNA origami platform for single-pair forster resonance energy transfer investigation of DNA-DNA interactions and ligation. *Journal of the American Chemical Society, 142*, 815–825.

Baumgartner, W., et al. (2000). Cadherin interaction probed by atomic force microscopy. *Proceedings of the National Academy of Sciences of the United States of America, 97*, 4005–4010.

Beamish, E., Tabard-Cossa, V., & Godin, M. (2019). Programmable DNA nanoswitch sensing with solid-state nanopores. *ACS Sensors, 4*, 2458–2464.

Bell, N. A. W., & Molloy, J. E. (2023). Single-molecule force spectroscopy reveals binding and bridging dynamics of PARP1 and PARP2 at DNA double-strand breaks. *Proceedings of the National Academy of Sciences of the United States of America, 120*, e2214209120.

Bevington, P. R., & Robinson, D. K. (2002). *Data reduction and error analysis for the physical sciences* (3rd ed.). Boston: McGraw-Hill,.

Brucale, M., et al. (2009). Pathogenic mutations shift the equilibria of α-synuclein single molecules towards structured conformers. *Chembiochem: A European Journal of Chemical Biology, 10*, 176–183.

Cao, Y., Balamurali, M. M., Sharma, D., & Li, H. B. (2007). A functional single-molecule binding assay via force spectroscopy. *Proceedings of the National Academy of Sciences of the United States of America, 104*, 15677–15681.

Chin, J. W., et al. (2002). Addition of p-azido-L-phenylalanine to the genetic code of *Escherichia coli*. *Journal of the American Chemical Society, 124*, 9026–9027.

Choi, H. K., Kim, H. G., Shon, M. J., & Yoon, T. Y. (2022). High-resolution single-molecule magnetic tweezers. *Annual Review of Biochemistry, 91*, 33–59.

Choi, J. W., Chen, J., Schreiber, S. L., & Clardy, J. (1996). Structure of the FKBP12-rapamycin complex interacting with the binding domain of human FRAP. *Science (New York, N. Y.), 273*, 239–242.

Colquhoun, D. (1971). *Lectures on biostatistics*. Oxford: Clarendon Press,.

Colquhoun, D., & Hawkes, A. G. (1981). On the stochastic properties of single ion channels. *Proceedings of the Royal Society B, 211*, 205–235.

Colquhoun, D. (2006). The quantitative analysis of drug-receptor interactions: A short history. *Trends in Pharmacological Sciences, 27*, 149–157.

Colquhoun, D. (2007). Why the Schild method is better than Schild realised. *Trends in Pharmacological Sciences, 28*, 608–614.

De Bragança, S., et al. (2023). APLF and long non-coding RNA NIHCOLE promote stable DNA synapsis in non-homologous end joining. *Cell Reports, 42*, 111917.

Duboc, C., Fan, J., Graves, E. T., & Strick, T. R. (2017). Preparation of DNA substrates and functionalized glass surfaces for correlative nanomanipulation and colocalization (nanoCOSM) of single molecules. *Methods in Enzymology, 582*, 275–296.

Dulin, D. (2024). An introduction to magnetic tweezers. *Methods in Molecular Biology, 2694*, 375–401.
Fuhrmann, A., et al. (2009). Quantitative analysis of single-molecule RNA-protein interaction. *Biophysical Journal, 96*, 5030–5039.
Funke, J. J., et al. (2016). Uncovering the forces between nucleosomes using DNA origami. *Science Advances, 2*, e1600974.
Galat, A. (2013). Functional diversity and pharmacological profiles of the FKBPs and their complexes with small natural ligands. *Cellular and Molecular Life Sciences, 70*, 3243–3275.
Gosse, C., & Croquette, V. (2002). Magnetic tweezers: Micromanipulation and force measurement at the molecular level. *Biophysical Journal, 82*, 3314–3329.
Geiger, T. M., et al. (2024). Discovery of a potent proteolysis targeting chimera enables targeting the scaffolding functions of FK506-binding protein 51 (FKBP51). *Angewandte Chemie International Edition, 63*, e202309706.
Gosse, C., et al. (2004). Micelles of lipid-oligonucleotide conjugates: Implications for membrane anchoring and base pairing. *The Journal of Physical Chemistry B, 108*, 6485–6497.
Gosse, C., Strick, T. R., & Kostrz, D. (2019). Molecular scaffolds: When DNA becomes the hardware for single-molecule investigations. *Current Opinion in Chemical Biology, 53*, 192–203.
Gosse, C., et al. (2020). A pressure-actuated flow cell for soft x-ray spectromicroscopy in liquid media. *Lab on a Chip, 20*, 3213–3229.
Halvorsen, K., Schaak, D., & Wong, W. P. (2011). Nanoengineering a single-molecule mechanical switch using DNA self-assembly. *Nanotechnology, 22*, 494005.
Han, L., et al. (2009). Concentration and length dependence of DNA looping in transcriptional regulation. *PLoS One, 4*, e5621.
Hinczewski, M., Hyeon, C., & Thirumalai, D. (2016). Directly measuring single-molecule heterogeneity using force spectroscopy. *Proceedings of the National Academy of Sciences of the United States of America, 113*, E3852–E3861.
Hu, C. P., Tahir, R., & Mao, H. B. (2022). Single-molecule mechanochemical sensing. *Accounts of Chemical Research, 55*, 1214–1225.
Johnson, K. C., & Thomas, W. E. (2018). How do we know when single-molecule force spectroscopy really tests single bonds? *Biophysical Journal, 114*, 2032–2039.
Joshi, D. C., Gosse, C., Huang, S.-Y., & Lin, J.-H. (2022). A curvilinear-path umbrella sampling approach to characterizing the interactions between rapamycin and three FKBP12 variants. *Frontiers in Molecular Biosciences, 9*, 879000.
Ke, Y. G., Meyer, T., Shih, W. M., & Bellot, G. (2016). Regulation at a distance of biomolecular interactions using a DNA origami nanoactuator. *Nature Communications, 7*, 10935.
Kilchherr, F., et al. (2016). Single-molecule dissection of stacking forces in DNA. *Science (New York, N. Y.), 353*, aaf5508.
Kim, B. H., et al. (2011). Single-molecule atomic force microscopy force spectroscopy study of Aβ-40 interactions. *Biochemistry, 50*, 5154–5162.
Kim, J., Zhang, C. Z., Zhang, X., & Springer, T. A. (2010). A mechanically stabilized receptor-ligand flex-bond important in the vasculature. *Nature, 466*, 992–995.
Kinz-Thompson, C. D., Bailey, N. A., & Gonzalez, R. L. (2016). Precisely and accurately inferring single-molecule rate constants. *Methods in Enzymology, 581*, 187–225.
Koirala, D., et al. (2011). A single-molecule platform for investigation of interactions between G-quadruplexes and small-molecule ligands. *Nature Chemistry, 3*, 782–787.
Kolos, J. M., Voll, A. M., Bauder, M., & Hausch, F. (2018). FKBP ligands – Where we are and where to go? *Frontiers in Pharmacology, 9*, 1425.
Kolos, J. M., et al. (2021). Picomolar FKBP inhibitors enabled by a single water-displacing methyl group in bicyclic [4.3.1] aza-amides. *Chemical Science, 12*, 14758–14765.

Kostrz, D., et al. (2019). A modular DNA scaffold to study protein-protein interactions at single-molecule resolution. *Nature Nanotechnology, 14*, 988–993.

Koussa, M. A., Halvorsen, K., Ward, A., & Wong, W. P. (2015). DNA nanoswitches: A quantitative platform for gel-based biomolecular interaction analysis. *Nature Methods, 12*, 123–126.

Kozany, C., Marz, A., Kress, C., & Hausch, F. (2009). Fluorescent probes to characterise FK506-binding proteins. *Chembiochem: A European Journal of Chemical Biology, 10*, 1402–1410.

Krasnoslobodtsev, A. V., et al. (2015). A flexible nanoarray approach for the assembly and probing of molecular complexes. *Biophysical Journal, 108*, 2333–2339.

Kuo, T. L., et al. (2010). Probing static disorder in Arrhenius kinetics by single-molecule force spectroscopy. *Proceedings of the National Academy of Sciences of the United States of America, 107*, 11336–11340.

Lajoie, M. J., et al. (2013). Genomically recoded organisms expand biological functions. *Science (New York, N. Y.), 342*, 357–360.

Ma, X. F., et al. (2019). Interactions between PHD3-bromo of MLL1 and H3K4me3 revealed by single-molecule magnetic tweezers in a parallel DNA circuit. *Bioconjugate Chemistry, 30*, 2998–3006.

Mak, S., Marszal, A., Matscheko, N., & Rant, U. (2023). Kinetic analysis of ternary and binary binding modes of the bispecific antibody emicizumab. *mAbs, 15*, e2149053.

Mickolajczyk, K. J., et al. (2022). The MIDAS domain of AAA mechanoenzyme Mdn1 forms catch bonds with two different substrates. *eLife, 11*, e73534.

Neuman, K. C., & Nagy, A. (2008). Single-molecule force spectroscopy: Optical tweezers, magnetic tweezers and atomic force microscopy. *Nature Methods, 5*, 491–505.

Nevo, R., et al. (2003). A molecular switch between alternative conformational states in the complex of Ran and importin β1. *Nature Structural Biology, 10*, 553–557.

Nickels, P. C., et al. (2016). Molecular force spectroscopy with a DNA origami-based nanoscopic force clamp. *Science (New York, N. Y.), 354*, 305–307.

Penth, M., Schellnhuber, K., Bennewitz, R., & Blass, J. (2021). Nanomechanics of self-assembled DNA building blocks. *Nanoscale, 13*, 9371–9380.

Pomplun, S., et al. (2018). Chemogenomic profiling of human and microbial FK506-binding proteins. *Journal of Medicinal Chemistry, 61*, 3660–3673.

Priest, D. G., et al. (2014). Quantitation of the DNA tethering effect in long-range DNA looping in vivo and in vitro using the Lac and lambda repressors. *Proceedings of the National Academy of Sciences of the United States of America, 111*, 349–354.

Purder, P. L., et al. (2023). Deconstructing protein binding of sulfonamides and sulfonamide analogues. *JACS Au, 3*, 2478–2486.

Raible, M., et al. (2006). Theoretical analysis of single-molecule force spectroscopy experiments: Heterogeneity of chemical bonds. *Biophysical Journal, 90*, 3851–3864.

Revyakin, A., Ebright, R. H., & Strick, T. R. (2005). Single-molecule DNA nanomanipulation: Improved resolution through use of shorter DNA fragments. *Nature Methods, 2*, 127–138.

Rico, F., et al. (2019). Heterogeneous and rate-dependent streptavidin-biotin unbinding revealed by high-speed force spectroscopy and atomistic simulations. *Proceedings of the National Academy of Sciences of the United States of America, 116*, 6594–6601.

Rieu, M., et al. (2021a). Single-molecule kinetic locking allows fluorescence-free quantification of protein/nucleic-acid binding. *Communications Biology, 4*, 1083.

Rieu, M., et al. (2021b). Parallel, linear, and subnanometric 3D tracking of microparticles with stereo darkfield interferometry. *Science Advances, 7*, eabe3902.

Rognoni, L., et al. (2012). Dynamic force sensing of filamin revealed in single-molecule experiments. *Proceedings of the National Academy of Sciences of the United States of America, 109*, 19679–19684.

Schickinger, M., Zacharias, M., & Dietz, H. (2018). Tethered multifluorophore motion reveals equilibrium transition kinetics of single DNA double helices. *Proceedings of the National Academy of Sciences of the United States of America, 115*, E7512–E7521.

Shelley, C., & Magleby, K. L. (2008). Linking exponential components to kinetic states in Markov models for single-channel gating. *The Journal of General Physiology, 132*, 295–312.

Shrestha, P., et al. (2021). Single-molecule mechanical fingerprinting with DNA nanoswitch calipers. *Nature Nanotechnology, 16*, 1362–1370.

Steinfield, J. I., Francisco, J. S., & Hase, W. L. (1998). *Chemical kinetics and dynamics*, (2nd ed.). Upper Saddle River: Prentice Hall,.

Strick, T. R., et al. (1996). The elasticity of a single supercoiled DNA molecule. *Science (New York, N. Y.), 271*, 1835–1837.

Vafabakhsh, R., & Ha, T. (2012). Extreme bendability of DNA less than 100 base pairs long revealed by single-molecule cyclization. *Science (New York, N. Y.), 337*, 1097–1101.

Vasilyeva, S. V., Filichev, V. V., & Boutorine, A. S. (2016). Application of Cu(I)-catalyzed azide-alkyne cycloaddition for the design and synthesis of sequence specific probes targeting double-stranded DNA. *Beilstein Journal of Organic Chemistry, 12*, 1348–1360.

Wang, J. L., et al. (2018). Dissection of DNA double-strand-break repair using novel single-molecule forceps. *Nature Structural & Molecular Biology, 25*, 482–487.

Wang, Y. J., et al. (2023). Combining DNA scaffolds and acoustic force spectroscopy to characterize individual protein bonds. *Biophysical Journal, 122*, 2518–2530.

Wang, Y. N., et al. (2019). Label-free single-molecule quantification of rapamycin-induced FKBP-FRB dimerization for direct control of cellular mechanotransduction. *Nano Letters, 19*, 7514–7525.

Xia, J. H., Zuo, J. C., & Li, H. B. (2022). Molecular homogeneity of GB1 revealed by single molecule force spectroscopy. *Nanoscale, 14*, 9925–9931.

Yang, D., Ward, A., Halvorsen, K., & Wong, W. P. (2016). Multiplexed single-molecule force spectroscopy using a centrifuge. *Nature Communications, 7*, 11026.

Young, T. S., Ahmad, I., Yin, J. A., & Schultz, P. G. (2010). An enhanced system for unnatural amino acid mutagenesis in *E. coli*. *Journal of Molecular Biology, 395*, 361–374.

Zhu, J. B., et al. (2023). Multiplexed nanopore-based nucleic acid sensing and bacterial identification using DNA dumbbell nanoswitches. *Journal of the American Chemical Society, 145*, 12115–12123.

CHAPTER THREE

Construction and operation of high-resolution magnetic tape head tweezers for measuring single-protein dynamics under force

Rafael Tapia-Rojo*
Physics Department, King's College London, London, United Kingdom
*Corresponding author. e-mail address: rafael.rojo@kcl.ac.uk

Contents

Abstract

Mechanical forces are critical to protein function across many biological contexts—from bacterial adhesion to muscle mechanics and mechanotransduction processes. Hence, understanding how mechanical forces govern protein activity has developed into a central scientific question. In this context, single-molecule magnetic tweezers has recently emerged as a valuable experimental tool, offering the capability to measure single proteins over physiologically relevant forces and timescales. In this chapter, we present a detailed protocol for the assembly and operation of our magnetic tape head tweezers instrument, specifically tailored to investigate protein dynamics. Our instrument boasts a simplified microscope design and incorporates a

Methods in Enzymology, Volume 694
ISSN 0076-6879, https://doi.org/10.1016/bs.mie.2023.12.010

magnetic tape head as the force-generating apparatus, facilitating precise force control and enhancing its temporal stability, enabling the study of single protein mechanics over extended timescales spanning several hours or even days. Moreover, its straightforward and cost-effective design ensures its accessibility to the wider scientific community. We anticipate that this technique will attract widespread interest within the growing field of mechanobiology and expect that this chapter will provide facilitated accessibility to this technology.

1. Introduction

In recent years, mounting evidence has emphasized the key role of protein mechanics in biology. Initial examples such as muscle contractility (Kellermayer, Smith, Granzier, & Bustamante, 1997; Rief, Gautel, Oesterhelt, Fernandez, & Gaub, 1997) or the bacterial ClpX proteolytic machinery (Maillard et al., 2011) soon made evident that the mechanical stability of proteins was a critical regulator of many physiological processes. These early instances promoted the implementation of single-molecule approaches to the study and manipulation of individual proteins with force, particularly focusing on their unfolding properties (Neuman & Nagy, 2008). More recently, fueled in part by the burgeoning field of mechanobiology and the pressing need to unveil the molecular underpinnings of cellular force-sensing pathways (Kechagia, Ivaska, & Roca-Cusachs, 2019; Vining & Mooney, 2017; Vogel & Sheetz, 2006; Yusko & Asbury, 2014), the study of protein nanomechanics has evolved into a broader and more pervasive problem, permeating many fundamental processes in biology. Consequently, understanding how protein activity is regulated by mechanical force has evolved into a central scientific question, naturally catalyzing the development of novel experimental methods to study protein dynamics under biologically relevant force conditions but also underscoring the imperative of making these approaches accessible to an ever-expanding scientific community (Bustamante, Alexander, Maciuba, & Kaiser, 2020; Choi, Kim, Shon, & Yoon, 2022; Mora, Stannard, & Garcia-Manyes, 2020).

Among the three primary single-molecule force spectroscopy techniques—atomic force microscopy (AFM), optical tweezers, and magnetic tweezers—, protein nanomechanics has been classically approached using AFM or optical tweezers. In contrast, magnetic tweezers, owing to its seeming lower resolution but also intrinsic ability to apply exceedingly low forces, was historically employed in the study of nucleic acids (Gosse & Croquette, 2002; Smith, Finzi, & Bustamante, 1992; Strick, Allemand, Bensimon, Bensimon, & Croquette, 1996). However, recent years have witnessed a transformative shift, driven by the confluence of technological advancements and the development of improved molecular tethering strategies, which have extended the utility of magnetic tweezers into the realm of single proteins (Choi et al., 2019; Löf et al., 2019; Popa et al., 2016; Tapia-Rojo, Eckels, & Fernández, 2019). Compared to AFM and optical tweezers, magnetic tweezers offer two distinct advantages: *(i)* intrinsic force-clamp conditions, affording direct manipulation of the pulling force over a broad spectrum (~0–100 pN); *(ii)* exceptional temporal stability, enabling prolonged measurements under equilibrium conditions. These inherent qualities render magnetic tweezers particularly well-suited to the study of protein dynamics under force. This proficiency has now been extensively showcased, for example, by resolving the folding pathway of membrane proteins (Choi et al., 2019), capturing the equilibrium dynamics of rare misfolded states (Tapia-Rojo et al., 2023), or revealing how protein-protein interactions are regulated by force (Tapia-Rojo, Alonso-Caballero, & Fernandez, 2020a; Le, Yu, & Yan, 2019; Franz et al., 2023), to name a few recent examples. However, while all magnetic tweezers instruments are based on the same fundamental concept (a bright-field microscope to track micron-sized beads and some means to apply and manipulate a magnetic field), there is an absence of major commercially available magnetic tweezers, and most laboratories employ homemade setups. This leads to a substantial variability in instrumental configurations and a lack of standardized instrumental designs, which has greatly hampered the wide dissemination of this technology, particularly among non-specialized laboratories.

Here, we aim to address this challenge by presenting a detailed protocol on how to assemble and set up a magnetic tweezers instrument based on our recent magnetic tape head tweezers design (Tapia-Rojo et al., 2019; Tapia-Rojo, Alonso-Caballero, Badilla, & Fernandez, 2021). Our instrument differs from other magnetic tweezers configurations in two main aspects: *(i)* we use a simplified bright-microscope assembly constructed with standard and inexpensive optomechanical components; *(ii)* we implement a tape head as the device to apply and manipulate the magnetic field (Tapia-Rojo et al., 2019). These particularities enhance its capability to apply and control pulling forces—over a broad range (0–40 pN), at high resolution (<0.1 pN), and with high bandwidth (>10 kHz)—and lead to its hallmark stability, allowing virtually driftless days-long measurements, making it particularly suitable to measure protein dynamics. Moreover, its relatively inexpensive cost and easy operation make it a highly accessible instrument to the scientific community. In this chapter, after a general description of the magnetic tape head tweezers instrument (Section 2), we detail how to assemble the microscope (Section 3.1), build and calibrate the electronics controlling the tape head (Section 3.2), and align the optical axis (Section 3.3). Next, we explain how to set up an experiment, focusing on the operation of the controlling software (Section 4), and finally, we detail how to calibrate the instrument (Section 5). All necessary material (software, circuit design, etc.) can be downloaded from (https://github.com/rafatrojo/Magnetic-Tweezers-Software). Of note, the focus of this chapter is on the instrumental aspects of our magnetic tweezers configuration; more general descriptions and protocols for conducting magnetic tweezers experiments (focusing on the fluid chamber functionalization, protein production, data interpretation, etc.) can be found elsewhere (Tapia-Rojo, Mora, & Garcia-Manyes, 2023).

2. Overview of single-molecule magnetic tweezers force spectroscopy

In magnetic tweezers, single molecules are tethered between a glass cover slide and a micron-sized superparamagnetic bead. A magnetic field is applied above the fluid chamber containing these beads, generating a pulling force on the beads and thereby stretching the tethered molecule. Using a bright field microscope, the conformational dynamics of the molecule (hallmarked by changes in its end-to-end length) are measured by tracking the superparamagnetic bead's vertical position relative to that of a non-magnetic reference bead, firmly attached to the glass cover slide.

Based on this fundamental notion, our magnetic tweezers instrument employs a custom-made bright-field microscope and implements a magnetic tape head to generate magnetic fields instead of the more common antiparallel permanent magnet arrangements. A tape head is simply a transducer that converts electrical signals into fluctuations of the magnetic field (Karlqvist, 1954). These electromagnet devices are ubiquitously used to read and record on magnetic tapes in different technologies like hard-drive disks, credit cards, or old VHS tapes. Magnetic tape heads consist of a core of magnetic material having a toroidal shape, tightly wound with wire, leaving a very narrow gap filled with a diamagnetic material. When an electric current is applied, a strong magnetic field is generated across the gap region, enabling the magnetization of a magnetic tape to store

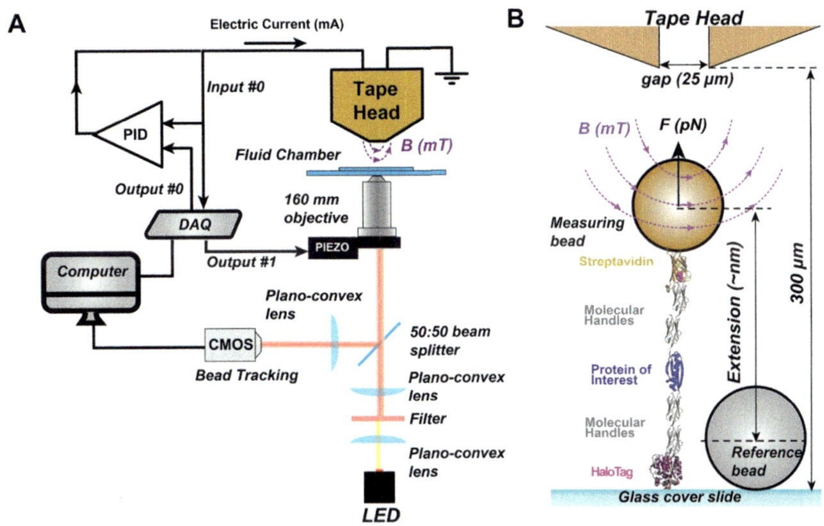

Fig. 1 Single-molecule magnetic tweezers for single protein experiments. (A) Schematics of a magnetic tweezers instrument, highlighting the bright field microscope and the computer control of the piezo scanner and magnetic tape head. (B) Detail of the single protein tether (not in scale) in a magnetic tweezers experiment. A fusion protein, containing the protein of interest flanked by some mechanically stiff proteins as molecular handles, an N-terminus HaloTag and a C-terminus biotinylated AviTag, is anchored between a glass cover slide and a superparamagnetic bead. The single protein is mechanically stretched under constant forces in the ~pN range, by magnetic fields of ~mT magnitude. For scale reference, the bead has a diameter of ~2.8 μm, while the protein tether has an extension of ~15 nm (when all proteins are folded).

information as the tape moves at constant speed over it. Over decades, this technology has been perfected to produce high magnetic fields with minimal heating and to modulate these fields at very high frequencies, a critical requirement for achieving high-density storage (Buschow, Long, & Grandjean, 2012). These capabilities render them exceptionally advantageous for magnetic tweezers applications, particularly as they permit swift force changes, hence overcoming a traditional limitation of magnetic tweezers instrumentation. Our magnetic tape head tweezers instrument specifically incorporates the #902836 tape head by Brush Industries, with a gap field ~5000 Gauss under a electric current of 1200 mA. Since the magnetic field reaching the superparamagnetic bead depends not only on the electric current but also on the distance between the tape head and the bead, we affix the tape head to a mounting piece that allows us to precisely position it at 300 μm over the magnetic beads and, hence, to exclusively control the pulling force through the electric current.

Fig. 1A presents a simplified diagram of our magnetic tweezers strategy, showing the optical pathway and electronic control of the instrument. The microscope is constructed around a 160 mm optical configuration, with illumination from the bottom and image collection using a high-speed CMOS camera. The objective lens is mounted on a nanopositioner piezo scanner. A data-acquisition (DAQ) card is responsible for digital-to-analog conversion and the management of the different input/output signals. Specifically, we use two analog output channels, one for controlling the position of the piezo and a second one for controlling the electric current (and hence force) supplied to the tape head, which is kept under electronic feedback using a Proportional-Integral (PI) circuit, ensuring precise force clamping. The instrument is controlled by a custom-made software program that communicates with the CMOS camera and DAQ card, conducts image processing, and collects the data throughout the experimentation process.

To measure the dynamics of individual proteins, we engineer and produce our protein constructs as fusion proteins, which are tethered between a superparamagnetic bead and a glass cover slide within home-made single-use fluid chambers (Popa et al., 2016). Fig. 1B shows a schematic of the molecular tethering strategy. The molecular construct comprises the protein of interest flanked by two mechanically rigid protein domains that serve as molecular handles (typically, the titin Ig32 domain or the pilin Spy0128 protein) (Tapia-Rojo et al., 2022). At the N-terminus end, a HaloTag allows for covalent and specific anchoring to functionalized glass cover slides, while a biotinylated C-terminal AviTag binds to the streptavidin-coated superparamagnetic bead. During an experiment, using a standard image analysis algorithm, we compare the vertical position of the measuring bead to that of a reference bead, firmly attached to the cover slide, enabling us to determine the molecular extension. In brief, the image analysis algorithm involves three main steps (Popa et al., 2016): *(i)* compute the Fourier transform of the bead's image; *(ii)* calculate the radial profile by integrating the intensity of the pixels at equal radial position; *(iii)* correlate the radial profile with a pre-acquired z-stack library (captured at the start of the experiment) and fit the correlation profiles to deduce the bead's position. Using this approach, our instrument offers the capability of measuring protein dynamics with spatial resolution at the ~nm scale and a temporal resolution <ms, while being able to maintain these measurements over very long times, up to several uninterrupted days.

3. Construction of the magnetic tape-head tweezers setup

In this section, we detail the steps involved in assembling and setting up the magnetic tape-head tweezers instrument. Table 1 lists the necessary parts, while Fig. 2 shows a three-dimensional model of the setup with each part tagged by the associated number indicated in Table 1.

3.1 Construction of the microscope

The magnetic tape head tweezers setup is based on a standard homemade bright field microscope built as a 30 mm cage system, illuminating the sample from the bottom, and using a 160 mm optical arrangement. Here, we describe the procedure to assemble the microscope; part numbers refer to Fig. 2 and Table 1.

3.1.1 Procedure (Time: 1–2 days)

1. Assemble the microscope scaffold (base and construction rail, parts #1 and #2). Secure it on top of a tapped optical table using four M6 screws.
2. Mount the first U-bench (#18) onto the rail carriage (#4). Attach the LED source (#16) to the U-bench using the threading adapter (the LED has an internal SM1 thread and the U-bench an internal SM1 thread). Insert the four cage rods (#3) through the four side holes of the U-bench.
3. Mount the first plano-convex lens (#19), filter (#20), and iris (#22) onto the U-bench using the four cage rods (#3). Tighten the locking screws to secure their positions temporarily (adjustments will be needed later). The use of the filter is optional (see **Note #1**).
4. Collimate the LED's beam by moving the position of the plano-convex lens (#19). This can be done by observing the shape of the beam over a white surface at different distances and adjusting the lens until a collimated beam is achieved. This would correspond to a position where the lens is ~35 mm away from the LED. Tighten the lens' locking screws to secure its position.
5. Mount the rail carriage (#4) onto the construction rail (#2) and bring it to a low position (~15 cm from the base).
6. Place the second plano-convex lens (#24) on top of the U-bench using the cage rods (#3). Its position will be adjusted later.
7. Mount the second U-bench (#25) onto a second rail carriage (#6) using three M6 screws and then on the microscope by inserting its

Table 1 List of components for the construction of the magnetic tape head tweezers.

#in Fig. 2	Company	Part Number	Part Description
1	Thorlabs	XT96P3	Base plate
2	Thorlabs	XT95-500	95 mm-Construction rail
3, 10,11	Thorlabs	ER8-4	Construction Rod (x4)
4, 6, 13, 16, 21	Newport	M-CXL95-120	Rail Carriage
5	Thorlabs	CP33/M – SMH39	Cage plate + thread adaptor
7	Ximea	MQ013MG-ON	CMOS camera
8	Thorlabs	CM1-DCH/M	30 mm cage cube
9	Thorlabs	LA1027 – CP35 (cage plate)	Mounted 35 mm plano-convex lens
12	Thorlabs	SM1ZA	Z flexor stage
14	Physik Instrumente	PI725. CDD	Piezo scanner
15	Zeiss	Different models	160 mm objective
17	Thorlabs	MCWHL5-C4-6500 K	LED source
18, 25	Thorlabs	CBB1/M	U-bench
19	Thorlabs	LA1509 – CP35 (cage plate)	Mounted 100 mm plano-convex lens
20	Thorlabs	FB6120-10 – CP35	Mounted filter
22	Thorlabs	CP20D	Iris diaphragm (optional)
23	Custom-made	*N/A*	Electronics box and circuit
24	Thorlabs	LA1509 – CP35 (cage plate)	Mounted 100 mm plano-convex lens

26	Thorlabs	BSW26R	50:50 beam splitter
27	Thorlabs	SM1S15 – SM1A39	Lens tube spacer + adaptor
28	Custom-made	*NA*	Mounting piece
29	Brush Industries	902836	Tape head
30	Newport	MS-125-XY	XY-linear stage
31	Custom made	*N/A*	Manipulation Fork
Not shown, 32	NI	USB-6341	DAQ Control Card
Not shown, 33	Different suppliers	*N/A*	12 V batteries 18 Ah
Not shown, 33	Different suppliers	*N/A*	Battery charger
Not shown, 34	PI	E-709. CHG	Piezo controller

The number associated with each part is indicated in Fig. 2. Some key components (#32–#34) are not shown in the diagram.

bottom part through the cage rods (#3) and tightening the locking screws. Mount the rail carriage (#6) onto the construction rail, approximately 10 cm above of the first rail carriage (#4).

8. Attach the 30 mm cage cube (#8) with the mounted 50:50 beam splitter (#26) to the free threaded end of the cage rods (#3). Note that the locking screws in the U-bench (#18) need to be loosened to do this, so the rods must be attached one by one not to change the position of the collimating lens.

9. Mount the cage rods (#10) on the free side of the cage cube (#8), following the same procedure as in the previous step. Insert the plano-convex lens (#9). Mount the CMOS camera (#7) on the cage plate (#5) using the c-mount thread adapter and then on the microscope using the cage rods. The position of the lens (#9) and camera (#7) will need to be adjusted later.

10. Attach the third set of rods (#11) to the top side of the cage cube, having previously inserted them through the U-bench (#25).

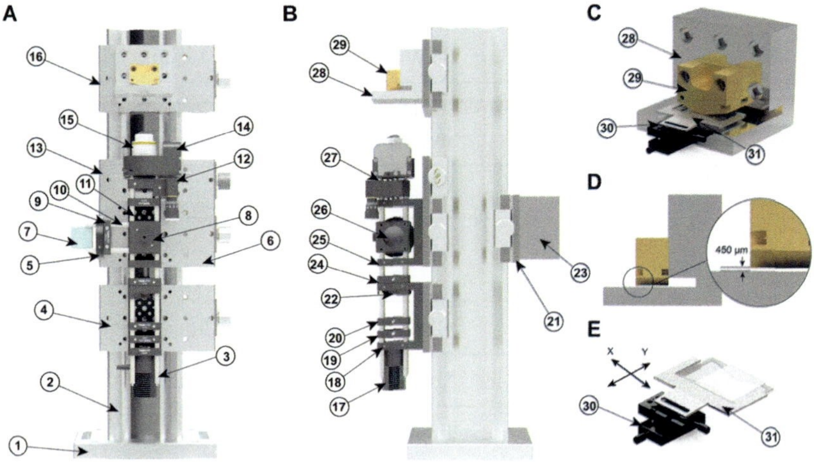

Fig. 2 **Three-dimensional model of the tape-head magnetic tweezers setup,** front (A) and side (B) views, with all parts indicated with arrows and numbers, following the details in Table 1. (C) Detail of the mounting piece with the XY-linear stage and how the fluid chamber is clamped to it. (D) Side view of the mounting piece highlighting the accurate positioning of the tape head (inset). (E) Detail of the XY linear stage and manipulation fork for displacing the fluid chamber when searching for beads.

11. Mount the z-translation mount (#12) on the microscope by inserting it through the rods (#10) and securing it in place using the locking screws.
12. Mount the piezo scanner (#14) on top of the z-translation mount by using a thread adaptor (#26). The z-mount has an internal SM1 thread, while the piezo can be purchased with different internal thread adapters, depending on the objective to be used. An SM1 to M25 has been used here.
13. Mount the 160 mm objective (#15) onto the piezo, following the piezo manufacturer's instructions.
14. Affix the mounting piece (#28) on a third rail carriage (#15) using five M6 screws. Do not mount the tape head at this stage.
15. Place a cover slide on top of the bottom opening in the mounting piece. Carefully move the rail carriage (#16) down to approach it to the objective lens and lock it in place. Ensure that the objective lens fits through the hole of the mounting piece, making direct contact with the glass cover slide and lifting it slightly when pushing. In case this is not possible, the bottom part of the aperture in the mounting piece needs to be sanded down until the objective fits comfortably. This step

is essential as different objective models might have different diameters and sizes, which require adapting the mounting piece to them.

16. Place a fluid chamber with reference beads on top of the mounting piece and try to form an image. Using the controlling software (see Section 4.2) or the camera software, visualize the image captured by the camera and find the focal plane by using the z-mount. It might be necessary to adjust manually the position of the rail carriage (#16) in case the z-mount does not have enough range. The microscope needs now to be adjusted until a clear image of the beads is formed. Note that lens #24 directs the light into the objective, while #9 forms the image into the camera. As a general guideline, lens (#24) should lie close to U-bench #18 (see Fig. 2), while lens #9 should be ~160 mm away from the objective lens.

17. Place the last rail carriage (#13) in a position so that the microscope is in focus when the rail carriage (#15) is on top of it. This rail carriage plays no role in the assembly and simply serves as a focus locker. The microscope is now ready.

Note #1: The filter is only required if using a white LED. Alternatively, a monochromatic LED can be used, which allows for sharper rings that improve the signal-to-noise ratio of the magnetic tweezers recordings. Moreover, using long-wavelength (~red) illumination minimizes the production of reactive oxygen species, improving the duration of the experiments.

Note #2: Alternative arrangements or modifications to the proposed microscope design can be introduced at the will of the user. The only critical components are the CMOS camera and the piezo scanner, as the controlling software is based on these specific parts. Using alternative components would require modifying the software code accordingly. The rest of the components can be modified without further implications.

Note #3: The mounting piece (#28) has been designed attending to the dimensions of the specific tape head model we employ (Fig. 2C). Upon fixing it to the piece, its gap is positioned precisely 450 μm away from the bottom surface, so that the beads are at a distance of 300 μm when using #1 cover slides (150 μm thick) to assemble the fluid chambers (Fig. 2D). Additionally, the XY-stage (#30), used to mount the manipulation fork (#31), is also fixed to the piece (Fig. 2E). The *.step* file for the fabrication of the mounting piece can also be found at https://github.com/rafatrojo/Magnetic-Tweezers-Software

3.2 Electronic circuit and calibration of the magnetic tape head

To achieve precise control of the electric current and allow for conducting swift force changes, we maintain the tape head under electronic feedback using a current-clamp Proportional-Integral (PI) feedback circuit (Fig. 3). It's worth noting that, due to the tape head's inductive properties, voltage clamping alone cannot guarantee control of the pulling force, as the tape head's resistance changes with the magnitude of the electric current. To address this, we incorporate a 2 Ω high-precision resistor connected in series with the tape head, clamping the voltage across it and, therefore, the electric current across the tape head (Fig. 3A).

3.2.1 Description of the PI feedback circuit

Fig. 3B depicts the schematic of the PI circuit, while Fig. 3C presents a photograph of the circuit, with specific components highlighted. In essence, the setpoint signal (*SIGNALS-3/4*) serves as the input to a summing amplifier (op amp *A0*), allowing for the superimposition of a second input signal (*SIGNALS-5/6*), which is optional unless we need to generate complex force signals, as discussed in (Tapia-Rojo, Alonso-Caballero, & Fernández, 2020b). The voltage drop across the high-precision resistor (*R17*) is measured by the differential amplifier *A1* (*SIGNALS-1/2*, going to the DAQ). An error amplifier (*A2*) calculates the difference between the setpoint and the measured voltage, serving as the input to the proportional-integral (PI) modules (*A3* and *A4*, respectively). The output of these PI modules is then amplified by an output gain amplified (*A5*). Lastly, a 60 W power amplifier (*A6*) generates the electric current supplied to the tape head.

When adequately calibrated, this feedback control allows for rapid changes of the electric current applied to the tape head, achieving response times faster than 50 μs, only limited by the slew rate of the power amplifier (*A6*). This capability enables swift changes in force or the generation of complex force signals, such as oscillating or randomly fluctuating forces (Tapia-Rojo et al., 2020b). A second setpoint containing the time-dependent component is superimposed using the summing amplifier (*SIGNALS-5/6*). This component is controlled by a second DAQ card, while the primary input (*SIGNALS-3/4*) defines the time-independent component. To ensure optimal time response, the circuit is powered by a pair of large-capacity 12 V batteries. This choice provides a cleaner source of electric current, overcoming the limitations of conventional power

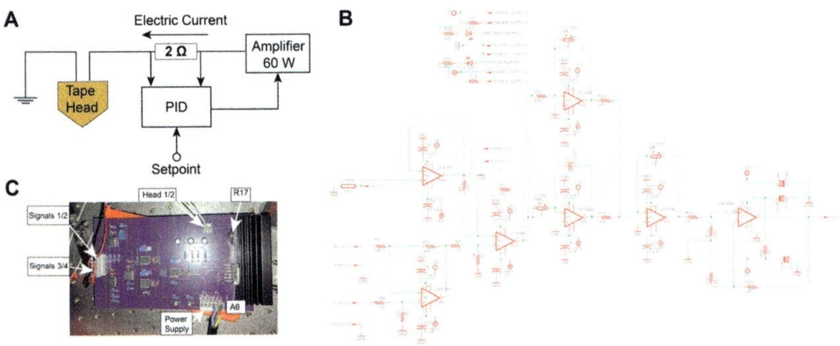

Fig. 3 Electronic design of the PI circuit controlling the tape head. (A) Block diagram showing the electronic design to control the force applied by the tape head. (B) Schematic of the PI circuit. (C) Photograph of the PI circuit, highlighting some of its components.

supplies, more evident when executing fast stepwise electric current changes. As proposed, employing large-capacity batteries (for example, 18 Ah) offers extended autonomy to the magnetic tweezers setup, allowing for long single-protein recordings.

3.2.2 Calibration of the PI circuit: Procedure (Time: 1–2 days)

1. Assemble the PI circuit according to the schematic shown in Fig. 3C. The Eagle (Autodesk) files for producing the PCB boards, including the parts list, can be downloaded from https://github.com/rafatrojo/Magnetic-Tweezers-Software
2. Solder an MTA connector to cable track-#1 of the tape head and connect it to the circuit (*HEAD-1/2*, Fig. 3B, C). Note that the tape head has two tracks, and only track-#1 is employed here.
3. Wire the batteries in series to set the +12 V, −12 V and ground channels.
4. Take two BNC cables and solder an MTA connector to one end of each cable. Use one of these cables to connect the input channel of the circuit (*SIGNALS-2/3*) to a function generator and the second one to connect the output of the circuit (*SIGNALS-3/4*) to an oscilloscope. Additionally, connect the output of the function generator to a different channel on the oscilloscope to be able to compare the input and the output signals.
5. Generate a square waveform with a frequency of 1 Hz and a peak-to-peak amplitude of 1 V as the function signal. While the exact amplitude

is not crucial, electric currents over ~1.5 A (corresponding to an amplitude of 3 V) could damage the tape head.

6. Calibrate the PI circuit by adjusting the output gain (*T_gain*), the gain of the P module (*P_gain*), and the I module (*I_tau*). Start by setting the *T_gain* until you observe an output signal on the oscilloscope that resembles the input square wave. Then, tune the *P_gain* to improve the quality and time-response of the output signal, and lastly, the *I_tau* gain, trying to optimize the time-response of the circuit. You can test the circuit's time response by increasing the input waveform's frequency, aiming for an optimal response time below 50 μs.
7. Once the circuit is calibrated, mount it on a rail carriage (Fig. 2, #23) to place it on the microscope. Connect the circuit to the DAQ card. The input (*SIGNALS-3/4*) should be connected to the DAQ's analog output #0, while the output (*Signals 1/2*) to the DAQ's analog input #0 (respectively, channels 15–16 and 1–2, when using the indicated DAQ model).
8. Test the connections by transmitting a constant voltage signal using DAQ's National Instruments software (*NI MAX*). For example, apply 1 V to the output channel #0 and read the input channel #0 using a multimeter, and ensure that they match.
9. The tape head is ready to use.

Note #4: The magnetization of the tape head saturates at ~1200 mA, which implies that the magnetic field-magnetization relationship is linear up to ~1000 mA, which is the range of electric currents implemented in the setup (the software program caps currents above 1200 mA).

3.3 Alignment of the tape head

To apply calibrated and reproducible forces, it is crucial to work precisely beneath the tape head's gap. Achieving this requires precise alignment of the optical axis of the microscope with the tape head. Although the mounting piece is designed to maintain track #1 of the tape head aligned with the optical axis, the narrow width of the head's gap (~25 μm) necessitates a fine alignment. This can be achieved by adjusting the position of the rail carriage (#16) and of the z-mount (#12).

3.3.1 Procedure (Time: ~1 h)

1. Mount the tape head on the mounting piece using the dowel-pin system. Note that 1/8-inch pins need to be used and that, when put in the mounting piece, they need to be tightly inserted; otherwise, this would lead to uncertainties in the tape head's position.

2. Lower the position of the rail carriage until the objective contacts the tape head and correct its position using the z-mount (#12) to find the focus until an image of the tape head is achieved.
3. The gap can be easily identified as a black stripe, in contrast to the silver-color surfaces at its sides. Upon assembling the microscope for the first time, the gap will likely be off by several microns. Find the gap's position by carefully pushing the z-mount left and right, hence displacing manually the optical axis. Align the gap's position with the microscope's optical axis by ensuring it is placed at the center of the field of view. To do this, first, tighten the rail carriage (#16) and the screws in the mounting piece (#28)—you should see how the gap shifts in position upon doing so. Then, slightly loosen the locking screws on the z-mount while displacing it laterally, aiming at setting the gap's position at the center of the field of view. Then, tighten the locking screws to fix the optical axis in place.
4. The setup is now ready to use. With this arrangement, the optical axis can remain stable over several weeks; however, check its position every week and correct it by repeating step **#4** if necessary.

4. Operation of the magnetic tape head tweezers software

Our magnetic tape head tweezers setup is controlled by a custom-developed software program (*MT_3* software) written in C++, using Qt for the graphic user interface (GUI) and OpenCV as a visual library. The software controls the position of the piezo, the input and output of the PI circuit, and carries out the image analysis algorithm to infer the relative height of the measuring bead and, hence, the molecular extension. In this section, we will describe how to operate the software to conduct a magnetic tweezers experiment. Further details on the protein production, fluid chamber preparation, and protein measurement can be found elsewhere (Popa et al., 2016; Tapia-Rojo et al., 2020a, 2023).

4.1 Installation of the *MT_3* software

The *MT_3* software has been developed as a Visual Studio 2019 project, working on Windows 10, and uses different libraries that need to be installed to compile the code, required if it needs to be modified, for example, to adjust the calibration parameters (see Section 5) or if alternative components (e.g. camera or tape head) are employed. These required software packages are: Visual Studio 2019 Community (environment for software development), Open CV 4.3 (image processing libraries), Ximea Software Package (software for the CMOS camera, including libraries for image acquisition, etc.), NI-DAQmx (software for the DAQ card, including drivers), and Qt 5.15 (language for the graphical user interface). The default version of the *MT_3* software can be readily installed, only requiring the Ximea and NI-DAQmx software for its operation.

4.2 Description and operation of the *MT_3* software

The program has three working modes: *(i)* initialize; *(ii)* region-of-interest; *(iii)* measure (Fig. 4). The program opens in mode *(i)* (Fig. 4A), which allows clicking on the button "ROI" to go to mode *(ii)*. Mode *(ii)* (Fig. 4B) shows the full field of view (~80 × 80 μm) with two white vertical lines that mark the region where the tape head's gap should be aligned (see Section 3.3). Mode *(ii)* allows selecting the measuring bead (left mouse click) and the reference bead (right mouse click), which define the ROI (central mouse click). Once the ROI is defined, the software goes into mode *(iii)*. Mode *(iii)* (Fig. 4C) is the measuring mode of the *MT_3* software, where the data is acquired and written to a binary file. The window "F(pN)" allows changing the force (sending an input voltage signal to the DAQ), and the window "OP (nm)" shows the position of the piezo (between 0 and 18,000 nm). Clicking the ticking box by the piezo position locks its position, so it is corrected every 1000 frames to keep the same focal plane, preventing long-term drift. On the left, the button "Stack" builds the z-stack library, and "Save off/on" allows writing the measured data to the binary file. If "Save on" is active, the *MT_3* program writes the measured data in a binary file (saved on the Desktop), which contains the time series of the recording structured in four columns: *1.* extension [nm]; *2.* force [pN]; *3.* time mark [μs]; *4.* electric current [mA], written as single-precision float numbers. The binary file can be opened to visualize the data in real time using a separate program (e.g. Igor Pro, Matlab, Python, ...).

The three additional windows are, from left to right, the radial profile of the measuring (blue) and reference (green) beads, the correlation profiles of each bead fitted to a Gaussian function to determine the relative height of the measuring bead as the distance between the peaks, and the ROI being recorded by the camera (see (Popa et al., 2016; Tapia-Rojo et al., 2023) for further details on the image analysis algorithm). Note that the larger the

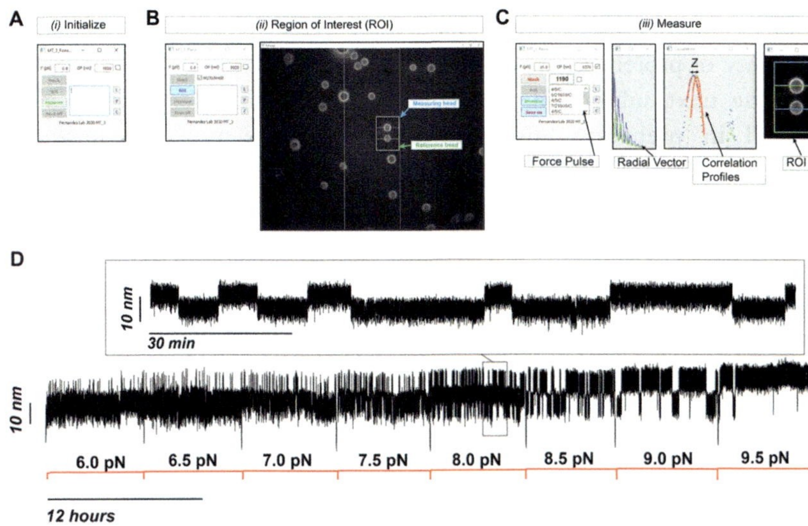

Fig. 4 The "*MT_3*" controlling software. (A) Operational mode "*(i)* Initialize", appearing as the software opens; (B) Operational mode "*(ii)* Region of Interest", showing the field of view and indicating examples reference (green) and measuring beads (blue), which define the ROI; (C) Operational mode "*(iii)* Measure", where data is acquired. The text box on the control panel allows defining a force pulse. The remaining windows, from left to right, show the real-time radial profiles for the magnetic (blue) and reference beads (green); correlation profiles, from where the protein extension "*Z*" is extracted as the distance between the Gaussian's (red lines) peaks; ROI, where the image analysis is done. (D) Example recording of a protein L monomer, measured over 48 h, exploring the range of forces where reversible folding transitions are observed.

ROI, the slower the data acquisition rate (indicated by the number in bold at the right of the "Stack" button). The ticking box at the right of the frames-per-second figure decreases the refreshing time of the images, therefore increasing the frame rate. Finally, the white text box allows for loading and using pre-defined force pulses, composed by combinations of constant force and force-ramp fragments, following the syntaxis: "*F/T/M*", where *F* is the desired force during the pulse (in pN), *T* the duration of that force-pulse (in seconds), and *M* the pulling mode, either *C* if a constant force is applied or *L* if a linear ramp is applied. Note that a linear ramp requires two lines, the first one indicating the starting force and the second one the final force value. The buttons at the right, "L", allow loading force pulses pre-written in a text file, "P" activates the pulse, and "C" closes the pulse (Tapia-Rojo et al., 2021). Fig. 4D shows a typical long recording (48 h) using a pre-defined force pulse to explore the folding probability of protein L.

In the next subsection, we will briefly explain the different steps involved in preparing and conducting a single-molecule magnetic tweezers experiment, employing the described instrument and controlling software.

5. Calibration of the setup

The magnetic tape head tweezers is designed to maintain the tape head at a fixed position so the pulling force is only controlled by the applied electric current. However, there are different factors that might lead to setup-to-setup calibration differences. Particularly, different tape head batches might have slightly different magnetic properties (this can be checked by measuring the inductance of the tape heads, which is around 0.4 mH, with a nominal error of ~20%); additionally, the manufacture of the mounting piece leads to small errors in the vertical positioning of the tape head (typically ±10 μm, although it can be manufactured up to different precisions). However, this can translate into significant force differences (see (Tapia-Rojo et al., 2019) for the full force-distance-current relationship). Hence, each setup must be calibrated individually to generate a reliable electric current-force relationship.

For our magnetic tape head tweezers, provided the tape head is fixed at a constant position so that the magnetic field (and force) can only be manipulated by the electric current, the pulling force F depends on the electric current I as (Tapia-Rojo et al., 2019):

$$F(I, z) = A \tan^{-1}\left(\frac{g/2}{z}\right)\frac{1}{1+\left(\frac{z}{g/2}\right)^2}I^2 + B\frac{1}{1+\left(\frac{z}{g/2}\right)^2}I \tag{1}$$

Where g is the width of the gap ($g = 25$ μm), z the distance between the tape head and the magnetic bead ($z = 300$ μm with the mounting piece), and A (with units of pN·mA^{-2}) and B (with units of pN·mA^{-1}), the calibration parameters to be determined. Calibrating the instrument requires some molecular quantity whose force dependence is well-described, measured for different electric current values (forces), allowing us to obtain A and B by fitting this quantity. As described in (Popa et al., 2016; Tapia-Rojo et al., 2019), our calibration strategy is based on the force dependency of the elastic properties of folding/unfolding proteins. Upon unfolding, a protein becomes an unstructured polypeptide chain, quickly equilibrating to an average end-to-end extension $\langle z \rangle$ imposed by the pulling force F, and accurately following standard polymer physics models, like the freely-jointed chain (FJC):

$$\langle z \rangle (F) = \Delta L_C \left[\coth\left(\frac{Fl_K}{kT}\right) - \frac{kT}{Fl_K} \right] \tag{2}$$

being ΔL_C is the change in contour length (the difference between the contour length of the protein and the size of its folded state), and l_K the Kuhn length, which, for an unfolded protein, typically, $l_K = 1.1$ nm.

We employ protein L (the B1 domain of protein L from *Peptostreptococcus Magnus*) as a molecular ruler, given its convenient folding/unfolding behavior, which shows measurable folding and unfolding events over a broad range of forces (~4–65 pN) (Popa et al., 2016; Tapia-Rojo et al., 2019; Valle-Orero et al., 2017). This contrasts with other common protein folding models, like the titin Ig27 (Rivas-Pardo et al., 2016), which only unfolds at very high forces (~100 pN) and folds at very low forces (~4 pN), or the talin $R3^{IVVI}$ domain (Tapia-Rojo et al., 2020a, 2020b), which only shows reversible folding/unfolding transitions over a narrow force range (~8–9 pN). To amplify our molecular signal, we use protein L octamers, providing more folding/unfolding events per unit time (Fig. 5A).

5.1 Procedure (Time: ~1 week)

1. Conduct magnetic tweezers experiments using protein L octamers exploring different force ranges, both exploring unfolding and folding forces, as illustrated in Fig. 5B. Detailed protocols for conducting experiments can be found elsewhere (Popa et al., 2016; Tapia-Rojo et al., 2019, 2023), but briefly:

- **i.** Incubate the protein L octamer construct (Fig. 5A) at a concentration of ~1 nM for 30–40 min in the fluid chamber (#31). Wash the fluid chamber with the working buffer to remove the non-anchored protein.
- **ii.** Stick two stripes of double-sided sticky tape to the manipulation fork and stick the fluid chamber to it by pressing gently (see Fig. 2E).
- **iii.** Spread a drop of immersion oil over the base of the mounting piece, and then carefully mount the fluid chamber by sliding it under the tape head while keeping the rail carriage (#28). The drop of oil keeps the fluid chamber adhered to the surface of the mounting piece.

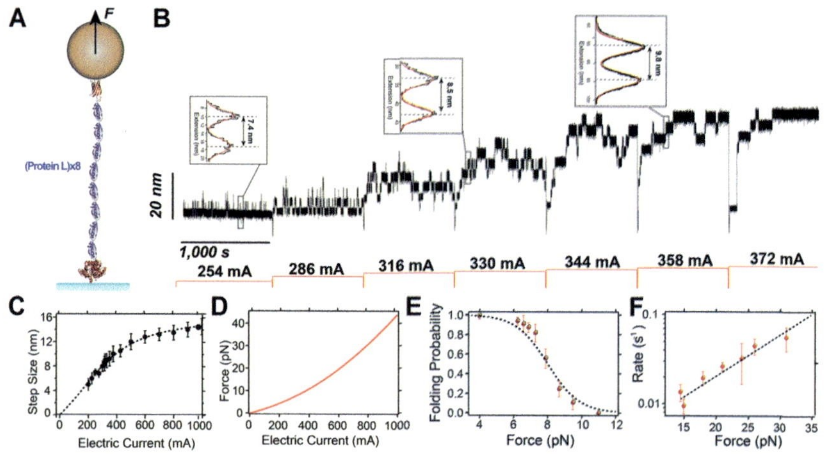

Fig. 5 **Calibration of the magnetic tape head tweezers setup.** (A) Schematics of the molecular construct employed for calibration of the instrument, containing eight repeats of protein L flanked by a HaloTag and a biotinylated AviTag. (B) Typical calibration experiment, wherein different force (electric current) pulses are probed, and the average step sizes (inset) corresponding to each force measured. (C) Step sizes of protein L as a function of the electric current. The dashed line represents a fit to Eq. 2 (using Eq. 1 for the force vs. current relationship), which determines the calibration parameters. (D) Calibration curve using the newly determined parameters. (E) Folding probability of protein L as measured with the newly determined calibration (red marks), compared to Eq. 3 with $F_{0.5} = 8.1$ pN, $R = 0.93$ pN (black dashed line). The experimental data accurately aligns with the previously described folding behavior of protein L, validating the calibration over the low force range. (F) Unfolding rates of protein L as measured with the newly determined calibration, compared to Eq. 4 with $k_U^0 = 0.0026\ \text{s}^{-1}$ and $x_U^{\dagger} = 0.44$ nm (black dashed line), validating the calibration at high force.

Screw the manipulation fork onto the XY-stage (#30) using a pair of 5/16–24 screws. Fasten the screws tight enough to have some grip but not too tight, or it will prevent displacing the chamber. Add immersion oil to the objective and lower the rail carriage until it lies on top of the focus locker (#13).

iv. Open the *MT_3* software and, using mode *(ii)*, adjust the position of the z-mount to focus the microscope. With force set to 0 pN, add your solution of magnetic beads to one of the wells of the fluid chamber and let incubate. Add then some force value (e.g. 4 pN) to pull away non-attached beads.

v. By displacing the fluid chamber laterally using the XY-stage (#30), find a measuring bead and a reference bead that is close by. Make sure the measuring bead is between the two white lines, which define the position of the tape head's gap. Define the ROI.

vi. Move the position of the piezo scanner down by 1280 nm (use the "OP" window or manually with the z-mount) and click "Stack" to acquire the z-stack library. The z-stack library is composed of 128 radial profiles for the measuring and reference beads, spaced by 20 nm steps. Bring the microscope back to focus and ensure that the correlation profiles are two narrow peaks with a well-fitted Gaussian (red solid line, Fig. 4C). Click the "Save Off" button to start acquiring data. Note that if there is a pre-existing binary file on the Desktop, the new data will be appended to it, so ensure there is not a pre-existing file so that a new one is created.

2. Measure the step sizes as a function of the electric current values. To do this, calculate a histogram of the protein's extension in the vicinity of a folding or unfolding event so that only a single event is analyzed, and fit a double Gaussian function to extract the distance between the peaks and, hence, the step size (Fig. 5B). To improve the signal-to-noise ratio, particularly necessary at low forces (<10 pN), smooth the recording (e.g., 151-points Savizky-Golay algorithm). Measure >100 steps at each electric current (force) value, evenly covering the full range between 200 mA and 1000 mA and use $N > 3$ different experiments.

3. Calculate the average step sizes $\langle z \rangle$ and plot them as a function of the electric current (Fig. 5C). Determine the fitting parameters A and B by fitting the data to Eq. 2, where the force-current relationship is given by Eq. 1. Use $\Delta L_C = 16.3$ nm and lK= 1.1 nm 17,28. This defines the calibration or Force Law (Fig. 5D).

4. Test the validity of the newly calculated force law by contrasting it against protein L's folding and unfolding properties:

4.1 *Folding equilibrium*: Over the range of forces where reversible folding/unfolding transitions are observed (should correspond to 4–10 pN, Fig. 5B), measure the folded fraction, or probability of populating the folded state P_F. To that aim, measure the dwell time in each folded state t_i being i the number of folded domains ($i = 0,\ldots,8$) and calculate the probability of populating state i as $p_i = t_i/t_t$, where t_t is the total time at a particular force, combining multiple proteins/force pulses; then, $P_F = \Sigma_i\, p_i/i$. The folded fraction should have a sigmoidal shape:

$$P_F(F) = \frac{1}{1 + e^{(F-F_{0.5})/w}} \tag{3}$$

Where, for protein L, $F_{0.5} = 8.1$ pN, and $w = 0.93$ pN. Plot the folded fraction as a function of force (using the newly determined force law) together with the analytical shape (Eq. 3), and check that the experimental data can be well-described with this expression (Fig. 5E).

4.2 *Unfolding dynamics*. Measure the unfolding rates over the range of forces where no refolding transitions are observed (should correspond to >10 pN). To that aim, for each unfolding cycle where the eight protein L domains are unfolded, measure the unfolding times t_U and calculate the unfolding rate as $r_U = 1/\langle t_U \rangle$. The unfolded rates should increase exponentially with force as:

$$r_U(F) = k_U^0 e^{Fx_U^\dagger/kT} \tag{4}$$

Where, for protein L, $k_U^0 = 0.0026\ \mathrm{s}^{-1}$ and $x_U^\dagger = 0.44$ nm. Plot the unfolding rates as a function of force (using the newly determined force law) together with the analytical shape (Eq. 4) and check that the experimental data is well-described by the previously determined expression (Fig. 5F).

5. The procedure described in step 4 ensures that the force law is correct over the full range of forces. If this is not the case, it indicates that the calibration is incorrect, likely due to poor statistics or some issue in the process. Repeat steps 1–3 in that case and check the new calibration curve following step 4.

Note #6: Our analytical description of the force dependency of the step sizes is based on the FJC model, but the worm-like chain (WLC) model can alternatively be employed; however, the FJC is usually preferred in magnetic tweezers, due to its more convenient *z(F)* form. Note that the FJC describes better the low force range (<10 pN), while the WLC works better at high forces. Typically, this implies that the FJC underestimates the contour length of a protein. For example, for protein L, $\Delta L_C = 18.6$ nm when using the WLC, being this value closer to the expected contour length than that predicted by the FJC model. Regardless of this, both models work equally well for calibration purposes when employing the right contour length value.

Note #7: Fig. 5 shows a calibration for a particular setup, where $A = 0.386$ pN mA^{-2} and $B = 9.462$ pN mA^{-1}. These are the parameters included in the *MT_3* software by default.

6. Summary and conclusions

In conclusion, this chapter has provided an in-depth exploration of our magnetic tape head tweezers instrument, focusing on its assembly and operation. Introduced in (Tapia-Rojo et al., 2021), this setup showcases the capability to manipulate very low forces at high resolution and over a large bandwidth (force range, 0–40 pN; force resolution <0.1 pN; force bandwidth >10 kHz), and, especially, with an exceptionally high stability, allowing continuous measurements of single molecules during several days. These characteristics make it particularly suitable for the study of single protein dynamics in equilibrium. While our primary focus and expertise revolve around single protein nanomechanics, this instrumentation can naturally be implemented in the study of other biomolecules like RNA, or DNA. However, certain capabilities inherent to conventional magnetic tweezers instrumentation, such as the application of torque or the facilitation of multiplexed measurements, which are particularly advantageous in the study of nucleic acids, cannot be integrated into this setup due to the employment of the tape head technology.

One notable benefit of the magnetic tape head tweezers instrument is its open design and relatively low cost, which we hope will make it accessible even to non-expert user. Moreover, this open architecture not only facilitates ease of use but also allows for flexibility in adapting and refining various aspects of the instrument for future enhancements. For instance, the possibility of replacing the specific tape head we currently employ with alternative devices featuring narrower gaps or stronger gap fields could expand the range of applicable forces if required. Furthermore, minor adjustments to the mounting piece can be made to approach the tape head's position and further expand the force range, albeit necessitating very narrow fluid chambers. Improving the time resolution, currently limited to ~1.6 kHz, could be achieved by the incorporation of faster cameras alongside improved image processing and computational methods. Recent advances in magnetic tweezers have incorporated temperature control capabilities through various approaches, which could be similarly integrated into this instrument (Seifert et al., 2020; Zhou et al., 2023). Finally, although necessitating more extensive modifications, the integration of single-molecule fluorescence with magnetic tweezers has proven successful, introducing an additional measuring coordinate to capture DNA-protein interactions or simply capture conformational changes that are orthogonal to the end-to-end length (Ivanov et al., 2018; Kemmerich et al., 2016). Similarly, this approach could be combined with our setup to investigate protein-protein interactions under force, opening exciting avenues to further delve into the molecular underpinnings of mechanotransduction processes.

Acknowledgements

We would like to thank Prof. Julio M. Fernandez for his invaluable contributions to the instrumental development and his legacy in the field of protein nanomechanics. We also thank Dr. Marc Mora for critical reading and constructive critique. RTR is supported by an EPSRC's Early Career Researcher International Collaboration grant (EP/Y001125/1) and a Royal Society Research Grant (RG\R2\232147).

References

Buschow, K. H. J., Long, G. J., & Grandjean, F. (2012). *High density digital recording*. Springer.

Bustamante, C., Alexander, L., Maciuba, K., & Kaiser, C. M. (2020). Single-molecule studies of protein folding with optical tweezers. *Annual Review of Biochemistry, 89*, 443–470.

Choi, H.-K., et al. (2019). Watching helical membrane proteins fold reveals a common N-to-C-terminal folding pathway. *Science (New York, N. Y.), 366*, 1150–1156.

Choi, H.-K., Kim, H. G., Shon, M. J., & Yoon, T.-Y. (2022). High-resolution single-molecule magnetic tweezers. *Annual Review of Biochemistry, 91*, 33–59.

Franz, F., et al. (2023). Allosteric activation of vinculin by talin. *Nature Communications, 14*, 4311.

Gosse, C., & Croquette, V. (2002). Magnetic tweezers: Micromanipulation and force measurement at the molecular level. *Biophysical Journal, 82*, 3314–3329.

Ivanov, I. E., et al. (2018). Multimodal measurements of single-molecule dynamics using FluoRBT. *Biophysical Journal, 114*, 278–282.

Karlqvist, O. (1954). *Calculation of the magnetic field in the ferromagnetic layer of a magnetic drum* (Elanders boktr; H. Lindståhls bokhandel i distribution).

Kechagia, J. Z., Ivaska, J., & Roca-Cusachs, P. (2019). Integrins as biomechanical sensors of the microenvironment. *Nature Reviews. Molecular Cell Biology, 20*, 457–473.

Kellermayer, M. S. Z., Smith, S. B., Granzier, H. L., & Bustamante, C. (1997). Folding-unfolding transitions in single titin molecules characterized with laser tweezers. *Science (New York, N. Y.), 276*, 1112–1116.

Kemmerich, F. E., et al. (2016). Simultaneous single-molecule force and fluorescence sampling of DNA nanostructure conformations using magnetic tweezers. *Nano Letters, 16*, 381–386.

Le, S., Yu, M., & Yan, J. (2019). Direct single-molecule quantification reveals unexpectedly high mechanical stability of vinculin—talin/α-catenin linkages. *Science Advances, 5*, eaav2720.

Löf, A., et al. (2019). Multiplexed protein force spectroscopy reveals equilibrium protein folding dynamics and the low-force response of von Willebrand factor. In: *Proceedings of the National Academy of Sciences, 116*, 18798–18807.

Maillard, R. A., et al. (2011). ClpX(P) generates mechanical force to unfold and translocate its protein substrates. *Cell, 145*, 459–469.

Mora, M., Stannard, A., & Garcia-Manyes, S. (2020). The nanomechanics of individual proteins. *Chemical Society Reviews, 49*, 6816–6832.

Neuman, K. C., & Nagy, A. (2008). Single-molecule force spectroscopy: Optical tweezers, magnetic tweezers and atomic force microscopy. *Nature Methods, 5*, 491–505.

Popa, I., et al. (2016). A HaloTag anchored ruler for week-long studies of protein dynamics. *Journal of the American Chemical Society, 138*, 10546–10553.

Rief, M., Gautel, M., Oesterhelt, F., Fernandez, J. M., & Gaub, H. E. (1997). Reversible unfolding of individual titin immunoglobulin domains by AFM. *Science (New York, N. Y.), 276*, 1109–1112.

Rivas-Pardo, J. A., et al. (2016). Work done by titin protein folding assists muscle contraction. *Cell Reports, 14*, 1339–1347.

Seifert, M., et al. (2020). Temperature controlled high-throughput magnetic tweezers show striking difference in activation energies of replicating viral RNA-dependent RNA polymerases. *Nucleic Acids Research, 48*, 5591–5602.

Smith, S. B., Finzi, L., & Bustamante, C. (1992). Direct mechanical measurements of the elasticity of single DNA molecules by using magnetic beads. *Science (New York, N. Y.), 258*, 1122–1126.

Strick, T. R., Allemand, J.-F., Bensimon, D., Bensimon, A., & Croquette, V. (1996). The elasticity of a single supercoiled DNA molecule. *Science (New York, N. Y.), 271*, 1835–1837.

Tapia-Rojo, R., Mora, M., Board, S., Walker, J., Boujemaa-Paterski, R., Medalia, O., & Garcia-Manyes, S. (2023). Enhanced statistical sampling reveals microscopic complexity in the talin mechanosensor folding energy landscapes. *Nature Physics, 19*, 52–60. https://www.nature.com/articles/s41567-022-01808-4.

Tapia-Rojo, R., Alonso-Caballero, A., & Fernandez, J. M. (2020a). Direct observation of a coil-to-helix contraction triggered by vinculin binding to talin. *Science Advances, 6*, eaaz4707.

Tapia-Rojo, R., Alonso-Caballero, Á., & Fernández, J. M. (2020b). Talin folding as the tuning fork of cellular mechanotransduction. In: *Proceedings of the National Academy of Sciences, 117*, 21346–21353.

Tapia-Rojo, R., Alonso-Caballero, A., Badilla, C. L., & Fernandez, J. M. (2021). Identical sequences, different behaviors: Protein diversity captured at the single-molecule level. *bioRxiv*. https://doi.org/10.1101/2021.02.24.432730.

Tapia-Rojo, R., Eckels, E. C., & Fernández, J. M. (2019). Ephemeral states in protein folding under force captured with a magnetic tweezers design. In: *Proceedings of the National Academy of Sciences, 116*, 7873–7878.

Tapia-Rojo, R., Mora, M., & Garcia-Manyes, S. (2023). Single-molecule magnetic tweezers to probe the equilibrium dynamics of individual proteins at physiologically-relevant forces and timescales. *Nature Protocols (In Press)*.

Valle-Orero, J., et al. (2017). Proteins breaking bad: A free energy perspective. *The Journal of Physical Chemistry Letters, 8*, 3642–3647.

Vining, K. H., & Mooney, D. J. (2017). Mechanical forces direct stem cell behaviour in development and regeneration. *Nature Reviews. Molecular Cell Biology, 18*, 728–742.

Vogel, V., & Sheetz, M. (2006). Local force and geometry sensing regulate cell functions. *Nature Reviews. Molecular Cell Biology, 7*, 265–275.

Yusko, E. C., & Asbury, C. L. (2014). Force is a signal that cells cannot ignore. *Molecular Biology of the Cell, 25*, 3717–3725.

Zhou, Y., et al. (2023). A novel design for magnetic tweezers with wide-range temperature control. *Biophysical Journal, 122*, 3860–3868.

CHAPTER FOUR

High-speed measurements of SNARE–complexin interactions using magnetic tweezers

Seokyun Hong[a,1], Taehyun Yang[a,1], Ara Go[b], Haesoo Kim[b], Tae-Young Yoon[c,d,*], and Min Ju Shon[a,e,*]

[a]Department of Physics, Pohang University of Science and Technology (POSTECH), Pohang, Republic of Korea
[b]Engitein Research Institute, Engitein, Siheung, Republic of Korea
[c]School of Biological Sciences and Institute for Molecular Biology and Genetics, Seoul National University, Seoul, Republic of Korea
[d]Department of Biomarker Discovery, PROTEINA Co., Ltd, Seoul, Republic of Korea
[e]School of Interdisciplinary Bioscience and Bioengineering, Pohang University of Science and Technology (POSTECH), Pohang, Republic of Korea
*Corresponding authors. e-mail address: tyyoon@snu.ac.kr; mjshon@postech.ac.kr

Contents

Abstract

In neuroscience, understanding the mechanics of synapses, especially the function of force-sensitive proteins at the molecular level, is essential. This need emphasizes the importance of precise measurement of synaptic protein interactions. Addressing this,

[1] These authors contributed equally to this work.

Methods in Enzymology, Volume 694
ISSN 0076-6879, https://doi.org/10.1016/bs.mie.2024.01.002

we introduce high-resolution magnetic tweezers (MT) as a novel method to probe the mechanics of synapse-related proteins with high precision. We demonstrate this technique through studying SNARE–complexin interactions, crucial for synaptic transmission, showcasing its capability to apply specific forces to individual molecules. Our results reveal that high-resolution MT provides in-depth insights into the stability and dynamic transitions of synaptic protein complexes. This method is a significant advancement in synapse biology, offering a new tool for researchers to investigate the impact of mechanical forces on synaptic functions and their implications for neurological disorders.

1. Introduction

The neuronal soluble *N*-ethylmaleimide-sensitive factor attachment protein receptor (SNARE) complexes are protein assemblies known to be involved in the exocytosis of synaptic vesicles (Söllner et al., 1993). These protein complex structures, attached to the lipid membranes of both the neuron and the synaptic vesicle, come together through a mechanism in which they intertwine, generating mechanical forces (Weber et al., 1998; Südhof & Rothman, 2009). It is through this force that the vesicle membranes fuse with the plasma membrane of the presynaptic neuron. This membrane fusion allows neurotransmitters to be released outside the cell, enabling the transmission of intercellular signals (Südhof, 2004).

Notably, experimental tools such as optical tweezers and magnetic tweezers (MT), capable of precisely applying forces at the sub-piconewton (pN) scale to individual molecules, have been employed to study the mechanical properties of SNARE proteins and their complexes. Using these techniques, researchers have discovered that 10–20 pN of forces applied to SNARE complexes can induce their gradual and reversible unzipping, during which distinct partially zippered states were observed (Gao et al., 2012). It is thought that the dynamic transition between these states might be responsible for the tight regulation of vesicle release at synapses, as well as in other places in cells (Zhang, Ma, & Bao, 2022; Zorman et al., 2014).

More recently, further research has shed light on how the nanomechanics of SNARE complexes can be influenced by their interactions with other proteins involved in the synaptic vesicle cycle (Zhang & Hughson, 2021). Representative SNARE interactors that have been investigated with force spectroscopy include αSNAP and NSF (Kim et al., 2021; Ma et al., 2016), chaperones Munc18 and Munc13 (Jiao et al., 2018; Kalyana Sundaram et al., 2021; Shu, Jin, Rothman, & Zhang, 2020), and complexin (Hao et al., 2023; Shon, Kim, & Yoon, 2018). In particular, complexin has been reported to exhibit seemingly opposing effects in supporting and inhibiting SNARE assembly and the resultant neurotransmission (Giraudo, Eng, Melia, & Rothman, 2006; Huntwork & Littleton, 2007; Maximov, Tang, Yang, Pang, & Südhof, 2009; Trimbuch & Rosenmund, 2016), but a direct evidence at the molecular level has long been lacking. The ability to conduct force spectroscopy analysis on more complex protein assemblies has recently been enhanced by the improved spatiotemporal resolution of single-molecule tweezers (Choi, Kim, Shon, & Yoon, 2022; Park et al., 2023). This approach allows for the comparison of the mechanical stability of distinct intermediate states of protein conformations and thus has proven to be a powerful method to test various models of force-dependent protein functions (Le, Liu, Lim, & Yan, 2016; Yang, Park, Rah, & Shon, 2022).

In this context, we describe the methodology of using high-resolution MT to investigate the mechanical properties of SNAREs and their association with complexin (Fig. 1A). It involves expressing and purifying SNARE proteins for experimentation, synthesizing DNA handles to facilitate single-molecule manipulation with MT, and applying controlled force schemes directly to protein molecules with MT, followed by analyzing the resulting changes. Given that SNARE complexes exhibit distinct extension

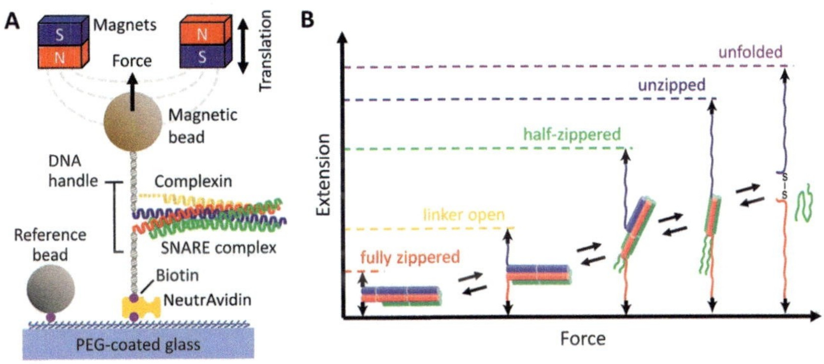

Fig. 1 Magnetic tweezers measurement scheme. (A) A single SNARE complex is attached to the glass surface and a magnetic bead via two DNA handles. SNARE–complexin interaction may lead to changes in SNARE complex conformation, which affect the vertical motion of the magnetic bead. To compensate for sample drift, magnetic bead coordinates are tracked relative to a reference bead (polystyrene microsphere) fixed on the same surface. Magnetic forces are applied by magnets above the sample chamber, controlled by translation motors. (B) Extension changes of a SNARE complex upon force application.

changes during gradual unfolding (Fig. 1B), it is pertinent to investigate how the introduction of an additional factor, like complexin, influences the distributions and kinetics of the intermediate states. Such methods can also be extended to study proteins other than SNAREs, particularly those that exhibit force-dependent changes in structure and protein–protein interactions (Kostrz et al., 2019). By leveraging the simplicity of the in vitro reconstituted systems, experiments can be specifically designed to explore the causal relationships between physiological cues and subsequent molecular changes that were previously challenging to address directly.

2. Preparation of SNARE–DNA construct

In this experiment, we utilize double-stranded DNA (dsDNA) fragments attached to SNARE complexes as molecular "handles". These DNA handles enable the attachment of protein samples to glass surfaces and magnetic beads, as well as the manipulation of target constructs (Fig. 1A). In addition, the characteristic force-extension curve (FEC) of the handles, following a worm-like chain model (Bouchiat et al., 1999), is essential for verifying the integrity of the target single-molecule construct. To achieve this, we prepare the dsDNA handles using polymerase chain reaction (PCR), then express and purify the proteins that make up the SNARE complex, and finally, conjugate the SNARE complexes with the DNA handles through the maleimide-thiol reaction (Fig. 2A–D).

2.1 Preparation of DNA handles

We begin by synthesizing two types of 510-base-pair (bp) dsDNA handles using PCR with 5′-labeled primers: biotin- and amine-labeled primers for the surface attachment with a lower handle, and azide- and amine-labeled primers for the bead attachment with an upper handle. Subsequently, the amine groups are converted to maleimide using the bifunctional crosslinker $SM(PEG)_2$.

2.1.1 Materials and equipment

- λ-DNA (NEB N3011S)
- Primers (IDT):
  - Primer B: 5′-[Biotin]TCGCCACCATCATTTCCA-3′
  - Primer Z: 5′-[Azide]TCGCCACCATCATTTCCA-3′
  - Primer N: 5′-[C6Amine]CATGTGGGTGACGCGAAA-3′

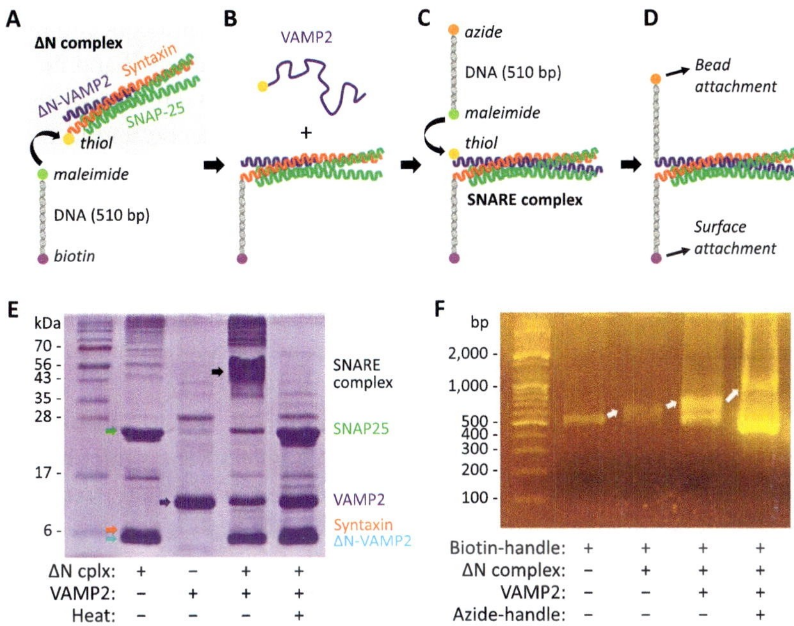

Fig. 2 Synthesis of the SNARE–DNA construct. (A) Biotin-labeled DNA handle (from PCR), functionalized with a maleimide group, reacts with a thiol group on the ΔN complex (syntaxin C-terminus cysteine). (B) VAMP2 addition completes SNARE complex assembly. (C) Azide-labeled DNA handle (from PCR), maleimide-functionalized, targets a thiol group on VAMP2 (C-terminus cysteine). (D) Full SNARE complex with two 510-bp DNA handles. (E) 15% SDS-PAGE of SNARE proteins and their complexes. (F) 1% agarose gel for verifying SNARE–DNA construct synthesis.

- nTaq DNA polymerase (Enzynomics P050A)
- nTaq buffer and dNTP (Enzynomics)
- Thermal cycler
- DNA purification kit
- Agarose gel electrophoresis setup
- UV spectrophotometer (Thermo NanoDrop)
- $SM(PEG)_2$ (Thermo 22102)
- Mixer (ELMI Intelli-Mixer™ RM-2M)
- Phosphate buffered saline (PBS) (pH 7.2) with 150 mM NaCl

2.1.2 Procedure

1. Prepare the three kinds of 5′-labeled primers listed above: Primers B, Z, and N. For long-term storage, make 100 μM solutions in water and keep them frozen.

2. Set up two sets of PCR reactions using the materials and concentrations listed above. For the biotin-labeled handle, use the Primers B and N. For the azide-labeled handle, use the Primers Z and N. Common materials include λ-DNA as the template, dNTPs, nTaq polymerase, and its buffer from the supplier. Use the following conditions: 1.25 μg/mL λ-DNA, 1.2 μM primers (each), 0.2 mM dNTP, 0.05 U/μL nTaq polymerase in 1X nTaq buffer, 80 μL reaction per tube, 40–50 tubes total
3. Run the reaction following a standard PCR protocol. A PCR thermal cycle consists of denaturation, annealing, and extension steps, each performed at 95 °C for 30 s, 53 °C for 30 s, and 72 °C for 5 min. Repeat the cycle 34 times.
4. Purify the PCR product using a purification kit. Then, measure the concentration by UV absorption at 260 nm. Collect the product handles in ~40 μL each; around 2 μM should be obtained. If necessary, verify the product size on a 1% agarose gel.
5. After adjusting the pH of the DNA handle solution to pH 6.5–7.5 with PBS, react the purified DNA handles with 5 mM SM(PEG)$_2$ for the conversion of amine groups to maleimide. Carry out the reaction for 1 h at room temperature with gentle rotation.
6. Purify the product once again using a DNA purification kit. If necessary, check the product size and concentration similarly to the PCR product.

2.1.3 Notes

- The final concentrations of DNA handle solutions should be at least 1 μM, particularly for the azide-labeled handle. This high concentration is necessary to improve the reaction yield when creating double-handled constructs (see Section 2.3).
- For the SM(PEG)$_2$ reaction, it is crucial to adjust the pH correctly to avoid cross-reactivity of the succinimide group.

2.2 Protein expression and purification

The SNARE complex we utilize comprises three rat SNARE proteins: VAMP2, syntaxin-1A, and SNAP-25 isoform b (SNAP-25b). The actual procedure of assembly involves combining VAMP2 with the ΔN complex, consisting of syntaxin, SNAP-25, and a short VAMP2 peptide (ΔN-VAMP2) without the N-terminal region. This ΔN complex aids in forming the ternary SNARE complex, especially in aligning incoming VAMP2 with syntaxin correctly (Pobbati, Stein, & Fasshauer, 2006). For the MT assay, we modify the native SNARE protein sequences as follows (Shon et al., 2018):

1. For improved solubility, the transmembrane domains of VAMP2 and syntaxin-1A are truncated, and the H_{abc} domain of syntaxin-1A is not included, for simplicity.
2. The C-termini of both VAMP2 and syntaxin-1A are modified with cysteine residues to enable DNA handle attachment via maleimide chemistry.
3. To prevent the SNARE complex from rupturing after unfolding, the residues at the −7th layer of the SNARE motifs in both VAMP2 and syntaxin-1A are substituted with cysteine. This modification creates an internally locked complex through a disulfide bond.
4. Four native cysteine residues in SNAP-25b are substituted with alanine to prevent unintended modifications.

In addition, we prepare free SNAP-25b and complexin-1 proteins, introduced separately into the sample chamber during the MT assay. The following procedure details standard recombinant protein expression and purification using *E. coli*, along with their verification through SDS-PAGE gel analysis. The purified proteins are then suitable for storage and subsequent attachment of DNA handles.

2.2.1 Materials and equipment

- Plasmids (coding sequences provided in the Supplementary Materials)
- – 6×His-tagged VAMP2 in pET28a
- – 6×His-tagged ΔN-VAMP2 and syntaxin-1A in pETDuet-1 (for ΔN complex)
- – Untagged SNAP-25b in pET28a (for ΔN complex)
- – 6×His-tagged SNAP-25b in pET28a
- – GST-tagged complexin in pGEX-4T-1
- Rosetta(DE3) competent *E. coli* cells (NEB C2528J)
- Agar plates and growth media (Luria-Bertani Broth, Terrific Broth, etc.) for colony selection and *E. coli* culture
- Antibiotics: ampicillin, chloramphenicol, and kanamycin
- Isopropyl β-d-1-thiogalactopyranoside (IPTG)
- General facilities for bacterial culture (culture bench, shaking incubator, centrifuge, etc.)
- Purification buffers (details provided in the Supplementary Materials)
- PBS (pH 7.2) with 150 mM NaCl
- Sonifier (Branson S-450D)

- Chromatography column (BIO-RAD 7321010)
- Ni-NTA agarose resin (Qiagen 30210)
- Glutathione agarose resin (Sigma G4510)
- Desalting column (Cytiva PD MiniTrap G-25)
- HRV 3 C protease
- Fast Protein Liquid Chromatography (FPLC) system
- Size-exclusion column: Superdex 75 (Cytiva)
- Centrifugal filters: Amicon Ultra-0.5, 10 kDa cutoff (Millipore UFC5010)
- SDS-PAGE setup

2.2.2 Procedure

1. Prepare the plasmids described above (available from the authors upon request).
2. Transform four aliquots of Rosetta cells with the corresponding plasmids for expressing VAMP2, the ΔN complex, free SNAP-25b, and complexin. For the ΔN complex specifically, perform a double transformation using both the duet vector (which contains 6×His-tagged ΔN-VAMP2 and syntaxin-1A in pETDuet-1) and the untagged SNAP-25b in the pET28a vector.
3. Express the proteins by following a general protocol for bacterial culture and harvesting. Prepare agar plates and use appropriate antibiotic combinations for each plasmid transformation.
4. Lyse the cells in Lysis Buffer, followed by sonication (15% amplitude with pulses of 5 s on and 5 s off for a total of 40 min) with cooling on ice.
5. Centrifuge the solution at 15,000 g for 30 min at 4 °C to remove unwanted precipitates.
6. Prepare gravity columns, loading them either with Ni-NTA resin (for SNAREs) or with glutathione-agarose resin (for complexin). Equilibrate them with PBS.
7. Pass the lysates through the columns, and then wash them with the specified Wash Buffers.
8. Elute the bound proteins either with the Elution Buffer containing imidazole (for His-tagged SNARE proteins) or by cleavage with HRV 3C protease (for GST-tagged complexin).
9. For SNARE proteins, use desalting columns to remove imidazole and for buffer exchange to PBS. For complexin, further purify the eluate using size exclusion chromatography (employ an FPLC system for this process and collect the peak fractions).

10. Determine protein concentrations by measuring UV absorption at 280 nm, and verify the product by SDS-PAGE (Fig. 2E). In particular, mix stoichiometric amounts of the ΔN complex and VAMP2 to assess the assembly of the SNARE complex and confirm the function of SNARE proteins. Store solutions with concentrations over several μM at −80 °C. For solutions with lower concentrations, concentrate them using centrifugal filters before storage.

2.3 Attachment of DNA handles to SNARE complex

This section outlines the preparation of the SNARE–DNA construct, a crucial component for MT experiments. The procedure begins with a maleimide-thiol reaction, which attaches the ΔN complex to the maleimide end of a biotin-labeled lower handle. Subsequently, VAMP2 is introduced to drive the formation of the complete SNARE complex and, after assembly, unbound proteins are removed using centrifugal filters. The process continues with another maleimide-thiol reaction, this time to attach an azide-labeled upper handle to the VAMP2 within the SNARE complex. To ensure proper progress, intermediates are verified through agarose gel electrophoresis. Based on our experience, the meticulous synthesis of this protein–DNA construct is critical for the successful execution of single-molecule analyses of the protein complex in MT experiments.

2.3.1 Materials and equipment

- Biotin-labeled and azide-labeled DNA handles prepared from Section 2.1
- Purified ΔN complex and VAMP2 prepared from Section 2.2
- PBS (pH 7.2) with 150 mM NaCl
- Centrifugal filters: Amicon Ultra-0.5, 100 kDa cutoff (Millipore UFC5100)

2.3.2 Procedure

1. Combine biotin-labeled DNA handles with 45 μM ΔN, maintaining a ~16-fold molar excess of ΔN over the DNA. In this reaction, the maleimide group on the DNA handle will covalently bond with the thiol located on the C-terminus of syntaxin within the ΔN complex. Incubate this mixture at room temperature for 3 h with gentle shaking.
2. Introduce a VAMP2 solution to establish a 1:2.5 molar ratio between ΔN and VAMP2. This addition will lead to the displacement of the ΔN-VAMP on syntaxin, forming the complete SNARE complex. Continue incubation at room temperature for 1 h, maintaining gentle shaking.

3. To remove unreacted proteins, transfer the reaction mixture into centrifugal filter devices and centrifuge at 13,000 g for 5 min at 4 °C. Repeat this centrifugation multiple times, replenishing with fresh PBS each time. Conclude with a 15-min final centrifugation to concentrate the product.
4. Assess the product concentration by measuring UV absorption at 260 nm and 280 nm using a spectrophotometer. As you remove free proteins, the A260/A280 ratio should approach that of pure DNA, since the UV absorption by a single SNARE complex attached to the 510-bp DNA handle is relatively weak.
5. Add a concentrated azide-labeled DNA handle solution to reach about a 10–20-fold molar excess over the biotin-labeled handle. In this step, the maleimide on the azide-labeled handle will react with the thiol on the C-terminus of VAMP2. Incubate the sample at room temperature overnight with constant agitation. Validate the final quality of the sample by performing 2% agarose gel electrophoresis and comparing it with intermediate samples (Fig. 2F).

2.3.3 Notes

- As the maleimide-thiol reaction occurs optimally between pH 6.5 and pH 7.5, it is essential to adjust the pH to around 7.2 using PBS.
- Given the substantial excess of SNARE proteins added to the first handle, effective removal of these proteins is crucial. This step ensures that the second handle can specifically react with the SNARE complex attached to the first handle.
- The excess amount of azide-labeled handle need not be removed before storage, as it can be washed away later in the sample chamber during the MT assay. Alternatively, the band around 1 kbp can be selectively extracted using gel electrophoresis to purify the double-handled product.

3. Magnetic tweezers assay

Single-molecule force spectroscopy experiments on the prepared SNARE–DNA constructs are conducted using a custom MT apparatus (Fig. 3A). For preparing sample slides, a low-cost flow-cell chamber is constructed using PEG-coated glass coverslips and double-sided tape, and then mounted onto the MT microscope. The sample solutions are introduced into the flow cell, allowing the SNARE–DNA constructs to tether

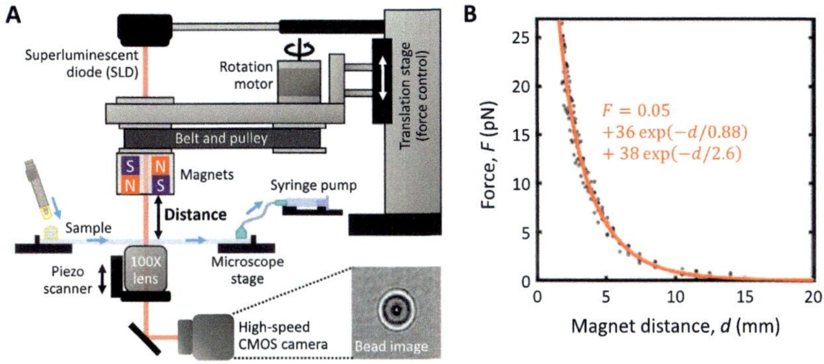

Fig. 3 Instrument setup. (A) Schematic of a high-resolution MT apparatus. (B) Representative force calibration results.

to the surface through NeutrAvidin–biotin linkage. Subsequently, magnetic beads functionalized with DBCO groups are added to interact with azide-labeled handles on the constructs.

Our in-house setup exemplifies the recent advancements in high-resolution MT apparatuses independently developed in various research labs. The technical specifications and capabilities of such setups have been thoroughly discussed in a recent review (Choi et al., 2022). Briefly, our apparatus includes a high-speed CMOS camera and a bright light source, paired with rapid bead-tracking software, enabling the recording of events at speeds up to 1.2 kHz or 830 μs. We have previously detailed the construction of this setup in a video protocol (Park et al., 2023). In this section, we provide a concise overview of the key components, as well as guidance on the assembly and calibration of the instrument.

3.1 Instrument

Fig. 3 presents an overview of our home-built MT system. A sample chamber is positioned on an inverted microscope stage, equipped with magnetic tweezing capabilities. A sample solution is drawn into the chamber using a syringe pump, precisely controlling the flow rate and volume. The magnetic force exerted on the magnetic beads within the chamber is determined by the adjustable distance between the magnets and the sample, managed by a linear translation stage. A rotation motor is utilized to spin the tethered magnetic beads before the actual measurements, aiding in identifying authentic single-molecule constructs.

Illumination in the system is provided by a superluminescent diode, whose coherent beam produces clear diffraction rings in the bead images (the bottom right inset of Fig. 3A). These images are captured by a high-speed CMOS camera through a 100X oil-immersion objective lens. The lens is mounted on a piezo nanopositioner, crucial for generating equidistant bead images as reference. These images form a look-up table, essential for analyzing the vertical movement of beads in experiments. Variations in the bead height result in changes to the diffraction pattern. This alteration allows for precise analysis of bead coordinates and the corresponding conformational changes in the target molecules.

After the instrument setup, a critical step is pre-calibrating the magnetic forces for the specific magnetic beads used in future experiments. This process involves gradually adjusting the distance between the magnet and the sample and observing the resulting changes in the thermal fluctuations of beads tethered by a reference construct, typically a long (5–20 kbp) dsDNA tether. The dsDNA tether is selected because the tethered bead motion can be effectively modeled as an inverted pendulum to precisely measure forces (Daldrop, Brutzer, Huhle, Kauert, & Seidel, 2015).

When magnets are set 1 mm apart, magnetic forces ranging from 0 to 30 pN can be consistently applied across ~15 mm of travel (Fig. 3B). This force-distance relationship is commonly modeled by a double exponential equation effective across a broad range of distances (Yu et al., 2014). These calibration measurements are then incorporated into the control software, allowing users to accurately adjust force levels during experiments.

3.1.1 Materials and equipment

- Inverted microscope (Olympus IX73)
- High-speed CMOS camera (Mikrotron EoSens 3CXP)
- Frame grabber (Active Silicon AS-FBD-4XCXP6-2PE8)
- 100X oil-immersion objective lens (Olympus UPLXAPO100XO)
- Superluminescent diode (Qphotonics QSDM-680-2)
- Optomechanical parts (Thorlabs)
- Translation stage (PI M-126. PD1)
- Stepper motor for rotation (Autonics A3K-S545W)
- Objective lens nanopositioner (Mad City Labs Nano-F100S)
- Neodymium magnets (LG magnet ND 10 × 10 × 12t)
- Custom magnet holder (drawing available upon request)

3.1.2 Procedure

1. Place an inverted microscope on an optical table. Attach a high-speed CMOS camera to the microscope.

2. Vertically mount a linear translation motor to control magnetic forces. Install a stepper motor with a belt and pulley system for magnet rotation control. Mount a nanopositioning lens scanner onto the microscope's nosepiece.
3. Design and fabricate a custom magnet holder that accommodates two magnets, spaced 1 mm apart. Create an aperture in the center of the holder for a collimated light beam to pass through. Mount the magnets along with their holder onto the motor assembly, positioning them a few millimeters above the sample stage. Ensure the magnets are aligned with the microscope's central field of view.
4. Position a superluminescent diode above the magnets. Align the light beam, magnets, and lens vertically at the center, starting with a low magnification lens and progressing to a 100X oil-immersion objective lens.
5. Perform force calibration using the designated magnetic beads (e.g. prepared from Section 3.2) and a reference construct. We typically employ a 5-kbp dsDNA to prepare a bead–tether construct to assess the lateral fluctuations of the beads. Analyze the power spectrum of these fluctuations, fitting it to an appropriate theoretical model to determine the applied force (Daldrop et al., 2015). This calibration should be repeated with several beads to ensure reliable averaging. The applied force versus magnet distance data should then be modeled using a double exponential equation (Yu et al., 2014). For further details and specific methodologies, refer to our previous work (Park et al., 2023).

3.1.3 Notes

- The procedure outlined above provides a concise overview of the setup process. For comprehensive guidance on establishing an MT microscope, including detailed video protocols, consult additional specialized sources (Lipfert, Lee, Ordu, Kerssemakers, & Dekker, 2014; Park et al., 2023; You, Le, Chen, Qin, & Yan, 2017).

3.2 Preparation of magnetic beads

The choice of magnetic beads in MT experiments is crucial, as it influences both the range of forces exerted and the measurement resolution (Choi et al., 2022). The two most commonly used types of magnetic beads are M-270 (2.8 μm) and MyOne (1 μm) (collectively Dynabeads from Thermo), which are highly homogeneous and extensively characterized superparamagnetic beads. Our preference is for M-270 beads due to their capacity to produce larger forces and clearer diffraction patterns, thereby enhancing resolution.

After selecting the appropriate beads, effective conjugation chemistry is essential for tethering them to the target, in this case, the SNARE–DNA construct. Various methods such as avidin–biotin interaction, antigen–antibody binding, or covalent coupling have been explored previously (Choi et al., 2022; Janissen et al., 2014). For our purposes, we have added an azide label on the upper handle of the construct (Section 2.3) to enable covalent bonding through copper-free click chemistry. Accordingly, we here functionalize the surface of M-270 epoxy beads with dibenzocyclooctyne (DBCO). This functionalization ensures that the beads readily react with the azide-labeled construct during sample assembly (Section 3.4), facilitating a stable attachment that can withstand high forces.

3.2.1 Materials and equipment

- Magnetic beads: Dynabeads™ M-270 Epoxy (Thermo 14301)
- Dibenzocyclooctyne-amine (DBCO-NH_2) (Sigma 761540)
- Dimethylformamide (DMF)
- Phosphate buffer (pH 7.4): 81 mM of sodium phosphate dibasic (Na_2HPO_4) and 19 mM of sodium phosphate monobasic (NaH_2PO_4)
- Ammonium sulfate ($(NH_4)_2SO_4$)
- Mixer (ELMI Intelli-Mixer™ RM-2M)
- Permanent magnet (for bead separation)

3.2.2 Procedure

1. Prepare a stock solution of M-270 epoxy beads by dissolving 60 mg of beads (as purchased) in 2 mL of DMF. Take 167.5 μL (equivalent to 5 mg) for immediate use and keep the rest at 4 °C.
2. Similarly, make a stock solution of DBCO-NH_2 by dissolving 10 mg of powder (as purchased) in 362 μL of DMF. Store frozen at −20 °C.
3. Prepare a 3 M solution of ammonium sulfate by dissolving 396 mg powder in 1 mL of phosphate buffer.
4. Use a magnet to sediment the beads at the bottom of the tube. Remove the supernatant (DMF) and resuspend the beads in fresh phosphate buffer. Repeat this buffer exchange process three times. Finally, adjust the volume to 194 μL using the phosphate buffer.
5. Initiate the reaction by adding 6 μL of DBCO-NH_2 solution to the beads, along with 100 μL of 3 M ammonium sulfate solution, making a total reaction volume of 300 μL. Incubate the mixture at room

temperature with gentle rotation for 3 h. After incubation, exchange the buffer to fresh phosphate buffer to quench the reaction. Repeat this buffer exchange process three times.

3.2.3 Notes

- The M-270 bead solutions can be stored at 4 °C for several weeks. Adding 0.1% BSA to the solution may help prevent bead aggregation.

3.3 Construction of sample chambers

A sample chamber featuring flow cells for MT assay is prepared by assembling two glass coverslips, spaced by a double-sided tape defining the channel width (a few millimeters) and height (~100 μm) (Fig. 4). To ensure the precise and controlled delivery of sample solutions within the channels, inlet and outlet ports are meticulously configured. The outlet is connected to syringe pump tubing for the regulated handling of liquids.

Before assembling the chamber, PEGylation of the coverslip surfaces is essential. This process involves the covalent modification of the surface with polyethylene glycol (PEG) molecules via aminosilane. The PEGylation serves a dual purpose: firstly, it prevents nonspecific adhesion of unwanted molecules, and secondly, about 1% of the PEG layer consists of biotin-modified PEG. This specific modification is what enables NeutrAvidin attachment for the subsequent capture of biotinylated SNARE–DNA constructs. This meticulous surface treatment is critical in significantly reducing nonspecific sample binding, thereby enhancing the quality and throughput of single-molecule results.

3.3.1 Materials

- Rectangular glass coverslips (No. 1.5): 50 × 24 mm for the top surface, 60 × 24 mm for the bottom surface

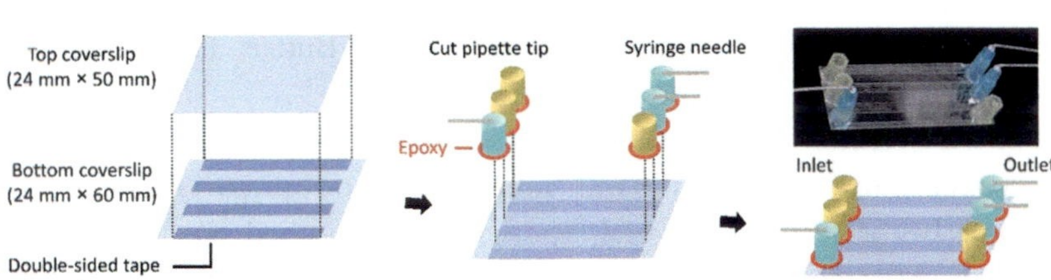

Fig. 4 Sample chamber assembly. A chamber with three flow channels is shown, created using double-sided tape between a bottom coverslip and a shorter top coverslip. Inlet and outlet ports are formed by gluing pipette tip segments and syringe needles with 5-minute epoxy.

- Staining jars to hold glass coverslips
- Piranha solution (3:1 mixture of concentrated sulfuric acid and 30% hydrogen peroxide)
- Methyl alcohol
- Acetic acid
- Aminosilane: N-[3-(trimethoxysilyl)propyl]ethylenediamine (Sigma #104884)
- Ultrasonic bath
- mPEG-succinimidyl valerate (mPEG-SVA) MW 5000 and MW 2000 (Laysan Bio)
- Biotin-PEG-SVA, MW 5000 (Laysan Bio)
- PEGylation Buffer (100 mM sodium bicarbonate (pH 9) with 400 mM potassium sulfate): dissolve 42 mg of sodium bicarbonate and 348.5 mg of potassium sulfate in 5 mL of ultrapure water.
- Syringe filters (0.20 μm pore, cellulose acetate) (Advantec 25CS020AS)
- Nitrogen gun
- Double-sided tape (3M)
- Cut pipette tips
- Syringe needles (22G)
- Epoxy glue (Devcon S-208)

3.3.2 Procedure

1. Clean glass coverslips in piranha solution for 20 min. Then rinse them thoroughly with ultrapure water, followed by a rinse with methanol.
2. Prepare a silane solution by mixing methanol, acetic acid, and aminosilane in a 200:10:3 volumetric ratio. Silanize the coverslips using this solution for 10 min at room temperature, then sonicate for 2 min. Allow to react for another 10 min, and finally rinse them with water and dry thoroughly.
3. Prepare a PEG solution by adding 99 mg of 5-kDa mPEG and 1 mg of 5-kDa biotin-PEG to 1 mL of the PEGylation Buffer. Then, filter the solution to remove potential aggregates.
4. For PEGylation of the coverslips, sandwich ~200 μL of PEG solution between a top coverslip (50 mm long) and a bottom coverslip (60 mm long). Let it react for several hours in a humidified box to prevent drying. After completion, rinse the coverslips with ultrapure water and blow dry them with a nitrogen gun.
5. Perform a second PEGylation step on the same glass, this time using 2-kDa mPEG instead of the 5-kDa used in the first step. This ensures

coverage of areas that may have been missed during the initial PEGylation. After completion, rinse them again thoroughly with ultrapure water and blow dry them with a nitrogen gun.

6. Store the PEGylated coverslips dry at −20 °C for up to several weeks.
7. Construct the chamber as illustrated in Fig. 4. Employ double-sided tape to demarcate the flow cell channel. Adapt cut segments of micropipette tips for the inlet reservoir and syringe needles for the outlet ports. Affix the ports with 5-minute epoxy, diligently inspecting for potential leaks to ensure integrity.

3.3.3 Notes

- Exercise extreme caution when handling piranha solution. Always add hydrogen peroxide to sulfuric acid, and not the reverse, as adding sulfuric acid to hydrogen peroxide can trigger a highly exothermic and potentially explosive reaction. For safety, use a larger container preheated to ~70 °C, which is capable of securely holding the staining jars with coverslips. This precaution helps to prevent the jars from cracking due to thermal shock.
- In the presence of moisture, the ester groups of mPEG-SVA and biotin-PEG-SVA can hydrolyze, rendering it inactive for coupling with amines. Therefore, the PEG powders should be handled in a way that limits its exposure to ambient moisture until it has reacted with the aminosilane.

3.4 Sample injection

In MT experiments, preparing surface-tethered single-molecule samples, such as the SNARE–DNA constructs used in this study, is vital. Once the sample chamber is loaded onto the microscope stage, a syringe pump connected to the chamber's flow cell is employed to sequentially introduce the required solutions. Controlled delivery, characterized by precise flow rate, volume, and timing, is crucial to ensure accurate and consistent preparation of the surface-tethered samples for investigation.

3.4.1 Materials

- Sample chamber (prepared from Section 3.3)
- 5 mL syringe (Henke-Ject)
- Polyethylene tubing (SCI, PE-60)
- PBS (pH 7.2) with 150 mM NaCl
- NeutrAvidin (Thermo 31000)

- SNARE–DNA construct (prepared from Section 2.3)
- 2,2′-Dithiodipyridine (DTDP) (Sigma D5767): 2 mM solution in 50 mM Tris-HCl (pH 8.0)
- Reference bead: Streptavidin-coated polystyrene particles (3.0–3.4 μm diameter) (Spherotech SVP-30-5)
- Magnetic bead: DBCO-coated M-270 beads (prepared from Section 3.2)
- SNAP-25b (prepared from Section 2.2)

3.4.2 Procedure

1. Place the sample chamber on the microscope stage and connect it to a syringe pump using polyethylene tubing. Equilibrate the chamber with fresh PBS.
2. Draw in 8 nM NeutrAvidin in PBS. After a 10 min incubation for binding, flush the chamber with 200 μL of PBS.
3. Draw in 5–10 nM SNARE–DNA construct in PBS. Incubate for 15 min to allow binding with NeutrAvidin, then flush the chamber with 200 μL of PBS.
4. Draw in the DTDP solution. After a 5 min incubation, flush the chamber with 200 μL of PBS.
5. Prepare a 25-fold dilution of the reference bead solution in PBS and introduce it into the chamber. After a 10 min incubation for surface attachment, flush the chamber with 200 μL of PBS.
6. Prior to introducing magnetic beads, remove the magnets. Spike 0.2 μL of DBCO-coated M-270 bead solution in 60 μL of 2.5 μM SNAP-25b in PBS, and introduce this mixture into the chamber. Allow a 10 min incubation for the reaction.
7. Insert the magnets into the holder to start the application of magnetic forces.
8. For experiments involving complexin, draw in a solution containing 3 μM complexin and 2.5 μM SNAP-25b.

3.4.3 Notes

- Optimal concentrations for the surface attachment of injected samples, such as NeutrAvidin and SNARE–DNA constructs, should be empirically determined to enhance productivity. While higher concentrations can increase surface sample density and throughput, they also risk the occurrence of doubly-tethered beads. Similarly, using high concentrations of beads increases the likelihood of finding a suitable reference bead near a magnetic bead-attached SNARE–DNA construct, but it may also cause complications in bead tracking.

- The flow channel we design (4 mm wide, 50 mm long, ~0.1 mm high) has a chamber volume of ~20 μL, so introducing ~30–50 μL of solutions is sufficient for all materials. When washing with 200 μL of PBS, this equates to ~10 times the chamber volume.
- For all steps, use a flow rate of ~200 μL/min for introducing the solutions.
- Avoid over-withdrawing the solution with the pump or introducing bubbles into the channel, which can potentially displace surface-tethered constructs and beads.

3.5 Force application procedure

Identification of single-molecule constructs is a crucial screening step, primarily based on their characteristic force-extension behavior. Our tether configuration consists of two 510-bp DNA handles, resulting in a total contour length of ~345 nm. Therefore, repetitively applying a force of 0 and 5 pN allows us to readily distinguish samples containing these handles. Additionally, we utilize rotating magnets to assess the single-tether behavior of the target construct, as doubly tethered samples restrict the free swiveling of the beads.

Once a suitable sample is selected, we record bead images while incrementally spacing the piezo lens scanner in small steps, typically 100 nm each. This process generates a "lookup table" of bead images. By documenting the changes in bead diffraction as a function of bead distance, we can estimate the actual vertical displacement of the beads when the lens remains stationary. This estimation is made after applying the necessary corrections for refractive index mismatches between the glass and the aqueous medium.

The investigation of SNARE complex properties begins by applying force ramps, continuously increasing the force to comprehensively analyze the mechanical unzipping of SNAREs. Following the identification of the force range that elicits responses from SNAREs, high-speed force-clamp measurements are conducted within this range, yielding extension distributions that indicate various intermediate conformations. By applying Hidden Markov Modeling (HMM) to the high-speed trajectories of extension changes, it becomes possible to accurately model both the population shifts (thermodynamics) and the transition rates (kinetics) simultaneously.

Additionally, considering the significant mechanical hysteresis in SNARE complexes, it is beneficial to examine the latency of unzipping events, particularly in cases involving abrupt structural changes (such as a 20–25 nm extension change in the unraveling of VAMP2 from the complex), through force-jump experiments with rapid force level transitions (e.g., from 10 to 15 pN). While optional, this analysis enhances the understanding of non-equilibrium behaviors, as exemplified by its application to the enzymatic disassembly of SNARE complexes by NSF (*N*-ethylmaleimide-sensitive factor) (Kim et al., 2021).

After establishing the baseline response of the SNARE complexes, we introduce free complexin into the solution to observe its effects on SNARE dynamics. This approach is adaptable for investigating interactions with other SNARE partners and can be broadly applied to various protein–protein and protein–ligand interactions.

3.5.1 Procedure

1. Scan the sample surface for target beads. Ensure they meet two criteria: free rotation upon magnet rotation (indicating single-tether geometry) and noticeable thermal fluctuations (indicating separation from the surface). Prefer beads with minimal footprint in rotation (see Notes). Finally, test vertical extensibility of the target construct with force variations between 0 and 5 pN. Make sure a reference bead is also in view.
2. For diffraction pattern vs. bead height calibration, create look-up tables of bead images using equal piezo steps. This calibration function is included in the software provided with our previous video article (Park et al., 2023).
3. To generate FECs, apply force ramps between 0 and 20 pN at 1 pN/s. The resulting curves typically resemble those of dsDNA tethers but with additional, superimposed features unique to the SNARE complexes.
4. For force-clamp measurements, calibrate and track bead images at 1.2 kHz. Avoid SNARE hysteresis by starting at 10 pN and swiftly moving to target forces (e.g., 12–17.4 pN as presented in Section 4). If a large (~20 nm) unzipping event occurs,. return to 10 pN for rezipping, then reapply higher forces. Record latency to unzipping events if needed.

3.5.2 Notes

- The off-centering of magnetic beads relative to their tethers is due to the alignment of the bead's magnetization with the external magnetic field (Klaue & Seidel, 2009). The degree of off-centering can be directly estimated from the radius of the magnetic bead's trajectory when the magnets are rotated (R_{off}). A large R_{off} indicates more significant distortions in the worm-like chain FEC of the SNARE–DNA construct. Although these distortions can be strategically corrected (Shon, Rah, & Yoon, 2019), they also tend to increase thermal noise in extension measurements, making the use of beads with smaller R_{off} values more desirable.

- The exact force levels at which proteins rupture may vary slightly due to several factors, such as buffer composition, tether chemistry and geometry, and the rate at which force is applied (Dudko, Hummer, & Szabo, 2008).

4. Representative results

In this section, we recapitulate key findings from the methods discussed in the chapter. Initially, the focus is on delineating the force-dependent characteristics of SNARE complexes. Subsequently, the discussion shifts to how the interaction with complexin modifies these behaviors. The experiments demonstrate notable alterations in the SNARE complex's stabilization and preferred states due to complexin, shedding light on its mechanical impact on SNARE behavior.

4.1 SNARE complex unzipping and intermediate conformations

The study begins by analyzing the SNARE complex's unzipping and intermediate states. Force-ramp experiments yield an FEC for the SNARE–DNA complex, displaying typical unzipping between 16 and 18 pN and confirming states like fully-zippered and half-zipped at lower forces (Fig. 5A). Force-clamp experiments, applying specific forces instantly, reveal changes in DNA length and enable identification of rapidly changing states, particularly at 16.7 pN (Fig. 5B, C). These tests confirm the mechanical properties of the SNARE complex.

4.2 Effect of complexin on SNARE complex conformation

After introducing complexin, the same force-ramp and force-clamp experiments are repeated. These experiments reveal that complexin increases the unzipping force of the SNARE complex to 18–21 pN, compared to the 16–18 pN range observed prior to complexin introduction, signifying its stabilizing effect (Fig. 6A). Force-clamp experiments at

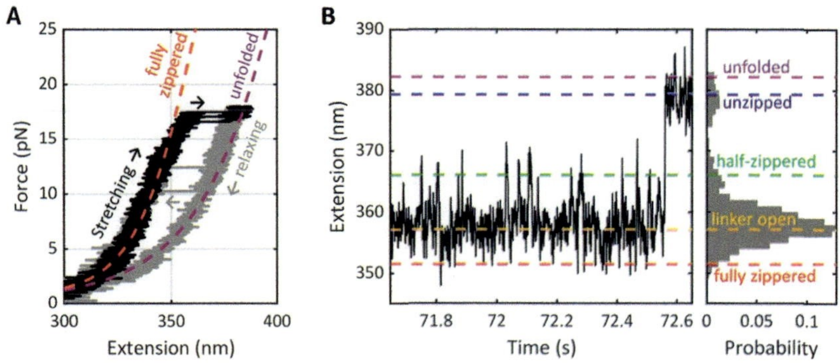

Fig. 5 Mechanical response of the SNARE complex. (A) Force-extension curve illustrating unzipping of the SNARE complex between 16 and 18 pN during increasing (red) and decreasing (blue) force ramps. (B) Extension changes over time at 16.7 pN, showing rapid and reversible transitions among fully zippered, linker-open, and half-zippered states, culminating in a 20-nm unzipping event (~72.6 s). The corresponding histogram of extension is shown on the right.

specific forces, such as 16.7 pN and 17.4 pN, further elucidate this effect, showing a disappearance of the half-zippered state and a preference for the linker-open state upon the addition of complexin (Fig. 6B).

In addition, investigations conducted at reduced force levels (12–14 pN) have revealed that the presence of complexin diminishes the frequency of the fully zippered state of the SNARE complex (Fig. 6C). This observation suggests a predilection for the linker-open state over the fully zippered state at lower forces (down to 12 pN) in the presence of complexin, thereby implying a destabilizing influence of complexin on the complete zippering of the SNARE complex. The observed dual effects of complexin can be attributed to the independent functions of its various segments (Shon et al., 2018), potentially resolving the paradoxical nature of its clamping activities observed in vivo (Trimbuch & Rosenmund, 2016). This comprehensive analysis enhances our understanding of the mechanical properties of the SNARE complex in the context of complexin interaction, shedding light on the intricate dynamics between these two pivotal proteins.

5. Discussion

In this chapter, we explored the methodological details for the study of SNARE–complexin interactions using magnetic tweezers. The

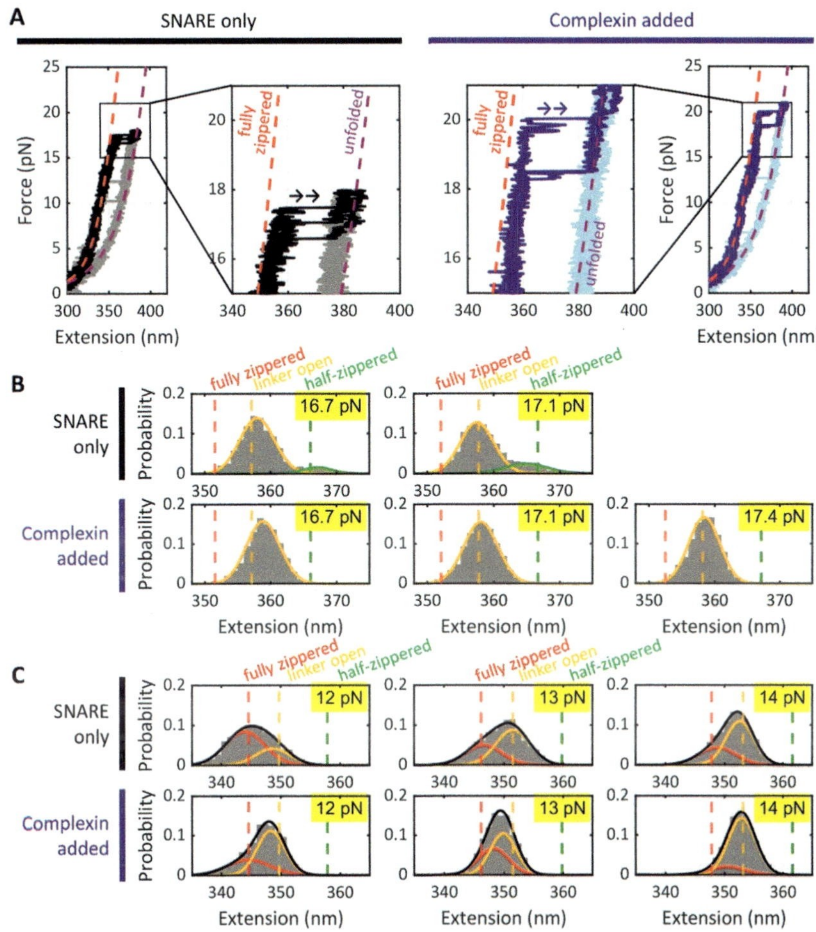

Fig. 6 Effect of complexin on SNARE complex conformation. (A) Force-extension curves display an increased unzipping force after complexin introduction, indicating its stabilizing effect. (B) Extension histograms before and after complexin addition, with force applications between 16 and 18 pN. Suppression of the half-zippered state by complexin suggests mechanical stabilization of the SNARE complex. (C) Extension histograms between 12 and 14 pN. A decreased population of the fully zippered state and a preference for the linker-open state indicates complexin's inhibitory role in full zippering of the SNARE complex.

application of single-molecule force spectroscopy has been instrumental in advancing our understanding of protein behaviors and interactions at the molecular level. By applying precise piconewton-scale forces to individual molecules, this method has shed light on the biophysical dynamics of the SNARE complex, a crucial component in physiological functions influenced by mechanical forces, and its modulation by complexin. The recent advancements in camera technology, enabling high-speed tracking, have further refined this method. The improved temporal resolution allows for detailed analyses of intermediate populations and their force-dependent shifts. They also permit investigations into rapid, short-timescale changes within biological molecules that were previously elusive using conventional methodologies. The integration of magnetic tweezers with high-speed tracking technology holds great promise for future research in understanding the properties and mechanisms of a wide array of biological molecules (Choi et al., 2022).

Despite these advancements, the chapter also highlights the challenges faced in sample preparation and the intricacies of conducting experiments with magnetic tweezers. Reflecting on our experience, we have frequently grappled with the issue of inconsistent sample quality, a predicament most apparent during the process of attaching DNA handles to the protein complex. Our electrophoresis results repeatedly underscored this challenge, revealing not only the formation of adequate samples but also highlighting the problems of suboptimal yields and significant losses of samples in purification stages. Addressing these challenges is essential for enhancing the accuracy and efficacy of experiments employing magnetic tweezers, which is likely to benefit a broader spectrum of research involving more complex biological samples, including membrane proteins and live cells (Cho et al., 2022; Choi et al., 2019).

Acknowledgment

This work was supported by the National Research Foundation of Korea (NRF) grant funded by the Korean government (MSIT) (NRF-2022R1C1C1012176, RS-2023–00218927, and NRF-2021R1A6A1A10042944).

Appendix A. Supporting information

Supplementary data associated with this article can be found in the online version at doi:10.1016/bs.mie.2024.01.002.

References

Bouchiat, C., Wang, M. D., Allemand, J., Strick, T., Block, S. M., & Croquette, V. (1999). Estimating the persistence length of a worm-like chain molecule from force-extension measurements. *Biophysical Journal, 76*(1 Pt 1), 409–413.

Cho, S., Shon, M. J., Son, B., Eun, G. S., Yoon, T.-Y., & Park, T. H. (2022). Tension exerted on cells by magnetic nanoparticles regulates differentiation of human mesenchymal stem cells. *Biomaterials Advances, 139*, 213028. https://doi.org/10.1016/j.bioadv.2022.213028.

Choi, H.-K., Kim, H. G., Shon, M. J., & Yoon, T.-Y. (2022). High-resolution single-molecule magnetic tweezers. *Annual Review of Biochemistry, 91*, 33–59. https://doi.org/10.1146/annurev-biochem-032620-104637.

Choi, H.-K., Min, D., Kang, H., Shon, M. J., Rah, S.-H., Kim, H. C., ... Yoon, T.-Y. (2019). Watching helical membrane proteins fold reveals a common N-to-C-terminal folding pathway. *Science (New York, N. Y.), 366*(6469), 1150–1156. https://doi.org/10.1126/science.aaw8208.

Daldrop, P., Brutzer, H., Huhle, A., Kauert, D. J., & Seidel, R. (2015). Extending the range for force calibration in magnetic tweezers. *Biophysical Journal, 108*(10), 2550–2561. https://doi.org/10.1016/j.bpj.2015.04.011.

Dudko, O. K., Hummer, G., & Szabo, A. (2008). Theory, analysis, and interpretation of single-molecule force spectroscopy experiments. *Proceedings of the National Academy of Sciences of the United States of America, 105*(41), 15755–15760. https://doi.org/10.1073/pnas.0806085105.

Gao, Y., Zorman, S., Gundersen, G., Xi, Z., Ma, L., Sirinakis, G., ... Zhang, Y. (2012). Single reconstituted neuronal SNARE complexes zipper in three distinct stages. *Science (New York, N. Y.), 337*(6100), 1340–1343. https://doi.org/10.1126/science.1224492.

Giraudo, C. G., Eng, W. S., Melia, T. J., & Rothman, J. E. (2006). A clamping mechanism involved in SNARE-dependent exocytosis. *Science (New York, N. Y.), 313*(5787), 676–680. https://doi.org/10.1126/science.1129450.

Hao, T., Feng, N., Gong, F., Yu, Y., Liu, J., & Ren, Y.-X. (2023). Complexin-1 regulated assembly of single neuronal SNARE complex revealed by single-molecule optical tweezers. *Communications Biology, 6*(1), https://doi.org/10.1038/s42003-023-04506-w Article 1.

Huntwork, S., & Littleton, J. T. (2007). A complexin fusion clamp regulates spontaneous neurotransmitter release and synaptic growth. *Nature Neuroscience, 10*(10), 1235–1237. https://doi.org/10.1038/nn1980.

Janissen, R., Berghuis, B. A., Dulin, D., Wink, M., van Laar, T., & Dekker, N. H. (2014). Invincible DNA tethers: Covalent DNA anchoring for enhanced temporal and force stability in magnetic tweezers experiments. *Nucleic Acids Research, 42*(18), e137. https://doi.org/10.1093/nar/gku677.

Jiao, J., He, M., Port, S. A., Baker, R. W., Xu, Y., Qu, H., ... Zhang, Y. (2018). Munc18-1 catalyzes neuronal SNARE assembly by templating SNARE association. *eLife, 7*, e41771. https://doi.org/10.7554/eLife.41771.

Kalyana Sundaram, R. V., Jin, H., Li, F., Shu, T., Coleman, J., Yang, J., ... Krishnakumar, S. S. (2021). Munc13 binds and recruits SNAP25 to chaperone SNARE complex assembly. *FEBS Letters, 595*(3), 297–309. https://doi.org/10.1002/1873-3468.14006.

Kim, C., Shon, M. J., Kim, S. H., Eun, G. S., Ryu, J.-K., Hyeon, C., ... Yoon, T.-Y. (2021). Extreme parsimony in ATP consumption by 20S complexes in the global disassembly of single SNARE complexes. *Nature Communications, 12*(1), https://doi.org/10.1038/s41467-021-23530-0 Article 1.

Klaue, D., & Seidel, R. (2009). Torsional stiffness of single superparamagnetic microspheres in an external magnetic field. *Physical Review Letters, 102*(2), 028302. https://doi.org/10.1103/PhysRevLett.102.028302.

Kostrz, D., Wayment-Steele, H. K., Wang, J. L., Follenfant, M., Pande, V. S., Strick, T. R., & Gosse, C. (2019). A modular DNA scaffold to study protein–protein interactions at single-molecule resolution. *Nature Nanotechnology, 14*(10), https://doi.org/10.1038/s41565-019-0542-7 Article 10.

Le, S., Liu, R., Lim, C. T., & Yan, J. (2016). Uncovering mechanosensing mechanisms at the single protein level using magnetic tweezers. *Methods (San Diego, Calif.), 94*, 13–18. https://doi.org/10.1016/j.ymeth.2015.08.020.

Lipfert, J., Lee, M., Ordu, O., Kerssemakers, J. W. J., & Dekker, N. H. (2014). Magnetic tweezers for the measurement of twist and torque. *JoVE (Journal of Visualized Experiments), 87*, e51503. https://doi.org/10.3791/51503.

Ma, L., Kang, Y., Jiao, J., Rebane, A. A., Cha, H. K., Xi, Z., ... Zhang, Y. (2016). α-SNAP enhances SNARE zippering by stabilizing the SNARE four-helix bundle. *Cell Reports, 15*(3), 531–539. https://doi.org/10.1016/j.celrep.2016.03.050.

Maximov, A., Tang, J., Yang, X., Pang, Z. P., & Südhof, T. C. (2009). Complexin controls the force transfer from SNARE complexes to membranes in fusion. *Science (New York, N. Y.), 323*(5913), 516–521. https://doi.org/10.1126/science.1166505.

Park, C., Yang, T., Rah, S.-H., Kim, H. G., Yoon, T.-Y., & Shon, M. J. (2023). High-speed magnetic tweezers for nanomechanical measurements on force-sensitive elements. *JoVE(195)*, e65137. https://doi.org/10.3791/65137.

Pobbati, A. V., Stein, A., & Fasshauer, D. (2006). N- to C-terminal SNARE complex assembly promotes rapid membrane fusion. *Science (New York, N. Y.), 313*(5787), 673–676. https://doi.org/10.1126/science.1129486.

Shon, M. J., Kim, H., & Yoon, T.-Y. (2018). Focused clamping of a single neuronal SNARE complex by complexin under high mechanical tension. *Nature Communications, 9*(1), https://doi.org/10.1038/s41467-018-06122-3 Article 1.

Shon, M. J., Rah, S.-H., & Yoon, T.-Y. (2019). Submicrometer elasticity of double-stranded DNA revealed by precision force-extension measurements with magnetic tweezers. *Science Advances, 5*(6), eaav1697. https://doi.org/10.1126/sciadv.aav1697.

Shu, T., Jin, H., Rothman, J. E., & Zhang, Y. (2020). Munc13-1 MUN domain and Munc18-1 cooperatively chaperone SNARE assembly through a tetrameric complex. *Proceedings of the National Academy of Sciences, 117*(2), 1036–1041. https://doi.org/10.1073/pnas.1914361117.

Söllner, T., Whiteheart, S. W., Brunner, M., Erdjument-Bromage, H., Geromanos, S., Tempst, P., & Rothman, J. E. (1993). SNAP receptors implicated in vesicle targeting and fusion. *Nature, 362*(6418), https://doi.org/10.1038/362318a0 Article 6418.

Südhof, T. C. (2004). The synaptic vesicle cycle. *Annual Review of Neuroscience, 27*(1), 509–547. https://doi.org/10.1146/annurev.neuro.26.041002.131412.

Südhof, T. C., & Rothman, J. E. (2009). Membrane fusion: Grappling with SNARE and SM proteins. *Science (New York, N. Y.), 323*(5913), 474–477. https://doi.org/10.1126/science.1161748.

Trimbuch, T., & Rosenmund, C. (2016). Should I stop or should I go? The role of complexin in neurotransmitter release. *Nature Reviews. Neuroscience, 17*(2), https://doi.org/10.1038/nrn.2015.16 Article 2.

Weber, T., Zemelman, B. V., McNew, J. A., Westermann, B., Gmachl, M., Parlati, F., ... Rothman, J. E. (1998). SNAREpins: Minimal machinery for membrane fusion. *Cell, 92*(6), 759–772. https://doi.org/10.1016/s0092-8674(00)81404-x.

Yang, T., Park, C., Rah, S.-H., & Shon, M. J. (2022). Nano-precision tweezers for mechanosensitive proteins and beyond. *Molecules and Cells, 45*(1), 16–25. https://doi.org/10.14348/molcells.2022.2026.

You, H., Le, S., Chen, H., Qin, L., & Yan, J. (2017). Single-molecule manipulation of G-quadruplexes by magnetic tweezers. *JoVE (Journal of Visualized Experiments), 127*, e56328. https://doi.org/10.3791/56328.

Yu, Z., Dulin, D., Cnossen, J., Köber, M., van Oene, M. M., Ordu, O., ... Dekker, N. H. (2014). A force calibration standard for magnetic tweezers. *Review of Scientific Instruments, 85*(12), 123114. https://doi.org/10.1063/1.4904148.

Zhang, Y., & Hughson, F. M. (2021). Chaperoning SNARE folding and assembly. *Annual Review of Biochemistry, 90*(1), 581–603. https://doi.org/10.1146/annurev-biochem-081820-103615.

Zhang, Y., Ma, L., & Bao, H. (2022). Energetics, kinetics, and pathways of SNARE assembly in membrane fusion. *Critical Reviews in Biochemistry and Molecular Biology, 57*(4), 443–460. https://doi.org/10.1080/10409238.2022.2121804.

Zorman, S., Rebane, A. A., Ma, L., Yang, G., Molski, M. A., Coleman, J., ... Zhang, Y. (2014). Common intermediates and kinetics, but different energetics, in the assembly of SNARE proteins. *eLife, 3*, e03348. https://doi.org/10.7554/eLife.03348.

CHAPTER FIVE

Force-fluorescence setup for observing protein–DNA interactions under load

Jaehun Jung[a], Subin Kim[b], Sang-Hyun Rah[a], Jayil Lee[b,c], and Min Ju Shon[a,d,*]

[a]Department of Physics, Pohang University of Science and Technology (POSTECH), Pohang, Republic of Korea
[b]Department of Biological Sciences, Ulsan National Institute of Science and Technology (UNIST), Ulsan, Republic of Korea
[c]Institute of Basic Science Center for Genomic Integrity, Ulsan, Republic of Korea
[d]School of Interdisciplinary Bioscience and Bioengineering, Pohang University of Science and Technology (POSTECH), Pohang, Republic of Korea
*Corresponding author. e-mail address: mjshon@postech.ac.kr

Contents

Abstract

This chapter explores advanced single-molecule techniques for studying protein–DNA interactions, particularly focusing on Replication Protein A (RPA) using a force-fluorescence setup. It combines magnetic tweezers (MT) with total internal

Methods in Enzymology, Volume 694
ISSN 0076-6879, https://doi.org/10.1016/bs.mie.2024.01.003

reflection fluorescence (TIRF) microscopy, enabling detailed observation of DNA behavior under mechanical stress. The chapter details the use of DNA hairpins and bare DNA to examine RPA's binding dynamics and its influence on DNA's mechanical properties. This approach provides deeper insights into RPA's role in DNA replication, repair, and recombination, highlighting its significance in maintaining genomic stability.

1. Introduction

In recent decades, significant advancements in single-molecule techniques have profoundly transformed molecular biology. These methods have been instrumental for in-depth analysis of molecular mechanisms in vitro, particularly for understanding structural changes and functional activities. They provide detailed insights into crucial biological processes involving nucleic acids and proteins (Ha, Kozlov, & Lohman, 2012).

Single-molecule imaging and manipulation techniques, standing as the two primary pillars of this area, offer unique and complementary insights. Imaging techniques, primarily fluorescence microscopy, are key for the sensitive detection and tracking of individual molecules. An example is the use of single-molecule fluorescence resonance energy transfer (FRET) combined with total internal reflection fluorescence microscopy (TIRF) (Roy, Hohng, & Ha, 2008). These techniques have been extensively employed to observe dynamic protein–DNA interactions at the single-molecule level in real-time. Recent developments like DNA curtain technology have also proved powerful in examining the movements of DNA-binding proteins on stretched DNA strands (Collins, Ye, Duzdevich, & Greene, 2014). On the other hand, single-molecule force spectroscopy techniques, such as optical tweezers, magnetic tweezers (MT), and atomic force microscopy, enable mechanical manipulation as well as measurements (Choi, Kim, Shon, & Yoon, 2022; Neuman & Nagy, 2008; Yang, Park, Rah, & Shon, 2022). These methods allow researchers to manipulate DNA substrates to study protein binding in relation to applied force (Chaurasiya, Paramanathan, McCauley, & Williams, 2010; Heller, Hoekstra, King, Peterman, & Wuite, 2014; Koch & Wang, 2003) or to directly probe the mechanics of protein–DNA complexes through purely mechanical means (Loenhout et al., 2013; Shon & Cohen, 2016).

The integration of force spectroscopy with fluorescence microscopy, humorously termed "fleezers", has gained interest over many years, showing great promise in molecular biology (Comstock, Ha, & Chemla, 2011; Ha, Kaiser, Myong, Wu, & Xiao, 2022; Hilario & Kowalczykowski, 2010; Jacobs & Blank, 2014). This multiplexed approach, correlating high-resolution fluorescence signals with mechanical load, empowers the study of nucleic acids and proteins, notably their structural dynamics and interactions (Comstock et al., 2015; Hohng et al., 2007; Ngo, Zhang, Zhou, Yodh, & Ha, 2015). Independent efforts to merge MT with TIRF microscopy have been reported by many labs (Bae et al., 2014; Brutzer, Schwarz, & Seidel, 2012; Graves et al., 2015; Kemmerich, Swoboda, et al., 2016; Kim, Jung, Lee, Lee, & Hong, 2021; Lebel, Basu, Oberstrass, Tretter, & Bryant, 2014; Madariaga-Marcos et al., 2018; Schwarz et al., 2013; Seol & Neuman, 2018; van Loenhout, de Grunt, & Dekker, 2012). We describe the use of this force-fluorescence method to study protein–DNA interactions, specifically involving human replication protein A (RPA) as a model system.

RPA, a major eukaryotic single-stranded DNA-binding (SSB) protein, plays a critical role in DNA replication, repair, and recombination (Iftode, Daniely, & Borowiec, 1999; Wold, 1997). Its ability to rapidly bind to single-stranded DNA (ssDNA) is key in preventing the formation of secondary structures, thereby maintaining genomic stability. Advances in single-molecule studies have greatly enhanced our understanding of SSB proteins, including RPA (Bain, Fischer, Chen, & Wold, 2018; Ha et al., 2012). Utilizing techniques like single-molecule imaging, these studies have elucidated the dynamic mechanisms of RPA (Gibb et al., 2014; Nguyen et al., 2014) and its interactions with other DNA-binding proteins (e.g. helicase and Rad51/52) (Awate, & Brosh, Jr., 2017; Grimme et al., 2010; Ma, Gibb, Kwon, Sung, & Greene, 2017), as well as with other proteins involved in DNA metabolism, (e.g. NSMF) (Kang et al., 2023). Furthermore, single-molecule tweezers have been pivotal in manipulating DNA substrates to study the correlation between DNA mechanics, structural changes, and RPA binding under varying force conditions. Different methods, such as the use of bare ssDNA and the creation of transient bubbles in dsDNA through unwinding torque (De Vlaminck et al., 2010), peeling via overstretching (Chen, Le, Basu, Chazin, & Yan, 2015; van Mameren et al., 2009), and the formation of DNA forks using hairpin structures (Kemmerich, Daldrop, et al., 2016), have provided unique mechanical insights into RPA's function.

This chapter provides methods for studying RPA–DNA interactions under various force conditions using a high-resolution MT setup combined

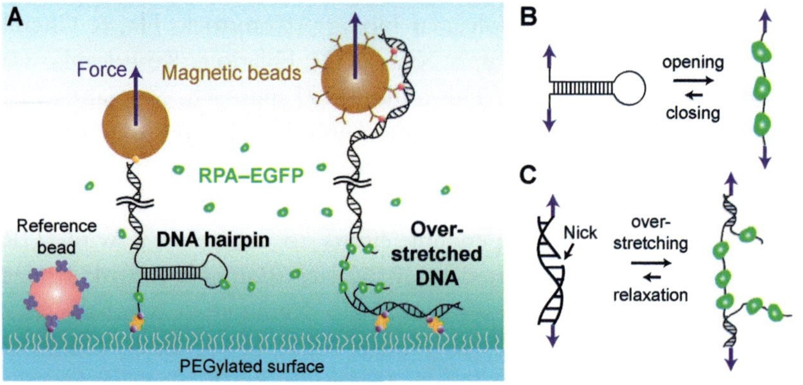

Fig. 1 Force-fluorescence study of protein–DNA interactions. (A) DNA constructs (hairpins and dsDNA for overstretching) are tethered between a PEGylated surface and a magnetic bead. Force application exposes ssDNA, enabling RPA binding. This process can be monitored by changes in DNA mechanics (using MT) and by fluorescence from RPA–EGFP (using TIRF). (B) A DNA hairpin undergoes a reversible transition between the open and the closed states. (C) An overstretched bare dsDNA fragment undergoes significant transformation, leading to the peeling of ssDNA strands, starting from single-stranded nicks. In both (B) and (C), RPA binding to ssDNA may hinder the reformation of double-stranded structure.

with TIRF microscopy (Fig. 1A). We employ two DNA structures: a DNA hairpin (Fig. 1B) and a bare 10-kbp dsDNA (Fig. 1C), leveraging their distinct force-dependent ssDNA exposure. The results indicate that RPA binding alters the force-extension behavior of DNA, markedly hindering its transition back to double-stranded form in both constructs. The direct involvement of RPA was confirmed through the visualization of its binding at the single-molecule level using TIRF imaging. This approach not only provides insights into the biological functions of RPA in modulating DNA structures within cell nuclei but also suggests the method's applicability to studying other proteins that remodel RPA–DNA interactions, thereby enriching our understanding of these complex molecular dynamics.

2. Force-fluorescence setup

2.1 MT instrument

Our MT setup is summarized here with a focus on its key elements (Fig. 2) (for a more detailed video protocol, refer to Park et al., 2023). The setup

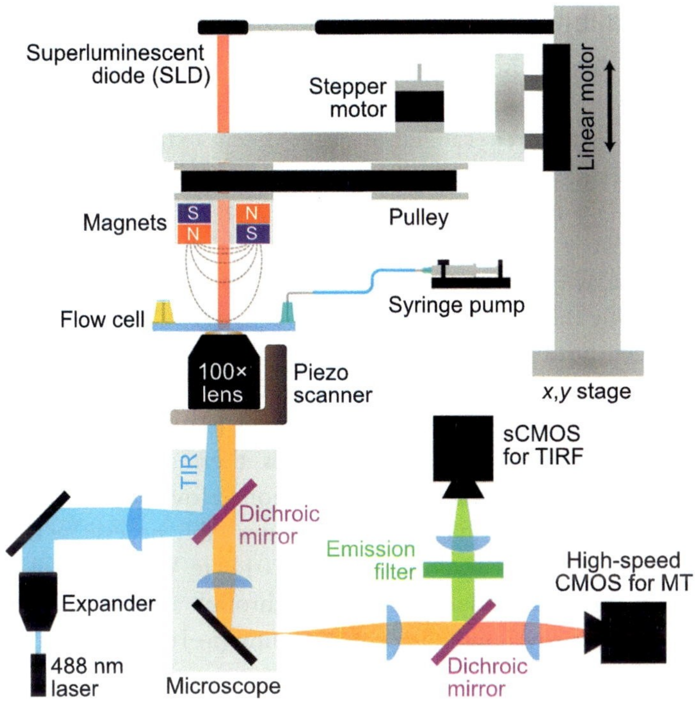

Fig. 2 MT-TIRF setup. A high-resolution MT instrument, built on an inverted microscope, is combined with a TIRF microscopy setup. The system features SLD transillumination for MT and a 488-nm blue laser for TIRF excitation. Imaging is conducted through a high-NA 100× oil-immersion objective lens, split into two channels by a 594 nm dichroic mirror: one for fluorescence (using an sCMOS camera) and the other for magnetic bead tracking (using a high-speed CMOS camera). The setup also includes motor-driven magnets, a nano-precision piezo scanner for lens distance control, and a syringe pump for sample delivery into the flow cell. To see this figure in color, go online.

includes an antiparallel pair of cubic (3/8″) neodymium magnets (N52), 1 mm apart, generating a strong magnetic field gradient that exerts distance-dependent upward forces on the magnetic beads. Force levels are adjusted by precisely controlling the magnet height from the sample surface using a motorized linear stage, vertically mounted on the optical table. Bead imaging is performed using a 100× magnification inverted microscope equipped with a high-speed CMOS camera, illuminated by a 680-nm beam from a super-luminescent diode (SLD). The nano-precision lens positioner is used for accurately controlling distances during pre-experiment calibration and for documenting bead images. It also enables real-time focus adjustments to correct for sample drift.

2.2 Force calibration

Interrogating the precise mechanical responses of biomolecules requires accurate calibration of the forces applied in an MT setup. Detailed methodologies for force calibration have been extensively covered in previous studies (Daldrop et al., 2015; te Velthuis, Kerssemakers, Lipfert, & Dekker 2010; Yu et al., 2014), which have also been integrated into our own prior research (Shon et al., 2019; Choi et al., 2019; Park et al., 2023). In this section, we will revisit these essential concepts to provide a brief overview of our calibration process.

For a given bead–tether construct, a simple force estimation based on inverted pendulum geometry ($F = lk_\mathrm{B}T/\langle\delta x^2\rangle$, where l is the tether length, $k_\mathrm{B}T$ is the thermal energy, and $\langle\delta x^2\rangle$ is the mean squared displacement that represents the extent of beads' thermal motion) provides a surprisingly effective first-order approximation. However, more accurate measurements across a broad range of forces invariably necessitate the use of power spectral analysis (Lansdorp & Saleh, 2012; te Velthuis et al., 2010). The work in this chapter demonstrates this through a case study involving both the low-force response of DNA hairpins (5–10 pN) and the high-force overstretching of dsDNA (>65 pN).

For the advanced force calibration, we typically synthesize a 5-kbp bare dsDNA via polymerase chain reaction (PCR) (similarly to Section 3.1 but excluding the hairpin motif). Subsequently, we assemble a bead–tether construct using the magnetic beads to be utilized in the following experiments (similar to Section 4.1). Primarily, we use M-270 beads (2.8-μm superparamagnetic beads), which have been a mainstay in MT experiments for many years. A tether length of 10–20 kbp is generally preferred to accurately capture the power spectra of thermal fluctuations. This range slows the bead–tether dynamics sufficiently, enabling more precise measurements, though it necessitates additional time for averaging out fluctuations at lower forces. Conversely, a 5-kbp tether, though shorter and less ideal, is made feasible by our high-speed camera. This camera effectively captures fluctuation dynamics even at high forces and decreases the time needed for averaging at lower forces. Additionally, we utilize the known behavior of dsDNA overstretching at 65 pN as a one-point validation for our high-force calibration.

The calibration process involves adjusting the magnet height and capturing bead images (Fig. 3A), ensuring sufficient duration at each force level. As the magnets approach the surface, the magnetic field gradient at

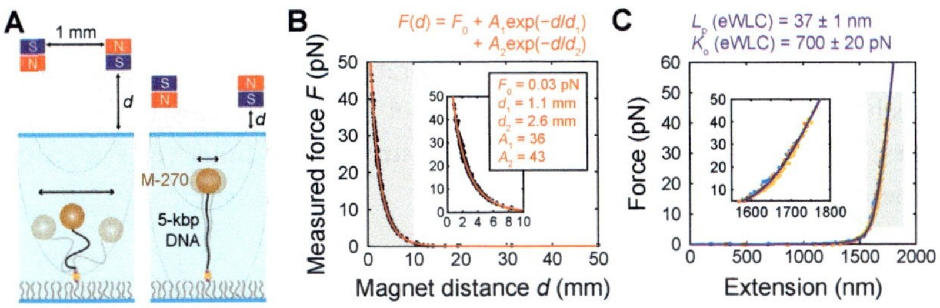

Fig. 3 Force calibration of MT. (A) Force calibration schematic showing forces measured from vertically arranged, antiparallel magnets (1 mm apart) through lateral displacements of a magnetic bead (M-270) tethered by 5-kbp dsDNA at varying distances (*d*). (B) Calibration results with measured force-fitted using a double-exponential model (*red*). (C) Verification of force-extension relationship for the 5-kbp calibration construct, with data ($n = 3$ beads) fitted to an eWLC model showing a persistence length (L_p) of 37 ± 1 nm and a stretch modulus (K_o) of 700 ± 20 pN. To see this figure in color, go online.

the sample plane increases, leading to a corresponding increase in the force exerted on the magnetic beads. This increased force reduces the fluctuation of the DNA-tethered beads. When the magnets are more than 20 mm away from the surface, the force changes slowly, and sparse sampling (about every 10 mm) is adequate. Closer distances within 10 mm require more frequent sampling, every ~1 mm. In the high-force regime, where the force increases rapidly, we perform nearly continuous measurements for each 0.1-mm movement of the magnets.

After collecting bead trajectory data, we analyze the power spectral density (PSD) of bead fluctuations at the magnet positions probed, noting that fluctuations are more pronounced in the *y*-direction, perpendicular to the field line. To determine the force value, a double-Lorentzian model is employed, as detailed in Daldrop, Brutzer, Huhle, Kauert, and Seidel (2015). The relationship between force (F) and magnet distance (d) is then plotted and fitted using a double-exponential equation (Yu et al., 2014):

$$F(d) = F_0 + A_1 \exp(-d/d_1) + A_2 \exp(-d/d_2) \quad (1)$$

where F_0 (baseline), A_1, A_2 (amplitudes), and d_1, d_2 (decay constants) the parameters we fit. A representative result is shown in Fig. 3B, with the force spanning 0–50 pN and showing a sharp increase in the last 10 mm of magnet distance.

To validate the accuracy of our force calibration, we check the force-extension curve (FEC) of the 5-kbp calibration construct, employing the extensible worm-like chain (eWLC) model (Fig. 3C). This model incorporates the persistence length (L_p) and stretch modulus (K_0) as fitting parameters, using the following equation:

$$\frac{FL_p}{k_B T} = \frac{1}{4}\left(1 - \frac{L}{L_0} + \frac{F}{K_0}\right)^{-2} - \frac{1}{4} + \frac{L}{L_0} - \frac{F}{K_0}. \tag{2}$$

Here, L_0 is the contour length (1.7 μm for 5 kbp), and $k_B T$ is the thermal energy (4.11 pN·nm at room temperature). Initially, we assess the FEC alignment with the eWLC model, which allows us to extract L_p and K_0 values. Our findings indicate L_p values around 42 nm and stretch modulus of ~700 pN, consistent with known DNA characteristics. These values were confirmed through repeated measurements on three distinct DNA constructs.

2.3 TIRF microscopy

Our MT instrument is combined with a total internal reflection fluorescence (TIRF) microscopy setup capable of single-molecule fluorescence detection (Fig. 2). The TIRF illumination is crucial for achieving low-background fluorescence imaging, particularly because magnetic beads near the surface often exhibit high autofluorescence. In our inverted microscope configuration, if the beam from the objective lens is directed straight up, it would directly illuminate the bead tethered by the target construct. This could obscure single-molecule fluorescence signals at the tether location, which is essential for precise biomolecular observations. Alternatively, confocal or highly Inclined and laminated optical sheet (HILO) illumination techniques can be employed to circumvent direct bead illumination to some degree.

The TIRF optics are configured according to standard protocols for through the objective illumination (Joo & Ha, 2012). A 488-nm laser beam, expanded 20 times, is directed onto the back focal plane of the objective lens via an achromatic lens. For TIR excitation, the beam is intentionally off-centered to the objective's back aperture, creating an incident angle surpassing the critical angle. The fluorescence emission, after passing through the microscope port, is separated from the 680-nm transillumination using a dichroic mirror. It is then imaged onto a high-quantum-yield scientific CMOS (sCMOS) camera.

3. Sample preparation

The success of single-molecule MT experiments critically depends on meticulous preparation of biomolecular samples. This process typically entails the preparation of specific target constructs and their subsequent attachment to surfaces within flow cells, followed by their coupling to magnetic beads. For probing the RPA–DNA interactions introduced here, we prepare two distinct types of DNA substrates (Sections 3.1 and 3.2, respectively), which are designed to reveal single-stranded regions for RPA binding under force. This feature makes them ideally suited for concurrent measurements of force and fluorescence from the interacting proteins. The DNA constructs are anchored onto PEGylated glass surfaces using biotin–avidin interactions. They are then coupled to magnetic beads, which are prepared to enable either click chemistry (Section 3.3) or antibody-mediated binding (Section 3.4). Additionally, we outline the protocols for expressing and purifying recombinant human RPA proteins fused to enhanced green fluorescent protein (EGFP) tag for single-molecule fluorescence imaging (Section 3.5).

3.1 DNA hairpin constructs

We first prepare a DNA construct incorporating a hairpin motif, consisting of a 15-bp stem and a 15-nt loop (Fig. 4A). The dynamics of such hairpin sequences have been extensively characterized using single-molecule tweezers (Woodside et al., 2006), demonstrating that they unfold at an applied force of 7–15 pN and refold upon relaxation. This two-state transition allows for the reversible transformation of the hairpin structure into a 54-nt ssDNA region, inclusive of the T9 linker at the 5′-end, and potentially induces the binding and unbinding of RPA (Fig. 1B). Considering the usual RPA occupancy of ~20–30 nt ssDNA (Wold, 1997), we estimate that 2–3 RPA trimers could bind to this region. Additionally, the construct includes a 5-kbp dsDNA handle, enhancing its detectability in single-molecule studies through force-extension response.

The DNA hairpin constructs are synthesized via one-step PCR amplification of a 5-kbp segment of λ-DNA using two specialized primers (Fig. 4A). The first primer, biotin-labeled at its 5′-end, is engineered to introduce the aforementioned hairpin motif at the surface end of the PCR product. Employed in our previous research (Shon, Rah, & Yoon, 2019), this built-in hairpin is also useful for correcting variations in the applied force by serving as an internal force gauge. The second primer, azide-labeled at the 5′-end, mediates subsequent attachment to DBCO-coated magnetic beads via copper-free click reaction.

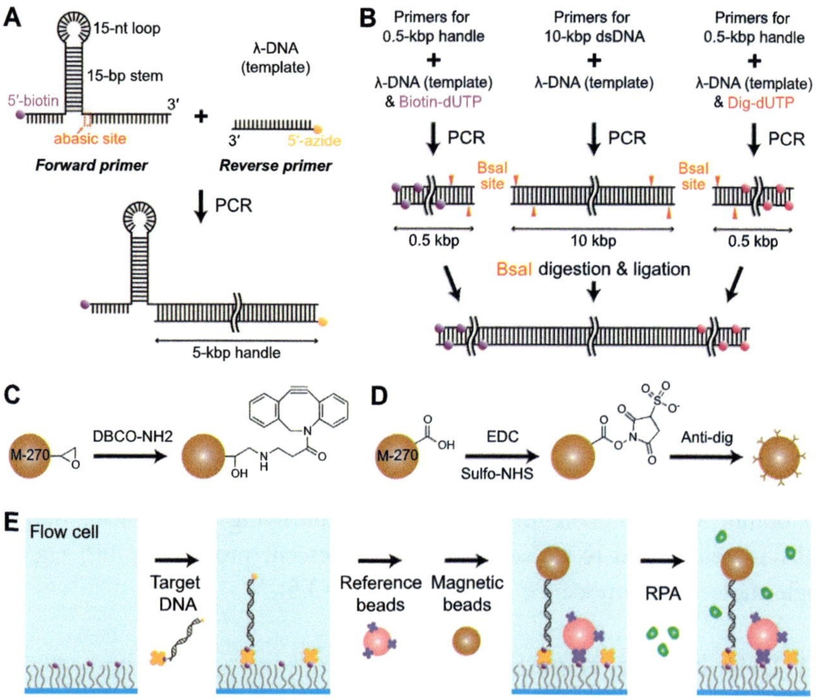

Fig. 4 Sample preparation. (A) 15-bp DNA hairpin constructs are prepared by PCR amplification of a 5-kbp region of λ-DNA. (B) 11-kbp dsDNA for overstretching is prepared by ligating biotin- and digoxigenin-labeled 0.5-kbp handles with a 10-kbp fragment after *BsaI* digestion. (C) Preparation of DBCO-labeled magnetic beads using M-270 epoxy beads, with click chemistry for coupling to DNA hairpin constructs. (D) Preparation of anti-dig-coated magnetic beads using M-270 carboxylic acid beads, employing carbodiimide crosslinking with EDC/sulfo-NHS for coupling to the overstretching construct. (E) Sample assembly via sequential introduction of solutions into a flow cell channel.

3.1.1 Materials and equipment

- λ-DNA (NEB N3011S) (PCR template).
- Primers (IDT):
  - Primer Bhp_5k-fwd: 5′-[Biotin] (T_9) GAG TCC TGG ATC CTG (T_{15}) CAG GAT CCA GGA CTC /idSp/ /idSp/ TCG CCA CCA TCA TTT CCA-3′ (hairpin underlined).
  - Primer Z_5k-rev: 5′-[Azide] TTA GAG AGT ATG GGT ATA TGA CAT CG-3′.
- dNTP (NEB).
- nTaq DNA polymerase with reaction buffer (Enzynomics P050A).

- Equipment for PCR and product verification: Thermal cycler, DNA purification kit, agarose gel electrophoresis setup, UV spectrophotometer (Thermo NanoDrop).

3.1.2 Procedure

1. Carry out PCR amplification:
   - 1.25 μg/mL λ-DNA, 1 μM of each primer, 0.2 mM dNTP, 0.05 U/μL nTaq polymerase in 1X nTaq buffer, adjust the final volume with ultrapure water.
   - Thermal cycle: Initial denaturation at 95 °C for 30 s; annealing at 52 °C for 30 s; extension at 72 °C for 5 min; repeated 34 cycles.
2. Purify the PCR products using a purification kit.
3. Measure the product concentration by UV absorption at 260 nm (A260) and perform agarose gel electrophoresis (1% gel) to verify the product size. A typical yield is ~35 μL of a ~200 nM solution.

3.2 dsDNA for overstretching

The second DNA substrate we create is a 10-kbp dsDNA fragment, with two terminal 0.5-kbp "handles" labeled with multiple biotin and digoxigenin molecules (Fig. 4B). These affinity-tailored termini facilitate attachment to glass surfaces and magnetic beads. The central 10-kbp region is initially double-stranded, not conducive to RPA binding. However, when subjected to forces above 65 pN, dsDNA undergoes an overstretching transition, resulting in a ~70% increase in its length (Smith, Cui, & Bustamante, 1996), potentially revealing RPA binding sites (Fig. 1C). The exact nature of this transition depends on the torsional constraint of the DNA ends: our construct's design restricts rotation, potentially raising the overstretching force to ~110 pN (Léger et al., 1999). However, the 65-pN overstretching can still occur in this DNA, possibly starting from nicks in the middle that allow torsional relaxation. Mechanistically, this change is driven by the gradual peeling of ssDNA (van Mameren et al., 2009; Zhang et al., 2013). The appearance of the 65-pN transition in our experiments would therefore indicate the peeling and exposure of ssDNA ends, likely originating from nicks, which are then available for RPA binding (Fig. 1C). Indeed, these interactions between RPA and DNA have been instrumental in elucidating the structural transitions of DNA during the overstretching process (Gross et al., 2011; King et al., 2013; van Mameren et al., 2009).

The 10-kbp construct for overstretching experiments is synthesized through the digestion and ligation of three PCR products (Fig. 4B): a central 10-kbp fragment and two 0.5-kbp handles, each multiply labeled with either biotin or digoxigenin. These labels, incorporated during the PCR as labeled nucleotides, mediate multi-point attachment to the coverslip surface and magnetic beads (see Section 3.4). For the digestion step, we use *Bsa*I, a restriction enzyme that targets asymmetric DNA sequences. This choice effectively prevents homo-dimerization of the cleaved products, thus improving the overall yield of the intended product. The final ligated product is then purified through gel electrophoresis to remove any unreacted end fragments, ensuring the integrity of the construct for subsequent experiments.

3.2.1 Materials and equipment

- λ-DNA (NEB N3011S) (PCR template).
- Primers for the 0.5-kbp handles (IDT) (*Bsa*I recognition sequence underlined):
  - Primer M-*Bsa*I_0.5k-fwd: 5′-TTT TTT GGT CTC TTG TGG TTC TGG CGT CGT TCT CG-3′.
  - Primer M_0.5k-rev: 5′-CGG AAC GCG ACA TCA CTC-3′.
- Primers for the 10-kbp central fragment (IDT) (*Bsa*I recognition sequence underlined):
  - Primer U-*Bsa*I_10k-fwd: 5′-TTT TTT GGT CTCTAG ACT GAA CTT AAC GGG GCA TCG TA-3′.
  - Primer U-*Bsa*I_10k-rev: 5′-TTT TTT GGT CTCTCA CAC AAA AAT AAA GGG AAA GAT AAG CGC-3′.
- dNTP (NEB).
- Biotin-16-dUTP (biotin-dUTP) (Sigma 11093070910).
- Digoxigenin-11-dUTP (dig-dUTP) (Sigma 11093088910).
- nTaq DNA polymerase with reaction buffer (Enzynomics P050A).
- Pfu-X DNA polymerase with reaction buffer (Solgent).
- Equipment for PCR and product verification: Thermal cycler, DNA purification kit, agarose gel electrophoresis setup, UV spectrophotometer (Thermo NanoDrop).
- Gel extraction kit.
- *Bsa*I with rCutSmart buffer (NEB R3733S).
- T4 DNA ligase with reaction buffer (NEB M0202S).

3.2.2 Procedure

1. Carry out PCR amplification of each construct:

- For the 0.5-kbp handle fragments: 0.6 μg/mL λ-DNA, 0.7 μM of each primer, 0.2 mM dNTP, 0.05 U/μL Pfu-X polymerase in 1X Pfu-X buffer, adjust the final volume with ultrapure water. Initial denaturation at 95 °C for 30 s; annealing at 52 °C for 30 s; extension at 72 °C for 30 s; repeated 34 cycles.
- For the 10-kbp central fragment: 1.25 μg/mL λ-DNA, 1 μM of each primer, 0.2 mM dNTP supplemented with 40 μM biotin-dUTP or dig-dUTP, 0.05 U/μL nTaq polymerase in 1X nTaq buffer, adjust the final volume with ultrapure water. Initial denaturation at 95 °C for 30 s; annealing at 63 °C for 30 s; extension at 72 °C for 5 min; repeated 34 cycles.

2. Purify the PCR products using a purification kit. Measure the product concentrations by A260 and run agarose gel (1%) electrophoresis to verity the product size. A typical yield is ~35 μL of ~3 μM handles and ~70 nM 10-kbp fragment.
3. Digest the purified fragments using *Bsa*I for 4 h at 37 °C.
4. Purify the digested products using a purification kit. Measure the product concentrations by A260 and run agarose gel (1%) electrophoresis to verity the digested size. A typical yield is ~35 μL of ~2 μM digested handles and ~50 nM digested 10-kbp fragment.
5. Ligate 20 nM of the digested 10-kbp fragment with 400 nM of the digested biotin/digoxigenin-labeled handles using T4 DNA ligase. Allow the ligation to proceed overnight at room temperature.
6. For gel purification, run the ligation product on a 1% agarose gel. Then, carefully excise the band corresponding to the 11-kbp product and proceed with gel extraction using a standard gel extraction kit.
7. Measure the product concentration by A260. A typical yield is ~35 μL of ~10 nM product (11 kbp dsDNA including the two 0.5-kbp terminal handles).

3.3 DBCO-functionalized magnetic beads

To attach magnetic beads to the DNA hairpin construct (as prepared in Section 3.1), we use a copper-free click reaction between the azide-labeled DNA and DBCO-functionalized M-270 beads, following the calibration described in Section 2.2. This method ensures rapid and highly selective covalent conjugation, producing bead–tether constructs durable under high forces for extended periods. Other methods, such as peptide-tag-based approaches for covalent bead coupling, are also viable (refer to Choi, Kim, Shon, and Yoon (2022) for a comprehensive review). The natural sedimentation of the magnetic beads to the bottom surface of the flow cell, close to the hairpin constructs, improves the reaction efficiency, typically taking only 20–30 min to complete. Importantly, the resulting bead–tether construct remains torsionally unconstrained owing to the single-chain nature of both NeutrAvidin–biotin and azide–DBCO linkages, thus allowing for unrestricted swiveling.

The DBCO functionalization of M-270 beads is performed by reacting M-270-epoxy beads with a bifunctional crosslinker, DBCO-NH_2 (Fig. 4C). A detailed protocol for this step is provided in the preceding chapter by our lab, titled "High-speed measurements of SNARE–complexin interactions using magnetic tweezers," specifically in Section 3.2. Please refer to this section for comprehensive guidance.

3.4 Anti-digoxigenin-coated magnetic beads

An alternative to click chemistry for bead coupling (Section 3.3) involves using digoxigenin and anti-digoxigenin (anti-dig) antibodies (Fig. 4D). This conventional method is popular due to the wide availability of digoxigenin-labeled oligos and the corresponding antibodies. Although specific and efficient, the antibody–antigen interaction is not covalent and may dissociate under applied force (Neuert, Albrecht, Pamir, & Gaub, 2006). Enhancing stability, multiple digoxigenin labels on DNA handles enable capture by several anti-dig molecules (Lipfert, Kerssemakers, Jager, & Dekker, 2010). The 10-kbp DNA, detailed in Section 3.2, is designed with multiple digoxigenin labels at one end for robust attachment to anti-dig-coated beads. A similar approach is used for the other end with multiple biotin labels, ensuring firm adherence to the glass surface through multiple NeutrAvidin molecules.

3.4.1 *Materials and equipment*

- Dynabeads™ M-270 carboxylic acid (Thermo 14305D).
- Anti-dig (Sigma 11333089001): Prepare a 1 mg/mL solution in phosphate-buffered saline (PBS).
- EDC (1-ethyl-3-(3-dimethylaminopropyl)carbodiimide hydrochloride) (Thermo 22980).
- Sulfo-NHS (N-hydroxysulfosuccinimide) (Thermo 24510).
- 2-Mercaptoethanol (Sigma).
- Phosphate buffer (pH 7.4): 81 mM of sodium phosphate dibasic (Na_2HPO_4) and 19 mM of sodium phosphate monobasic (NaH_2PO_4).
- Activation buffer: 0.1 M MES (pH 6.0) with 500 mM NaCl and 0.1% Tween-20.

- Coupling buffer: PB with 150 mM NaCl and 0.1% Tween-20.
- Quenching buffer: 0.1 M Tris-HCl (pH 8.5) with 0.1% Tween-20.
- Storage buffer: PB with 150 mM NaCl and 0.01% Tween-20.
- Mixer (ELMI Intelli-Mixer™ RM-2 M).
- Permanent magnet (for bead separation).

3.4.2 Procedure

1. Take 1 mg of M-270 beads from a stock solution (~30 mg/mL) and add it to 1 mL of activation buffer. Use magnets to replace the medium with fresh activation buffer. Repeat this washing process twice and remove the activation buffer in the final wash.
2. Dissolve 10 mg of EDC and 22 mg of sulfo-NHS in 1 mL of activation buffer by vortexing. Immediately add this solution to the collected magnetic beads and mix by vortexing. Incubate for 15 min at room temperature with slow tilt rotation.
3. Add 2.8 μL of 2-mercaptoethanol to quench the EDC reaction. Incubate for 10 min at room temperature with slow rotation. After incubation, wash twice with 1 mL of coupling buffer.
4. Add 60 μL of anti-dig solution to the bead mixture. Incubate for 3 h at room temperature with slow tilt rotation.
5. Wash the beads twice with 1 mL of quenching buffer, then incubate for an additional 10 min at room temperature with slow tilt rotation.
6. Wash the beads three times with 1 mL of storage buffer. Finally, resuspend the beads in 50 μL of storage buffer and store the solution at 4 °C.

3.5 RPA–EGFP expression and purification

This section provides the protocol for purifying the enhanced green fluorescent protein (EGFP)-tagged human RPA (RPA–EGFP), a heterotrimeric complex composed of RPA70, RPA32, and RPA14 subunits. The RPA70 subunit is fused with an 8xHis and EGFP tag at its carboxyl terminus. This protocol, an adaptation of the previously published methods (Kang et al., 2023; Kim et al., 2024), involves the expression of RPA subunits in *Escherichia coli* (*E. coli*), followed by a series of chromatographic steps: Ni-nitrilotriacetic acid (Ni-NTA) affinity chromatography, anion exchange chromatography, and size-exclusion chromatography, for sequential purification.

3.5.1 Expression of RPA

3.5.1.1 Materials and equipment

- BL21(DE3) chemically competent *E. coli* (Enzynomics).
- LB broth (Difco).

- Ampicillin (MBcell).
- IPTG (Isopropyl β-D-1 thiogalactopyranoside) (Glentham).
- T-Buffer: 12.5 mM Tris-HCl (pH 7.5) with 0.25 mM EDTA, 0.5 mM DTT, and 5% glycerol.
- Resuspension Buffer: T-Buffer supplemented with 100 mM NaCl.
- Tip sonicator (Qsonica).

3.5.1.2 Procedure

1. The coding sequences of all human RPA subunits (RPA70–EGFP, RPA32, and RPA14) are subcloned into the pET11d vector.
2. Transform 200 ng of the plasmids into 50 μL of *E. coli* BL21(DE3) competent cells.
3. Inoculate a single colony of *E. coli* into 50 mL of LB broth supplemented with 0.1 mg/mL ampicillin for pre-culture.
4. Transfer 40 mL of the pre-cultured *E. coli* cells into 4 L of LB broth with 0.1 mg/mL ampicillin. Incubate the culture in a shaking incubator at 37 °C and 200 rpm until the optical density at 600 nm (OD_{600}) reaches ~0.4, then cool the cells to 16 °C.
5. Induce protein expression by adding 1 mM IPTG when OD_{600} is around 0.6. Continue the incubation at 16 °C overnight.
6. Harvest the cells by centrifugation at 4 °C and 4000*g* for 20 min. Resuspend the cell pellets in 40 mL of Resuspension Buffer. The cells can be stored at −80 °C until use.
7. Lyse the cells using sonication at 47% amplitude, with 15 s pulses on and 100 s off, for a total active sonication time of 120 s. Keep the cells on ice to mitigate temperature rise and, if necessary, increase the duration of the pulse-off time.
8. Ultracentrifuge the lysates at 4 °C and 100,000*g* for 1 h. Collect the supernatant.

3.5.2 Column chromatography

All chromatography processes are conducted at 4 °C.

3.5.2.1 Materials and equipment

- HisPur™ Ni-NTA resin (Thermo Fisher scientific).
- Econo-Column chromatography column, 2.5 × 20 cm (Bio-Rad).
- Ni-NTA Wash Buffer: Resuspension Buffer supplemented with 20 mM imidazole.
- Ni-NTA Elution Buffer: Resuspension Buffer supplemented with 300 mM imidazole.

- Hitrap Q HP anion exchange chromatography column, 1 mL (Cytiva).
- ÄKTA pure chromatography system (Cytiva).
- Size-exclusion column: Superdex 200 Increase 10/300 GL (Cytiva).
- Amicon Ultra centrifugal filters, 30 kDa MWCO (Merck).

3.5.2.2 Procedure

1. Add 10 mL of Ni-NTA resin to an Econo-Column, creating a 5 mL bed volume. Equilibrate the column with 25 mL (5 × bed volume) of Ni-NTA Wash Buffer.
2. Load the clarified lysates from Section 3.5.1 onto the column. To enhance protein binding, reload the flow-through onto the column. Keep 5 μL of both the lysates and the flow-through for SDS-PAGE analysis.
3. Wash the column with 15 mL (3× bed volume) of Ni-NTA Wash Buffer. Keep 5 μL of the wash solution for SDS-PAGE.
4. Elute the protein using 15 mL (3× bed volume) of Ni-NTA Elution Buffer, collecting 1 mL fractions.
5. Analyze the fractions with 10% SDS-PAGE to identify those containing RPA, along with the clarified lysates, flow-through, and wash samples. Pool the fractions with RPA.
6. For ion exchange chromatography, equilibrate the Hitrap Q HP anion exchange column (1 mL) with 5 mL (5× bed volume) of Resuspension Buffer at a flow rate of 1 mL/min.
7. Load the pooled sample onto the column using Resuspension Buffer at 0.5 mL/min.
8. Wash the column with 5 mL (5× bed volume) of Resuspension Buffer at 1 mL/min.
9. Elute the protein with a salt gradient from Resuspension Buffer to T-buffer supplemented with 1 M NaCl, collecting 0.5 mL fractions. Run the salt gradient for 60 min at a flow rate of 0.5 mL/min to improve the separation of the RPA trimers from other proteins.
10. Use 10% SDS-PAGE to identify and pool the fractions containing RPA.
11. Concentrate the pooled fractions to 500 μL using an Amicon filter, centrifuging at 7500*g*.
12. For size-exclusion chromatography, equilibrate the Superdex column with T-buffer supplemented with 40 mM NaCl using the ÄKTA pure chromatography system.

13. Load the protein sample onto the column at a flow rate of 0.4 mL/min, collecting 0.5 mL fractions.
14. Analyze the fractions with 10% SDS-PAGE to pool those containing RPA.
15. Concentrate the pooled fractions as necessary using centrifugal filters. Aliquot the sample and store at −80 °C.

4. Force-fluorescence measurements

Using the materials prepared above, bead–DNA constructs are assembled in custom flow cells for microscopy. We first characterize the mechanical response of the DNA samples alone, and then introduce EGFP-tagged RPA (RPA–EGFP) molecules to examine changes in DNA mechanics. After that, the colocalization of RPA molecules with the bead–DNA constructs is checked via fluorescence. Finally, the dynamic exchange of RPA is followed by measuring the fluorescence recovery after photobleaching (FRAP) at the single-tether level.

4.1 Preparation of bead–tether constructs

Flow cells for force-fluorescence experiments are constructed by bonding two glass coverslips with double-sided tape. The glass surfaces are coated with PEG, of which a small fraction contains a terminal biotin group. This design enables specific tethering of biotin-containing target DNA constructs via NeutrAvidin, while simultaneously suppressing the non-specific adsorption of other molecules.

Details for the PEGylation of glass coverslips and sample chamber fabrication are the same as introduced in the preceding chapter by our lab, "High-speed measurements of SNARE–complexin interactions using magnetic tweezers", specifically in Section 3.3. Please refer to this section for comprehensive guidance, or our previous research introducing the same (Cho et al., 2022; Kim, Shon, et al., 2021; Park et al., 2023; Shon, Kim, & Yoon, 2018). In this section, we focus on delivering samples into the flow cells designated for RPA experiments. Once the flow cell is assembled, DNA constructs are attached to the surface and linked to magnetic beads for conducting MT experiments (Fig. 4E). Once a suitable DNA construct is identified, a solution of RPA is introduced to initiate its binding to the DNA substrate. All liquid handling tasks are performed using a syringe pump for precise control of flow rate and volume.

4.1.1 Materials and equipment

- Sample chamber: Refer to the previous Chapter: "High-speed measurements of SNARE–complexin interactions using magnetic tweezers" (Section 3.3).
- Syringe pump (KD Scientific Legato 210).
- Plastic syringe (5 mL) (Henke-Ject).
- PE-60 polyethylene tubing (SCI).
- Tris buffer: 50 mM Tris-HCl (pH 8.0) with 50 mM NaCl.
- BSA (Thermo): 1% in Tris buffer.
- NeutrAvidin (Thermo 31000): 2 μM in Tris buffer.
- DNA constructs (from Sections 3.1 and 3.2).
- Reference bead: Streptavidin-coated polystyrene particles (3.0–3.4 μm diameter) (Spherotech SVP-30-5).
- Magnetic beads (from Sections 3.3 and 3.4).
- RPA binding buffer: 1 mM DTT and 0.1% BSA in Tris buffer.
- Purified RPA–EGFP (from Section 3.5) in the RPA binding buffer.

4.1.2 Procedure

1. Prepare a mixture of 1 nM DNA (either hairpin or 10-kbp dsDNA) with 1 μM NeutrAvidin at a 1:1000 molar ratio. Incubate for 1 h with gentle rotation.
2. Connect the outlet syringe needle of the flow cell to the tubing leading to the syringe pump. Place the chamber on the microscope stage and equilibrate the channel with Tris buffer.
3. Introduce the BSA solution into the channel for passivation. Incubate for 10 min and then wash the chamber with 200 μL of Tris buffer.
4. Introduce the NeutrAvidin/DNA mixture. Incubate for 15 min and then wash the chamber with 200 μL of Tris buffer.
5. Dilute the reference bead solution 25 times in Tris buffer and introduce it into the channel. Incubate for 10 min and then wash the chamber with 200 μL of Tris buffer.
6. Dilute the magnetic bead solution 100 times in Tris buffer and introduce it into the channel. Incubate for 20 min and then wash the chamber with 200 μL of Tris buffer.

4.1.3 Notes

- Since the flow channel volume is typically ~20 μL, it is recommended to use ~50 μL of each solution for complete exchange. Also, use at least 200 μL for washing between steps.
- Maintain a flow rate of 200 μL/min for all steps.

4.2 RPA binding on DNA hairpins

With DNA hairpins, we first characterize their force-dependent opening/closing dynamics using MT (Fig. 5A). This involves profiling the overall FECs under force ramps (Fig. 5B) and investigating the telegraph-like length variations under constant forces (Fig. 5C). Then, after adding RPA, we test whether the exposed ssDNA regions upon hairpin opening are available for RPA binding. This is assessed both through mechanical changes in hairpin response observed by MT (Fig. 5D–F) and via fluorescence signal increase in TIRF imaging (Fig. 5G), which is capable of detecting single RPA–EGFP complexes. Finally, we aim to observe the exchange of RPA between the bound and unbound free molecules in the solution through FRAP (Fig. 5H).

4.2.1 Procedure

1. To identify a genuine DNA hairpin construct, search for a bead whose height changes abruptly when transitioning between 0 and 5 pN, indicating a large shift comparable to its expected tether length (1.7 μm). Additionally, rotate the magnets at 5 pN and check if the radius of the bead's trajectory is less than ~500 nm (the smaller the better, see Notes).
2. Apply force ramps from 0 to 15 pN, moving up and down at 1 pN/s. Verify that the FECs align with an eWLC model with reasonable values of persistence length and stretch modulus.
3. Check if the FEC displays a break due to the hairpin dynamics in the 5–10 pN regime, transitioning from one eWLC FEC to another, more extended curve.
4. With the previously identified force-responsive regime in mind, perform force-jump measurements within the same range, alternating between two force levels that respectively induce hairpin opening and closing. Generally, starting with 5 and 10 pN is advisable. Use the maximum motor speed possible to swiftly switch between the two levels. Confirm that the observed opening distance corresponds to the length change expected from the hairpin sequence.
5. While applying 1 pN, gently introduce 30 nM RPA into the channel (20 μL/min).
6. Repeat Steps 2–4 to look for any changes resulting from RPA binding.
7. For fluorescence imaging, switch off the SLD and illuminate the field with a 488-nm laser configured for TIRF excitation.

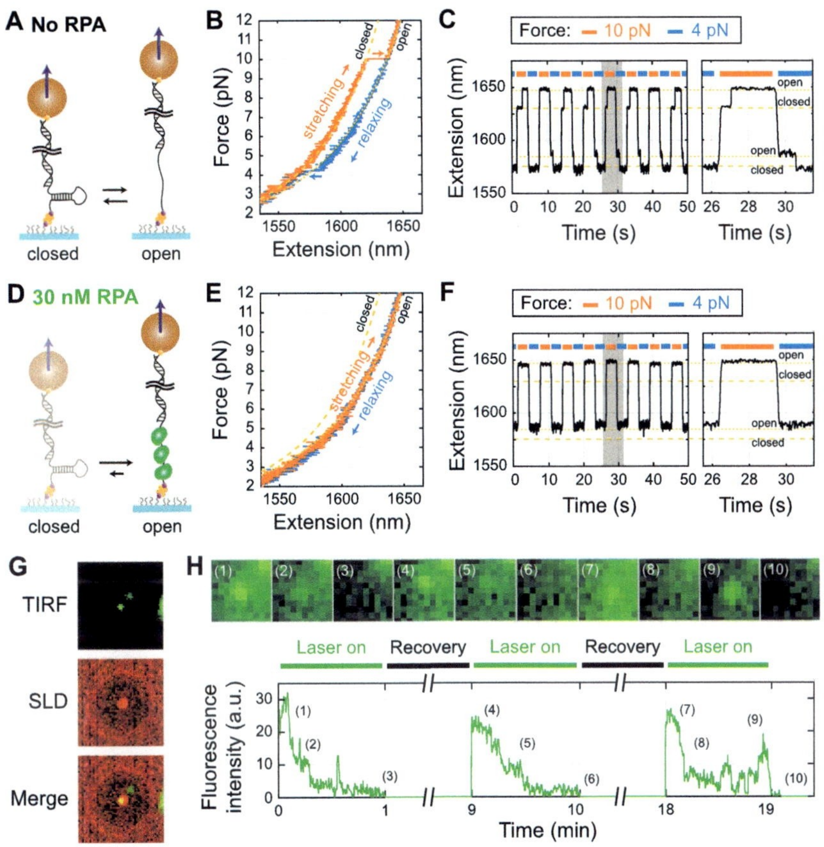

Fig. 5 Probing RPA–ssDNA interactions with DNA hairpins. (A, D) Hairpin two-state transition without (A) and with RPA (D). (B, E) FECs before (B) and after (E) RPA introduction, with yellow dashed and dotted lines representing eWLC models for closed/open states (with ssDNA region by a FJC model). (C, F) Force-jump experiment results without (C) and with (F) RPA, recorded at 100 Hz. (G) RPA–EGFP colocalization with magnetic bead shown in TIRF and SLD images. (H) FRAP experiments on ssDNA-bound RPA at 10 Hz, using three-point median filter for intensity data smoothing. To see this figure in color, go online.

8. Merge the bead image obtained under SLD illumination with the TIRF image to verify the colocalization of RPA molecules with the bead–DNA construct.
9. To investigate the dynamic exchange of RPA via FRAP, bleach the fluorescent molecules in the field of view, then shut off the laser for 8 min. Subsequently, image the same region to observe fluorescence recovery.

4.2.2 Notes

- The radius of bead's trajectory during magnet rotation indicates the degree of bead tilt from its tether. A larger radius of rotation correlates with significant distortion in the FECs deviating from model curves, as well as a poorer signal-to-noise ratio in length measurements using MT (Shon et al., 2019). Consequently, employing beads with a smaller radius is preferable for more accurate results.
- The hairpin may display force hysteresis, which is the difference in force levels during opening and closing, primarily determined by the loop length (Bercy & Bockelmann, 2015).
- The opening distance of hairpins can be estimated based on the force-dependent extension of the hairpin sequence. In our case, opening the stem (2×15 nt) and the loop (15 nt) exposes a total of 45 nt of ssDNA. Accounting for the 2-nm width of the hairpin in the closed state, this corresponds to an extension change of 11 nm at 5 pN and 17 nm at 10 pN.

4.3 RPA binding on overstretched DNA

Using the 10-kbp bare dsDNA construct, we follow a similar sequence of objectives and measurements as outlined for DNA hairpins in Section 4.2. The key difference lies in the structural changes that expose ssDNA, which occur via overstretching at a significantly higher force regime, above 60 pN. We first document the overstretching response by recording FECs, which likely lead to the peeling of the ssDNA strand from nicks (Fig. 6A). Upon introducing RPA, we then examine changes in the FECs and detect fluorescence via TIRF imaging (Fig. 6B). Similar to our approach with hairpins, we also observe the dynamic exchange of RPA with the freely available RPA in the bulk solution, using FRAP (Fig. 6C).

4.3.1 Procedure

- Follow the Procedure described in Section 4.2 with these modifications:
  - In Step 1, note that the contour length of the 10-kbp construct is 3.4 μm, which should result in more distinct bead height changes when transitioning from 0 to 5 pN.
  - For Steps 2 and 3, apply force ramps ranging from 0 to 70 pN. Pay attention to overstretching at ~65 pN, marked by a sudden, 1.7-times increase in extension.
  - Omit Step 4 (force-jump experiments) in this protocol.
  - In Step 9 (FRAP experiments), shorten the waiting time in darkness to 20 s

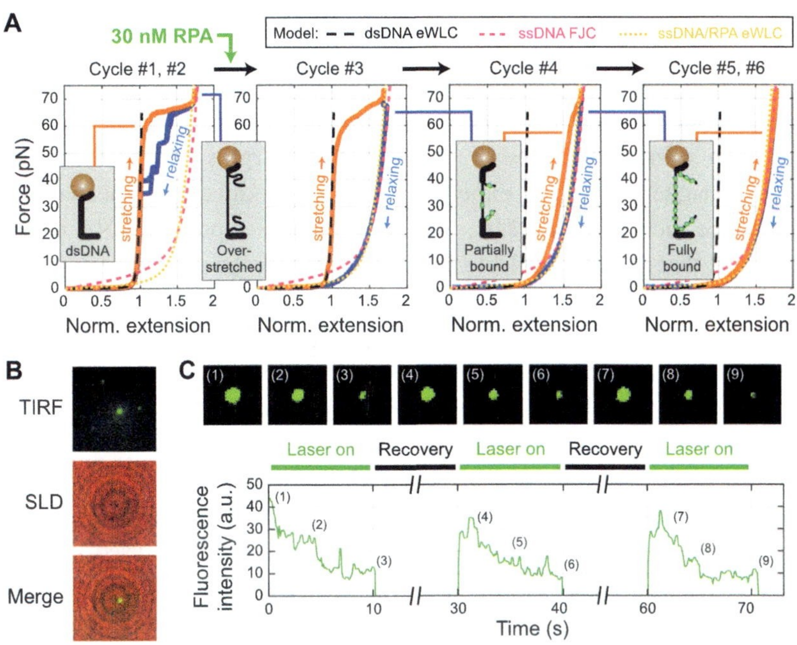

Fig. 6 Probing RPA–ssDNA interactions with overstretched DNA. (A) Experimental design and results with representative FECs for each force ramp cycle. Cycles #1 and #2 represent DNA without RPA, showing overstretching and hysteresis during relaxation due to ssDNA peeling. Cycles #3 and #4 display a shift to the RPA-bound state, inhibiting the return to the double-stranded form. Cycles #5 and #6 indicate a stable RPA-bound state without a further change. Model curves include eWLC for dsDNA (*black long-dashed*), extensible FJC for ssDNA (*magenta dashed*), and eWLC for RPA-bound ssDNA with $L_p = 2.7$ nm (*yellow dotted*). (B) RPA–EGFP colocalization with the magnetic bead in TIRF and SLD images. (C) FRAP experiments on ssDNA-bound RPA at 10 Hz, with intensity data smoothed using a three-point median filter. To see this figure in color, go online.

4.3.2 Notes

- After overstretching and peeling occur, DNA constructs become extremely unstable, leading to possible rupture. Therefore, changing the ramp rate during the overstretched region to 2–4 pN/s may enhance the repeatability of experiments. Alternatively, adjusting the overall ramp rate to 2–4 pN/s may be beneficial.
- After RPA coats the ssDNA region, DNA becomes stable but sticky, potentially attaching to the surface at ~0 pN. To prevent this, introducing an offset force of ~0.5 pN may help avoid attachment.

5. Discussion

In this chapter, we have presented detailed methodologies for conducting force-fluorescence measurements to observe the RPA–DNA interaction using two types of DNA: DNA hairpins and bare dsDNA. Initially, magnetic tweezer experiments were conducted to mechanically characterize each type of DNA, which potentially exposes ssDNA regions under load that may be available for RPA binding. After introducing RPA into the solution, we examined the binding through both MT (observing mechanical changes in FECs) and TIRF (detecting single-molecule fluorescence) modalities setup on the same microscope, thereby providing a comprehensive analysis of the RPA–DNA interaction.

For the DNA hairpin, prior to the introduction of RPA, clear two-state transitions were observed (Fig. 5A–C), each corresponding to the open and closed conformations of a hairpin motif. The measured transition closely matched the theoretical predictions for ssDNA region extension, modeled as a freely jointed chain (FJC) (Dessinges et al., 2002). Upon adding RPA, the hairpin remained almost exclusively in the open state, indicating that bound RPA significantly hinders the closing transition (Fig. 5D–F). This effect is likely pronounced in the loop region, where closing requires the zippering of proximal nucleotides from opposite sides of the loop. The presence of RPA on the probed construct was confirmed by single-molecule fluorescence of RPA–EGFP (Fig. 5G,H), revealing up to 3 molecules simultaneously on the same strand, likely occupying the open ssDNA region. Notably, in the 30 nM RPA solution, a significant exchange reaction was observed between the ssDNA-bound and free RPA (Fig. 5H), consistent with a previous result (Gibb et al., 2014).

In the experiments with bare dsDNA, the well-known overstretching transition at 65 pN was observed (Fig. 6A), even under torsional constraints imposed by 0.5-kbp multi-tethering handles. This observation suggests the presence of at least one ssDNA nick in the construct. Consistently, significant hysteresis in the stretching/relaxation FECs (Fig. 6A: Cycles #1 and #2) indicated ssDNA peeling (Smith et al., 1996; van Mameren et al., 2009). Intriguingly, upon introducing RPA, immediate changes in the force-extension behavior were observed after the first overstretching during relaxation (Fig. 6A: Cycle #3). This suggests extensive DNA peeling and melting, followed by the rapid and contiguous coating and stabilization of RPA on the exposed ssDNA. Although RPA binding slightly decreased after the first relaxation in the presence of RPA, evidenced by a minor reduction in extension (Fig. 6A: Cycle #4), subsequent cycles conformed to the fully bound ssDNA extension curve (Fig. 6A: Cycles #5 and #6). The FECs in the RPA-bound state, while similar to predictions from an extensible FJC model for ssDNA after overstretching (magenta dashed curve in Fig. 6A) (Smith et al., 1996), deviate distinctly from this model in the force regime below 10 pN. This deviation indicates unique mechanical properties of RPA-bound ssDNA, better represented by an eWLC model with an L_p of 2.7 nm (yellow dotted curve in Fig. 6A). This conforms to other observations of RPA-bound ssDNA FECs showing increased rigidity (Chen et al., 2015; De Vlaminck et al., 2010). Finally, akin to the hairpin substrate, a rapid exchange between DNA-bound and free RPA was observed (Fig. 6B,C). Note that while this substrate can host more RPA molecules than the hairpin, TIRF excitation limits the illumination depth to ~200 nm, visualizing RPA trimers near the surface only.

In this chapter, we detailed the use of a force-fluorescence setup combining MT and TIRF microscopy to investigate RPA–DNA interactions. This technique offers real-time, single-molecule insights into RPA binding under mechanical stress, enhancing our understanding of its role in DNA repair and replication. This approach paves the way for more comprehensive studies into protein–DNA interactions, particularly in dynamic, phase-separated RPA condensates (Spegg et al., 2023), underscoring its potential in elucidating complex cellular processes.

Acknowledgments

This work was supported by the National Research Foundation of Korea (NRF) grant funded by the Korean government (MSIT) (NRF-2022R1C1C1012176, RS-2023-00218927, and NRF-2021R1A6A1A10042944).

Appendix A. Supporting information

Supplementary data associated with this article can be found in the online version at https://doi.org/10.1016/bs.mie.2024.01.003.

References

Awate, S., & Brosh, R., Jr. (2017). Interactive roles of DNA helicases and translocases with the single-stranded DNA binding protein RPA in nucleic acid metabolism. *International Journal of Molecular Sciences, 18*(6), 1233. https://doi.org/10.3390/ijms18061233.

Bae, W., Kim, K., Min, D., Ryu, J.-K., Hyeon, C., & Yoon, T.-Y. (2014). Programmed folding of DNA origami structures through single-molecule force control. *Nature Communications, 5*(1), 5654. https://doi.org/10.1038/ncomms6654.

Bain, F. E., Fischer, L. A., Chen, R., & Wold, M. S. (2018). Chapter seventeen—Single-molecule analysis of replication protein A–DNA interactions. In M. Spies, & A. Malkova (Eds.), *Methods in enzymology: 600*, (pp. 439–461). Academic Press. ⟨https://doi.org/10.1016/bs.mie.2017.11.016⟩.

Bercy, M., & Bockelmann, U. (2015). Hairpins under tension: RNA versus DNA. *Nucleic Acids Research, 43*(20), 9928–9936. https://doi.org/10.1093/nar/gkv860.

Brutzer, H., Schwarz, F. W., & Seidel, R. (2012). Scanning evanescent fields using a pointlike light source and a nanomechanical DNA gear. *Nano Letters, 12*(1), 473–478. https://doi.org/10.1021/nl203876w.

Chaurasiya, K. R., Paramanathan, T., McCauley, M. J., & Williams, M. C. (2010). Biophysical characterization of DNA binding from single molecule force measurements. *Physics of Life Reviews,* 7(3), 299–341. https://doi.org/10.1016/j.plrev.2010.06.001.

Chen, J., Le, S., Basu, A., Chazin, W. J., & Yan, J. (2015). Mechanochemical regulations of RPA's binding to ssDNA. *Scientific Reports, 5*(1), 1. https://doi.org/10.1038/srep09296.

Cho, S., Shon, M. J., Son, B., Eun, G. S., Yoon, T.-Y., & Park, T. H. (2022). Tension exerted on cells by magnetic nanoparticles regulates differentiation of human mesenchymal stem cells. *Biomaterials Advances, 139*, 213028. https://doi.org/10.1016/j.bioadv.2022.213028.

Choi, H.-K., Kim, H. G., Shon, M. J., & Yoon, T.-Y. (2022). High-resolution single-molecule magnetic tweezers. *Annual Review of Biochemistry, 91*(1), 33–59. https://doi.org/10.1146/annurev-biochem-032620-104637.

Choi, H.-K., Min, D., Kang, H., Shon, M. J., Rah, S.-H., Kim, H. C., ... Yoon, T.-Y. (2019). Watching helical membrane proteins fold reveals a common N-to-C-terminal folding pathway. *Science (New York, N. Y.), 366*(6469), 1150–1156. https://doi.org/10.1126/science.aaw8208.

Collins, B. E., Ye, L. F., Duzdevich, D., & Greene, E. C. (2014). Chapter 12 - DNA curtains: Novel tools for imaging protein–nucleic acid interactions at the single-molecule level. In J. C. Waters, & T. Wittman (Eds.), *Methods in cell biology: 123*, (pp. 217–234). Academic Press. ⟨https://doi.org/10.1016/B978-0-12-420138-5.00012-4⟩.

Comstock, M. J., Ha, T., & Chemla, Y. R. (2011). Ultrahigh-resolution optical trap with single-fluorophore sensitivity. *Nature Methods, 8*(4), 4. https://doi.org/10.1038/nmeth.1574.

Comstock, M. J., Whitley, K. D., Jia, H., Sokoloski, J., Lohman, T. M., Ha, T., & Chemla, Y. R. (2015). Direct observation of structure-function relationship in a nucleic acid–processing enzyme. *Science (New York, N. Y.), 348*(6232), 352–354. https://doi.org/10.1126/science.aaa0130.

Daldrop, P., Brutzer, H., Huhle, A., Kauert, D. J., & Seidel, R. (2015). Extending the range for force calibration in magnetic tweezers. *Biophysical Journal, 108*(10), 2550–2561. https://doi.org/10.1016/j.bpj.2015.04.011.

De Vlaminck, I., Vidic, I., van Loenhout, M. T. J., Kanaar, R., Lebbink, J. H. G., & Dekker, C. (2010). Torsional regulation of hRPA-induced unwinding of double-stranded DNA. *Nucleic Acids Research, 38*(12), 4133–4142. https://doi.org/10.1093/nar/gkq067.

Dessinges, M.-N., Maier, B., Zhang, Y., Peliti, M., Bensimon, D., & Croquette, V. (2002). Stretching single stranded DNA, a model polyelectrolyte. *Physical Review Letters, 89*(24), 248102. https://doi.org/10.1103/PhysRevLett.89.248102.

Gibb, B., Ye, L. F., Gergoudis, S. C., Kwon, Y., Niu, H., Sung, P., & Greene, E. C. (2014). Concentration-dependent exchange of replication protein A on single-stranded DNA revealed by single-molecule imaging. *PLoS One, 9*(2), e87922. https://doi.org/10.1371/journal.pone.0087922.

Graves, E. T., Duboc, C., Fan, J., Stransky, F., Leroux-Coyau, M., & Strick, T. R. (2015). A dynamic DNA-repair complex observed by correlative single-molecule nanomanipulation and fluorescence. *Nature Structural & Molecular Biology, 22*(6), 6. https://doi.org/10.1038/nsmb.3019.

Grimme, J. M., Honda, M., Wright, R., Okuno, Y., Rothenberg, E., Mazin, A. V., ... Spies, M. (2010). Human Rad52 binds and wraps single-stranded DNA and mediates annealing via two hRad52–ssDNA complexes. *Nucleic Acids Research, 38*(9), 2917–2930. https://doi.org/10.1093/nar/gkp1249.

Gross, P., Laurens, N., Oddershede, L. B., Bockelmann, U., Peterman, E. J. G., & Wuite, G. J. L. (2011). Quantifying how DNA stretches, melts and changes twist under tension. *Nature Physics, 7*(9), 9. https://doi.org/10.1038/nphys2002.

Ha, T., Kaiser, C., Myong, S., Wu, B., & Xiao, J. (2022). Next generation single-molecule techniques: Imaging, labeling, and manipulation in vitro and in cellulo. *Molecular Cell, 82*(2), 304–314. https://doi.org/10.1016/j.molcel.2021.12.019.

Ha, T., Kozlov, A. G., & Lohman, T. M. (2012). Single-molecule views of protein movement on single-stranded DNA. *Annual Review of Biophysics, 41*(1), 295–319. https://doi.org/10.1146/annurev-biophys-042910-155351.

Heller, I., Hoekstra, T. P., King, G. A., Peterman, E. J. G., & Wuite, G. J. L. (2014). Optical tweezers analysis of DNA–protein complexes. *Chemical Reviews, 114*(6), 3087–3119. https://doi.org/10.1021/cr4003006.

Hilario, J., & Kowalczykowski, S. C. (2010). Visualizing protein–DNA interactions at the single-molecule level. *Current Opinion in Chemical Biology, 14*(1), 15–22. https://doi.org/10.1016/j.cbpa.2009.10.035.

Hohng, S., Zhou, R., Nahas, M. K., Yu, J., Schulten, K., Lilley, D. M. J., & Ha, T. (2007). Fluorescence-force spectroscopy maps two-dimensional reaction landscape of the holliday junction. *Science (New York, N. Y.), 318*(5848), 279–283. https://doi.org/10.1126/science.1146113.

Iftode, C., Daniely, Y., & Borowiec, J. A. (1999). Replication protein A (RPA): The eukaryotic SSB. *Critical Reviews in Biochemistry and Molecular Biology, 34*(3), 141–180. https://doi.org/10.1080/10409239991209255.

Jacobs, M. J., & Blank, K. (2014). Joining forces: Integrating the mechanical and optical single molecule toolkits. *Chemical Science, 5*(5), 1680–1697. https://doi.org/10.1039/C3SC52502C.

Joo, C., & Ha, T. (2012). Single-molecule FRET with total internal reflection microscopy. *Cold Spring Harbor Protocols, 2012*(12), pdb.top072058. https://doi.org/10.1101/pdb.top072058.

Kang, Y., Han, Y. G., Khim, K. W., Choi, W. G., Ju, M. K., Park, K., ... Lee, J. Y. (2023). Alteration of replication protein A binding mode on single-stranded DNA by NSMF potentiates RPA phosphorylation by ATR kinase. *Nucleic Acids Research, 51*(15), 7936–7950. https://doi.org/10.1093/nar/gkad543.

Kemmerich, F. E., Daldrop, P., Pinto, C., Levikova, M., Cejka, P., & Seidel, R. (2016). Force regulated dynamics of RPA on a DNA fork. *Nucleic Acids Research, 44*(12), 5837–5848. https://doi.org/10.1093/nar/gkw187.

Kemmerich, F. E., Swoboda, M., Kauert, D. J., Grieb, M. S., Hahn, S., Schwarz, F. W., ... Schlierf, M. (2016). Simultaneous single-molecule force and fluorescence sampling of DNA nanostructure conformations using magnetic tweezers. *Nano Letters, 16*(1), 381–386. https://doi.org/10.1021/acs.nanolett.5b03956.

Kim, C., Shon, M. J., Kim, S. H., Eun, G. S., Ryu, J.-K., Hyeon, C., ... Yoon, T.-Y. (2021). Extreme parsimony in ATP consumption by 20S complexes in the global disassembly of single SNARE complexes. *Nature Communications, 12*(1), 3206. https://doi.org/10.1038/s41467-021-23530-0.

Kim, S., Shin, W. H., Kang, Y., Kim, H., & Lee, J. Y. (2024). Direct visualization of replication and R-loop collision using single-molecule imaging. *Nucleic Acids Research, 52*, 259–273. https://doi.org/10.1093/nar/gkad1101.

Kim, S. H., Jung, H. J., Lee, I.-B., Lee, N.-K., & Hong, S.-C. (2021). Sequence-dependent cost for Z-form shapes the torsion-driven B–Z transition via close interplay of Z-DNA and DNA bubble. *Nucleic Acids Research, 49*(7), 3651–3660. https://doi.org/10.1093/nar/gkab153.

King, G. A., Gross, P., Bockelmann, U., Modesti, M., Wuite, G. J. L., & Peterman, E. J. G. (2013). Revealing the competition between peeled ssDNA, melting bubbles, and S-DNA during DNA overstretching using fluorescence microscopy. *Proceedings of the National Academy of Sciences, 110*(10), 3859–3864. https://doi.org/10.1073/pnas.1213676110.

Koch, S. J., & Wang, M. D. (2003). Dynamic force spectroscopy of protein-DNA interactions by unzipping DNA. *Physical Review Letters, 91*(2), 028103. https://doi.org/10.1103/PhysRevLett.91.028103.

Lansdorp, B. M., & Saleh, O. A. (2012). Power spectrum and Allan variance methods for calibrating single-molecule video-tracking instruments. *Review of Scientific Instruments, 83*(2), 025115. https://doi.org/10.1063/1.3687431.

Lebel, P., Basu, A., Oberstrass, F. C., Tretter, E. M., & Bryant, Z. (2014). Gold rotor bead tracking for high-speed measurements of DNA twist, torque and extension. *Nature Methods, 11*(4), 4. https://doi.org/10.1038/nmeth.2854.

Léger, J. F., Romano, G., Sarkar, A., Robert, J., Bourdieu, L., Chatenay, D., & Marko, J. F. (1999). Structural transitions of a twisted and stretched DNA molecule. *Physical Review Letters, 83*(5), 1066–1069. https://doi.org/10.1103/PhysRevLett.83.1066.

Lipfert, J., Kerssemakers, J. W. J., Jager, T., & Dekker, N. H. (2010). Magnetic torque tweezers: Measuring torsional stiffness in DNA and RecA-DNA filaments. *Nature Methods,* 7(12), 12. https://doi.org/10.1038/nmeth.1520.

Loenhout, M. T. J., van, Vlaminck, I. D., Flebus, B., Blanken, J. F., den, Zweifel, L. P., Hooning, K. M., ... Dekker, C. (2013). Scanning a DNA molecule for bound proteins using hybrid magnetic and optical tweezers. *PLoS One, 8*(6), e65329. https://doi.org/10.1371/journal.pone.0065329.

Ma, C. J., Gibb, B., Kwon, Y., Sung, P., & Greene, E. C. (2017). Protein dynamics of human RPA and RAD51 on ssDNA during assembly and disassembly of the RAD51 filament. *Nucleic Acids Research, 45*(2), 749–761. https://doi.org/10.1093/nar/gkw1125.

Madariaga-Marcos, J., Hormeño, S., Pastrana, C. L., Fisher, G. L. M., Dillingham, M. S., & Moreno-Herrero, F. (2018). Force determination in lateral magnetic tweezers combined with TIRF microscopy. *Nanoscale, 10*(9), 4579–4590. https://doi.org/10.1039/C7NR07344E.

Neuert, G., Albrecht, C., Pamir, E., & Gaub, H. E. (2006). Dynamic force spectroscopy of the digoxigenin–antibody complex. *FEBS Letters, 580*(2), 505–509. https://doi.org/10.1016/j.febslet.2005.12.052.

Neuman, K. C., & Nagy, A. (2008). Single-molecule force spectroscopy: Optical tweezers, magnetic tweezers and atomic force microscopy. *Nature Methods, 5*(6), 6. https://doi.org/10.1038/nmeth.1218.

Ngo, T. T. M., Zhang, Q., Zhou, R., Yodh, J. G., & Ha, T. (2015). Asymmetric unwrapping of nucleosomes under tension directed by DNA local flexibility. *Cell, 160*(6), 1135–1144. https://doi.org/10.1016/j.cell.2015.02.001.

Nguyen, B., Sokoloski, J., Galletto, R., Elson, E. L., Wold, M. S., & Lohman, T. M. (2014). Diffusion of human replication protein A along single-stranded DNA. *Journal of Molecular Biology, 426*(19), 3246–3261. https://doi.org/10.1016/j.jmb.2014.07.014.

Park, C., Yang, T., Rah, S.-H., Kim, H. G., Yoon, T.-Y., & Shon, M. J. (2023). High-speed magnetic tweezers for nanomechanical measurements on force-sensitive elements. *JoVE, 195*, e65137. https://doi.org/10.3791/65137.

Roy, R., Hohng, S., & Ha, T. (2008). A practical guide to single-molecule FRET. *Nature Methods, 5*(6), 6. https://doi.org/10.1038/nmeth.1208.

Schwarz, F. W., Tóth, J., van Aelst, K., Cui, G., Clausing, S., Szczelkun, M. D., & Seidel, R. (2013). The helicase-like domains of type III restriction enzymes trigger long-range diffusion along DNA. *Science (New York, N. Y.), 340*(6130), 353–356. https://doi.org/10.1126/science.1231122.

Seol, Y., & Neuman, K. C. (2018). Combined magnetic tweezers and micro-mirror total internal reflection fluorescence microscope for single-molecule manipulation and visualization. In E. J. G. Peterman (Vol. Ed.), *Single molecule analysis: 1665*, (pp. 297–316). Springer New York. ⟨https://doi.org/10.1007/978-1-4939-7271-5_16⟩.

Shon, M. J., & Cohen, A. E. (2016). Nano-mechanical measurements of protein-DNA interactions with a silicon nitride pulley. *Nucleic Acids Research, 44*(1), e7. https://doi.org/10.1093/nar/gkv866.

Shon, M. J., Kim, H., & Yoon, T.-Y. (2018). Focused clamping of a single neuronal SNARE complex by complexin under high mechanical tension. *Nature Communications, 9*(1), 3639. https://doi.org/10.1038/s41467-018-06122-3.

Shon, M. J., Rah, S.-H., & Yoon, T.-Y. (2019). Submicrometer elasticity of double-stranded DNA revealed by precision force-extension measurements with magnetic tweezers. *Science Advances, 5*(6), eaav1697. https://doi.org/10.1126/sciadv.aav1697.

Smith, S. B., Cui, Y., & Bustamante, C. (1996). Overstretching B-DNA: The elastic response of individual double-stranded and single-stranded DNA molecules. *Science (New York, N. Y.), 271*(5250), 795–799. https://doi.org/10.1126/science.271.5250.795.

Spegg, V., Panagopoulos, A., Stout, M., Krishnan, A., Reginato, G., Imhof, R., ... Altmeyer, M. (2023). Phase separation properties of RPA combine high-affinity ssDNA binding with dynamic condensate functions at telomeres. *Nature Structural & Molecular Biology, 30*(4), 4. https://doi.org/10.1038/s41594-023-00932-w.

te Velthuis, A. J. W., Kerssemakers, J. W. J., Lipfert, J., & Dekker, N. H. (2010). Quantitative guidelines for force calibration through spectral analysis of magnetic tweezers data. *Biophysical Journal, 99*(4), 1292–1302. https://doi.org/10.1016/j.bpj.2010.06.008.

van Loenhout, M. T. J., de Grunt, M. V., & Dekker, C. (2012). Dynamics of DNA supercoils. *Science (New York, N. Y.), 338*(6103), 94–97. https://doi.org/10.1126/science.1225810.

van Mameren, J., Gross, P., Farge, G., Hooijman, P., Modesti, M., Falkenberg, M., ... Peterman, E. J. G. (2009). Unraveling the structure of DNA during overstretching by using multicolor, single-molecule fluorescence imaging. *Proceedings of the National Academy of Sciences, 106*(43), 18231–18236. https://doi.org/10.1073/pnas.0904322106.

Wold, M. S. (1997). Replication protein A: A heterotrimeric, single-stranded DNA-binding protein required for eukaryotic DNA metabolism. *Annual Review of Biochemistry, 66*(1), 61–92. https://doi.org/10.1146/annurev.biochem.66.1.61.

Woodside, M. T., Behnke-Parks, W. M., Larizadeh, K., Travers, K., Herschlag, D., & Block, S. M. (2006). Nanomechanical measurements of the sequence-dependent folding landscapes of single nucleic acid hairpins. *Proceedings of the National Academy of Sciences of the United States of America, 103*(16), 6190–6195. https://doi.org/10.1073/pnas.0511048103.

Yang, T., Park, C., Rah, S.-H., & Shon, M. J. (2022). Nano-precision tweezers for mechanosensitive proteins and beyond. *Molecules and Cells, 45*(1), 16–25. https://doi.org/10.14348/molcells.2022.2026.

Yu, Z., Dulin, D., Cnossen, J., Köber, M., van Oene, M. M., Ordu, O., ... Dekker, N. H. (2014). A force calibration standard for magnetic tweezers. *Review of Scientific Instruments, 85*(12), 123114. https://doi.org/10.1063/1.4904148.

Zhang, X., Chen, H., Le, S., Rouzina, I., Doyle, P. S., & Yan, J. (2013). Revealing the competition between peeled ssDNA, melting bubbles, and S-DNA during DNA overstretching by single-molecule calorimetry. *Proceedings of the National Academy of Sciences, 110*(10), 3865–3870. https://doi.org/10.1073/pnas.1213740110.

CHAPTER SIX

Integrating magnetic tweezers and single-molecule FRET: A comprehensive approach to studying local and global conformational changes simultaneously

Hae Jun Jung[a], Beom-Hyeon Park[a], Sook Ho Kim[a,b], and Seok-Cheol Hong[a,*]

[a]Center for Molecular Spectroscopy and Dynamics, Institute for Basic Science, Department of Physics, Korea University, Seoul, Korea

[b]College of Veterinary Medicine, Chungbuk National University, Cheongju, Korea

*Corresponding author. e-mail address: hongsc@korea.ac.kr

Contents

Methods in Enzymology, Volume 694
ISSN 0076-6879, https://doi.org/10.1016/bs.mie.2024.01.007

Abstract

This chapter presents the integration of magnetic tweezers with single-molecule FRET technology, a significant advancement in the study of nucleic acids and other biological systems. We detail the technical aspects, challenges, and current status of this hybrid technique, which combines the global manipulation and observation capabilities of magnetic tweezers with the local conformational detection of smFRET. This innovative approach enhances our ability to analyze and understand the molecular mechanics of biological systems. The chapter serves as our first formal documentation of this method, offering insights and methodologies developed in our laboratory over the past decade.

1. Introduction

Magnetic tweezers (MT) have been a key tool for studying the mechanics of biological systems as well as the biochemical and enzymatic reactions in individual DNA molecules. This technique applies a specific level of tension and a controlled number of twists to a tethered DNA molecule. It delivers highly accurate information on the position of the tether's end, which is then used to determine the overall end-to-end distance of the molecule. However, this information is generally indirect and does not provide details about local conformational changes within the molecule. In this aspect, the single-molecule FRET (smFRET) technique, which acts as a nanoscale ruler for tracking local conformational changes, has gained prominence. More than a decade ago, we successfully and for the first time integrated magnetic tweezers with smFRET, preserving the ability to apply torque. Since we have not formally published this combined method, we outline here the technical details, challenges, and recent enhancements of this hybrid technique developed in our laboratory (Kim, Jung, Lee, Lee, & Hong, 2021; Lee, Kim, & Hong, 2010).

2. Materials and methods

In this section, we introduce the hybrid device that combines magnetic tweezers and smFRET and address the issues that should be considered during the process of setup. These two techniques use different light sources and separate cameras for measurements. This combination adds additional issues which are not present in individual techniques. To deal with these issues, we pay attention to several points of consideration such as the selection of light sources and cameras, camera synchronization, focal planes (conjugate planes of the cameras), and the vertical dimension of the DNA tether.

2.1 Magnetic tweezers

Our magnetic tweezers are built on a commercial inverted microscope (ECLIPSE Ti-U, Nikon).

To manipulate the force or torque applied to a magnetic microbead (e.g. Dynabead M280), either permanent magnets or electromagnets (Gosse & Croquette, 2002; Smith, Finzi, & Bustamante, 1992) can be used with them installed on a custom-made mount. This mount is composed of two DC motors, each of which controls a different degree of freedom of magnets, orientational and vertical, and are controlled via a motor controller (C-863, PI) connected to a PC. For permanent magnets, the controls of their height and orientation are essential to vary the force and torque applied to a bead. On the other hand, electromagnets can be used to directly manipulate the force to a bead by varying the input voltage to magnetic poles through a DAQ board (DT9854, Data Translation). In order to generate a rotating magnetic field with electromagnets, one can use electromagnets with multipoles (more than two poles facing the sample, typically six poles) and alternate sequentially the polarity of each pole to rotate the direction of the horizontal component of the magnetic field at the sample plane. In this case, the rotational motor in the mount is not for full rotation of the magnets but for fine adjustment of the initial orientation of the magnets.

The setup uses an LED (850 nm, M850L2, Thorlabs) to illuminate the sample from above, and the light passes through a 60X oil immersion objective lens (CFI Apochromat 60X TIRF, Nikon). The image of the sample chamber is captured at a speed of 300 Hz by a CMOS camera (STC-CMB401 PCL, Sentech). This speed can be acquired by using the trigger signal provided by an external waveform generator (33500B, Keysight). The x and y positions of the sample chamber are controlled by a micro-positioning stage (PI-Line M-686, PI), and the z-axis position of the chamber is adjusted by a piezoelectric positioner (P-725, PI).

2.1.1 Hybrid technique of magnetic tweezers with TIRF-based single molecule FRET

Fluorescence resonance energy transfer (FRET) detects fluorescence signals from donor and acceptor dyes in the manner depending on their relative distance. By measuring the intensity of signal from each dye and calculate the FRET efficiency, we can deduce their relative distance and therefore the structural information of the sample at nanometer scales. If magnetic tweezers are combined with FRET, one can track the conformational change in a specific region of a tethered DNA under

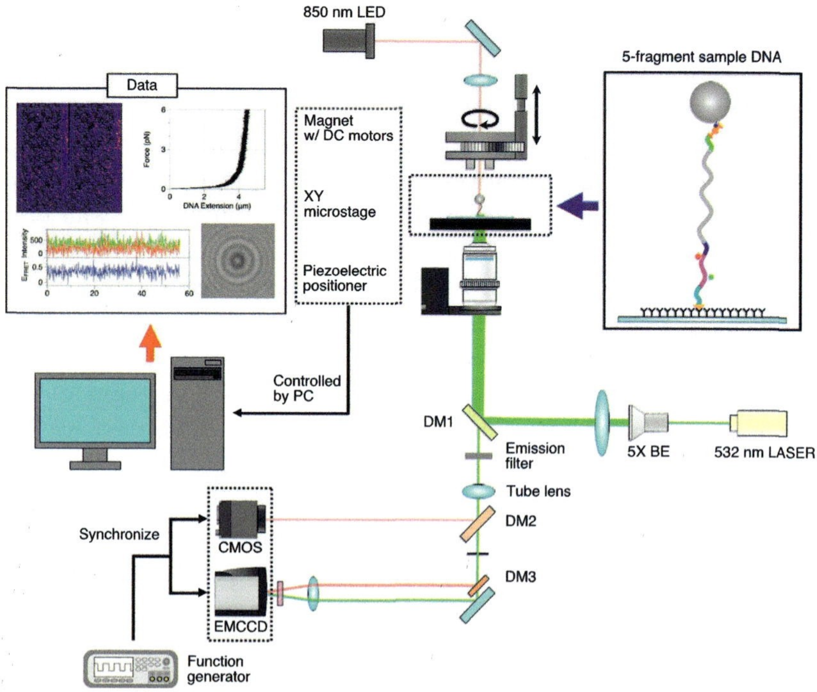

Fig. 1 Schematics of the hybrid MT-FRET set-up (BE: beam expander; DM: dichroic mirror).

varying force or torque. As a platform for smFRET assay, which enables us to measure the fluorescence from single dye molecules and thus provides the FRET efficiency value from each individual dye-labeled molecule, we chose the TIRF-type microscopy setup, which allows us to monitor molecules attached to the sample surface with low background (Joo & Ha, 2012). An additional advantage of a TIRF-based microscope is that such a setup allows us to have the space above the sample chamber free and available for magnets. We use 532 nm CW laser (Coherent, USA) for Cy3 dye excitation, and, if necessary, a 633 nm laser can be used for direct Cy5 dye excitation. To filter out the input beam from detection, an emission filter is installed. The path for fluorescence signal is split out from the path for IR beam used as an imaging light source in magnetic tweezers by a dichroic mirror (DM2 in Fig. 1) and it is further separated into two using a slit and another dichroic mirror (DM3 in Fig. 1). The separated beams, one for Cy3 and the other for Cy5 fluorescence, are observed simultaneously on an EMCCD camera (iXon Ultra, Andor) typically at 10 Hz. For simultaneous detection of the two fluorescent signals with a single CCD camera, as usually in the case for most smFRET assays, these signals are directed to the same CCD with a small lateral displacement so that the donor signal illuminates one half of the CCD chip and the acceptor signal beam the other half. That path that collects donor and acceptor signal is called donor and acceptor channel, respectively.

2.1.2 Issues arising from incorporation of MT with TIRF smFRET and feasible remedies

Different from a single modal setup, magnetic tweezers only or smFRET only, there are several issues to be solved in their combined setup, in particular, to measure their signals simultaneously.

First, it is essential to accurately synchronize the acquisition of images in the two different cameras. To solve this, we introduced a waveform generator (33500B, Keysight) that can generate two triggering signals at 10 Hz and 300 Hz simultaneously through individual channels, which are to be fed to the two cameras. Each camera records an array of images in synchronization with the supplied trigger signal.

Next, commercial magnetic beads (e.g. Dynabead) are auto-fluorescent so they forbid weak fluorescent signal from being detected by emitting strong background noise when exposed to light excitation. This happens even in the TIRF configuration if the bead sinks to the glass surface (buffer-glass interface) and enters the region where the evanescent field exists. This situation makes it hard to detect feeble fluorescent signals from single fluorophore molecules attached to a DNA tether. An approach to circumvent this issue would be to spatially separate the bead from the region fluorescent dyes are located. In the vertical geometry of typical magnetic tweezers, one can use a long DNA tether so that the bead is located above the region of evanescent field. We usually use DNA molecules, the full extension of which is longer than 4 μm. For example, in our 5-fragment tether configuration, a 12-kbp digest product from a 15-kbp vector (pPIA2–6) is used as the insert (longest part of the 5-fragment ligated tether). The end-to-end distance of a DNA tether is affected by the tension to the tether so that in our case with a 12-kbp tether, the tension to a torsion-free DNA in smFRET measurement is kept above ~0.2 pN and the tension to a negatively supercoiled DNA (sigma ~−0.04) above ~0.5 pN to prevent the bead from approaching too close to the interface.

Lastly, the LED illumination used for magnetic tweezers can also contribute to some noise in FRET channels, depending on the intensity and the spectral range of LED. To minimize such effects, we chose an LED, the center wavelength of which is 850 nm (in the near-IR range) and a dichroic mirror (DM2 in Fig. 1), the cut-off wavelength of which is 745 nm to exclude the LED light from the beam path to EMCCD. For MT imaging, we also selected a CMOS camera that can sensitively capture this spectral range.

3. Protocols and procedures

3.1 Solution

Here, we introduce buffer solutions used in the experimental procedures described in this chapter.

- T50 solution (10 mM Tris-HCl, 50 mM Na^+, pH 8.0)
- 80 mM PCA (protocatechuic acid) solution (308.24 mg of PCA in 25 mL of distilled water)
- CC (click chemistry) buffer (20 mM Tris-HCl, 5 mM EDTA in 50 mL of distilled water, see Note 1)
- PCA-PCD oxygen scavenging system (24 μL of 80 mM PCA solution, 2 μL of rPCO (Bacterial Protocatechuate 3,4-Dioxygenase from OYC), 774 μL of imaging buffer)

For more information about oxygen scavenging systems and imaging buffers, see Note 2 and Note 3, respectively.

3.2 DNA sample

The DNA tether molecule used in our assays typically consists of five fragments. These are arranged from glass to bead as follows: a linker that anchors the molecule to the bottom of the chamber; the DNA core, part that contains the sequence of our research; a connector that prevents undesirable multiple inclusion of the DNA cores; an insert that provides sufficient length for the DNA molecule needed in the experiment; and finally, a linker that binds to a magnetic bead. The following text explains how each DNA fragment is produced and assembled.

3.2.1 Transformation

Mix 1 μL plasmid, either pUC 19 (2686 bp) or pPIA 2–6 (15070 bp) with 9 μL *E. coli* (HIT Competent Cells-JM109 High 10^8 from RBC) and incubate the mixture for 10 min. Then, pour 5 μL of this mixture onto the center of an LB agar dish, spread gently, cover it with a lid, label, seal with parafilm, and incubate at 37 °C for 8–16 h.

3.2.2 Pre-culture

Add 5 mL LB medium and 5.5 μL ampicillin to a 15 mL round tube (Falcon® 14 mL round bottom polystyrene test tube). Use a sterilized micropipette to transfer an *E. coli* colony from a petri dish to the tube. Sterilize the tube opening with flame, cap, label, and incubate at 37 °C and 250 rpm for 16 h.

3.2.3 Miniprep

From this step, we collect tens of μL of purified plasmid in each tube, which would be used as the insert or PCR templates in the following reactions. Follow the instructions provided in the QIAprep Spin Miniprep Kit manual (QIAGEN).

3.2.4 PCR

This PCR step is conducted to create 'linker' fragments. A linker is a specialized region of the tether designed to attach either end of the molecule to the bottom of a chamber or to a magnetic bead. We use three types of linkers: functionalized with digoxigenin, biotin, or DBCO. In our case, the digoxigenin and DBCO linkers are used to anchor the assembled DNA to the bottom of a chamber while the biotin linker is specifically designed for binding with streptavidin-coated magnetic beads.

3.2.4.1 Reagents

- Primer1: 5′-GTG CCA CCT GAC GTC TAA GA-3′
- Primer2: 5′-GTT CCA CTG AGC GTC AGA CC-3′
- dNTP (dATP, dGTP, dCTP, dTTP) (Takara)
- Digoxigenin-11-dUTP, alkali-stable (Roche)
- Biotin-16-dUTP (Roche)
- 5-DBCO-PEG$_4$-dUTP, 10 mM Aqueous Solution (Merck)
- *Taq* DNA polymerase (Solgent)

3.2.4.2 Procedure

1. To prepare a 1 mM dNTP mixture, mix 3 μL each of 10 mM dATP, dGTP, and dCTP, and 2.1 μL of 10 mM dTTP in a tube. For a digoxigenin-labeled dUTP mixture, add 9 μL of 1 mM Digoxigenin-11-dUTP; for a biotin-labeled dUTP mixture, use 9 μL of 1 mM Biotin-16-dUTP instead. In both cases, add distilled water to bring the final volume to 30 μL.

2. To prepare a 2.5 mM dNTP mixture containing DBCO-labeled dUTP, mix the following components: 10 μL each of 10 mM dATP, dGTP, and dCTP, 7 μL of 10 mM dTTP, and 3 μL of 10 mM 5-DBCO-PEG$_4$-dUTP.
3. To prepare for either digoxigenin or biotin linker PCR, mix the following components: 2 μL of 10 μM Primer1, 2 μL of 10 μM Primer2, and 1 ng of pUC19. Add 5 μL of 10X *Taq* buffer and 0.25 μL of *Taq* DNA polymerase (5 units/μL). For the digoxigenin linker PCR, include 10 μL of 1 mM dNTP mixture with digoxigenin; for the biotin linker PCR, use 10 μL of 1 mM dNTP mixture with biotin instead. In both cases, add distilled water to bring the final volume to 50 μL.
4. To prepare for DBCO linker PCR, mix the following components: 2 μL of 10 μM Primer1, 2 μL of 10 μM Primer2, 4 μL of 2.5 mM dNTP mixture with DBCO, 1 ng of pUC19, 5 μL of 10X *Taq* buffer, and 0.25 μL of *Taq* DNA polymerase (5 units/μL). Add distilled water to bring the final volume up to 50 μL.
5. Repeat the following PCR reaction cycle 40 times: denaturation step at 95 °C for 30 s, annealing step at 52 °C for 40 s, and extension step at 72 °C for 95 s

3.2.5 PCR purification

From this step, we collect purified PCR products, which would be used to generate DNA linkers by restriction digestion. Follow the instructions provided in the QIAquick PCR Purification Kit manual. (QIAGEN).

3.2.6 Hybridization

This hybridization step is designed to form a DNA core and a DNA connector by annealing two single stranded DNA oligonucleotides to double-stranded DNA with cohesive overhangs at both ends. The complementary overhang of core and connector features a sequence where the first four nucleotides from the 5′ end are non-palindromic (Table 1).

3.2.6.1 Procedure

In this procedure, Strand-1 and Strand-2 represent Core-1 or Connector-1 and Core-2 and Connector-2, respectively.

1. Mix 1 μL of 100 μM Strand-1 stock with 99 μL of T50 to make 1 μM Strand-1 solution (100 μL total).
2. Mix 1 μL of 100 μM Strand-2 stock with 99 μL of T50 to make 1 μM Strand-2 solution (100 μL total).

3. Combine 10 μL of 1 μM Strand-1 solution, 10 μL of 1 μM Strand-2 solution, and 80 μL of T50 to make a 100 nM strand mixture.
4. Place the 100 nM strand mixture in a thermal cycler and heat it to 85 °C. After reaching this temperature, gradually decrease the heat by 0.5 °C every 5 min.

3.2.7 Enzyme cut

To generate complementary cohesive overhangs for ligation, PCR products and plasmid molecules are digested with appropriate restriction enzymes. We use enzymes all from New England Biolabs in this enzyme cutting step.

3.2.7.1 Procedure

1. To make a linker (digoxigenin, biotin, or DBCO), mix 25 μL of the respective linker PCR product (digoxigenin, biotin, or DBCO) with 5 μL of CutSmart® buffer and 2 μL of a restriction enzyme. Then, add distilled water to bring the total volume up to 50 μL. Choose the appropriate restriction enzyme based on your specific needs. In our case, we use enzymes like *Sal*I, *Sac*I, SbfI and *Bam*HI to create linkers of around 500 bp and 1000 bp in length.
2. To make an insert, mix 41 μL of pPIA 2-6 solution (prepared from the preceding miniprep step), 5 μL of CutSmart® buffer, and 2 μL of each of two different restriction enzymes. Since the insert is to be ligated with a connector and a biotin linker, two different restriction enzymes are required. Choose restriction enzymes that are suitable for your specific needs. In our experiments, we use enzymes like SbfI, *Bam*HI, and *Eag*I to create the insert of approximately 6000 bp, 8000 bp, and 14,000 bp in length.
3. Incubate the enzyme-DNA mixture at 37 °C for 4–12 h.
4. Perform heat inactivation, typically around 65 °C for 10 min. Consult the specific enzyme's user guide for the exact temperature.

3.2.8 Ligation

1. Before initiating the ligation step, measure the concentrations of all the fragments. The volume of a fragment to be added in the ligation mixture varies based on its concentration (see Note 4).
2. To prepare the ligation mixture, combine 2 μL of 5 nM insert, 0.75 μL of either 40 nM digoxigenin or DBCO linker, 0.5 μL of 60 nM biotin linker, and 0.9 μL each of 100 nM DNA core and connector. Add 3 μL of 10X T4 DNA ligase buffer and 3.5 μL of T4 DNA ligase. Finally, add

distilled water to bring total volume to 30 μL. With different concentrations of fragments, adjust their volumes accordingly.

3.3 Sample chamber

In magnetic tweezers assays, DNA molecules of interest need to be immobilized in the sample chamber to manipulate them and observe their response to mechanical control for an extended period of time. To prepare a sample chamber is often a critical step for successful and efficient assays because suitably prepared sample chamber would allow us to have and find a sufficient number (sometimes just one) of 'good' molecules (fulfilling all the molecular conditions required for the intended assay, for example, of suitable length, singly tethered (or doubly tethered), torsionally constrained or free, presence of donor and acceptor dyes, etc) within a reasonable length of time of molecule search. The quality of surface coating/passivation also affects the cleanness of the sample surface/view field and thus the quality of image. The cleanness of the sample surface is also important to avoid unidentified source of fluorescence background.

3.3.1 Flow chamber assembly

1. We use a laser engraver to make holes in 24 × 60 mm cover glasses (Marienfeld). The two holes in each cover glass are spaced about 50 mm apart.
2. Sonicate cover glasses with and without holes in acetone for 30 min.
3. Apply plasma treatment to the glass surface for 15 min to clean them and enhance their hydrophilicity.
4. Prepare pieces of cover glass for digoxigenin-based DNA immobilization as follows: first, mix 44.1 mL of 99.8% ethanol, 4.9 mL of distilled water, and 1 mL of (3-glycidyloxypropyl) tri-methoxy-silane (GPTMS, see Note 5); then, place pieces of cleaned cover glass in a jar with this mixture and incubate them in a shaking incubator for 1 h; after incubation, gently wash them first with ethanol and then with flowing distilled water.
5. Prepare pieces of cover glass for DBCO-based DNA immobilization as follows: first, mix 48 mL of acetone, 1 mL of distilled water, and 1 mL of (3-aminopropyl) tri-ethoxy-silane. (APTES); then, place pieces of cover glass in a jar with this mixture and incubate them in a shaking incubator for 1 h; after incubation, gently wash them first with acetone and then with flowing distilled water.
6. Place a layer of parafilm with an aperture that is approximately 10 mm wide and 50 mm long between two pieces of cover glass, one with holes

and the other without. Then, have them stuck by melting the film by applying pressure and heat on a heat block at a temperature of 100 °C or higher for 30 s. The parafilm layer contains a fluidic channel shaped by the aperture for buffer exchange and also serves as a sealing gasket for the sample chamber.

3.3.2 Immobilizing DNA tether molecules in the sample chamber

3.3.2.1 Reagents

- NeutrAvidin (Thermo Fisher)
- Anti-digoxigenin (Roche)
- BlockAid (Invitrogen)
- BSA (Merck)
- NHS-PEG4-Azide (Thermo Fisher)
- MS(PEG)4 Methyl-PEG-NHS-Ester (Thermo Fisher)
- Dynabeads™ M-280 Streptavidin (Invitrogen)

3.3.2.2 Procedure

1. The sample chamber, designed to support the immobilization of digoxigenin-functionalized DNA, is coated or treated in the following manner: first, add 100 μL of 100 μg/mL anti-digoxigenin and incubate at 37 °C for 1 h; then, wash with 1X PBS. For passivation, introduce 800 μL of either 100 μg/mL BSA or 1:1 mixture of 1X PBS and BlockAid. After ensuring the passivation solution fully saturates the surface, allow it to sit for 30 min. After incubation, wash the chamber with 1 mL of 1X PBS.
2. The sample chamber, designed to support the immobilization of DBCO-functionalized DNA, is coated or treated in the following manner: first, mix 40 μL of 250 mM NHS-PEG4-Azide, which binds to DBCO by click chemistry, with 20 μL of 10X PBS, add 100 μL of distilled water, and vortex this mixture to dissolve; next, add 40 μL of 250 mM MS(PEG)4 Methyl-PEG-NHS-Ester used for passivation and vortex again to ensure it is fully dissolved; then, inject this solution into the chamber and incubate for 1 h; lastly, wash the chamber using 1 mL of CC buffer without Tween. (see Note 6)
3. To prepare a 1/10X DNA solution, mix 9 μL of 1X PBS with 1 μL of 1X DNA solution. Next, to attach beads to DNA, add 10 μL of 1X PBS (for DNA bearing a digoxigenin linker) or CC buffer with Tween 20 (for DNA bearing a DBCO linker) to a new tube. Then, add 5 μL of 2.8 μm beads and 5 μL of the just prepared 1/10X DNA

solution. Gently agitate the mixture for 20 s to prevent the magnetic beads from settling, and then let it stand undisturbed for 40 s. Repeat this shaking and resting cycle 10 times for a total duration of 10 min.

4. Add 300 μL of 1X PBS (for DNA bearing a digoxigenin linker) or CC buffer with Tween 20 (for DNA bearing a DBCO linker) to the above solution and magnetically pull down DNA for 10 min in a dark environment. After 10 min, decant only the above solution. Repeat this process twice (see Note 7).
5. Add 500 μL of 1X PBS (for DNA bearing a digoxigenin linker) or CC buffer with Tween 20 (for DNA bearing a DBCO linker) to the above solution, gently mix and transfer to the chamber using a syringe. Incubate for 1 h.
6. Inject 1 mL of buffer into the chamber after washing it with 3 mL of 1X PBS (see Note 8): use a buffer with the appropriate salt concentration for experiments exclusively using magnetic tweezers, or an imaging buffer with the right salt concentration and an oxygen scavenger for any experiments that include smFRET assays.

3.4 MT-FRET experiments

3.4.1 Identification of suitable DNA molecules

Depending on the type of assay of interest, certain conditions should be met for molecules under assay. Here, we shall mention the following general issues of quality check: length of DNA; number of DNA molecules tethered to a bead of interest; torsional constraint; presence of donor and acceptor dyes; presence of a reference bead. Below, the quality check is described in the typical order of inquiry.

3.4.1.1 Presence of donor and acceptor dyes by checking fluorescent signal

1. Bring a fluctuating DNA molecule to the center of the screen.
2. Briefly turn on the laser and examine the signals in both donor and acceptor channels. If the signals are present at the bead's position on the screen with their relative intensities indicating the appropriate level of FRET efficiency (high FRET: stronger in the acceptor channel than in the donor channel; low FRET: stronger in the donor channel), turn off the laser and proceed to assess other qualities of the molecule. If these conditions are not met, search for another molecule.

3.4.1.2 Multiplicity of DNA molecules tethered to the bead tested via braid check

1. Underwind the DNA molecule by 100 turns while applying a force greater than 1 pN.
2. Measure the length of the molecule after underwinding or continue to monitor its length during underwinding.
3. If the DNA molecule is shortened considerably after a few turns of winding, it indicates that the molecule is multiply tethered and forms a braid upon winding. If a singly tethered molecule is needed, search for another molecule. If the length of the molecule remains unchanged, the molecule is likely singly tethered and go to 4.
4. Check if the length of the molecule changes significantly by reducing applied force (by raising the permanent magnets).
5. If the length of the molecule changes significantly, indicating the formation of plectonemes, it assures that the molecule is singly tethered.

3.4.1.3 Torsional constraint of a DNA molecule attached to the bead tested via nick check: in some cases, the state of torsional constraint may not be an issue; otherwise, one of the two states, torsionally constrained or free, should be used for assay (see Note 9)

1. Overwind the DNA molecule by 100 turns while applying a force greater than 1 pN.
2. Measure the length of the molecule after overwinding or continue to monitor its length during overwinding.
3. If the DNA molecule exhibits a noticeable reduction in length, it indicates that the molecule is not nicked but torsionally constrained, permitting (positive) supercoiling-induced buckling.
4. Check if the length of the molecule changes significantly by applying stronger force (by lowering the permanent magnets).
5. If the length of the molecule changes significantly, suggesting the dissolution of plectonemes, it further confirms that the molecule is not nicked.

3.4.1.4 Length of the DNA tether in a suitable range

1. In this step, we basically record the profile of diffraction patterns of the bead (see Fig. 2) attached to the DNA molecule at various magnet-chamber distances (MD) and a background image for background subtraction (see 4.1.1).
2. Set the range of scan of magnet-chamber distance, which is typically from 1 mm to 15 mm. In our current setup, the force exerted to the

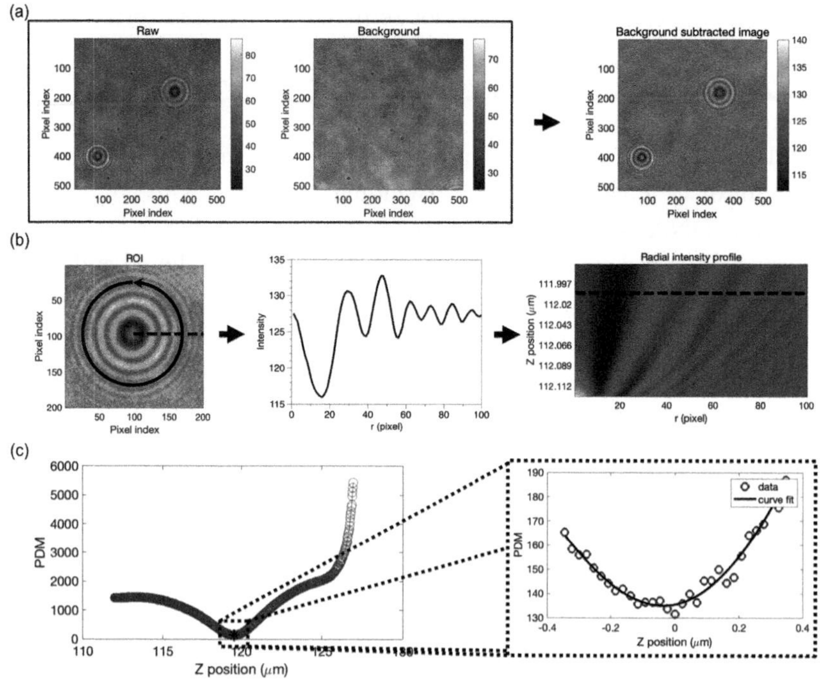

Fig. 2 Schematics of image analysis. (A) Background subtraction in the image analysis for magnetic tweezers assay. By subtracting background (center) from a raw image (left), we can get clear and cleaner image (right). (B) Generation of the radial intensity profile look-up table (LUT). A radial intensity profile is calculated by collecting the mean intensity value of the points of equal distance from the center (left, center). The profiles are stacked to form the profile LUT (right). The y-axis of LUT indicates the vertical position of the piezoelectric positioner that enables vertical movement of the objective. (C) Evaluation of the z-position of a magnetic bead by least-square fitting of the profile discrepancy measure (PDM) to a polynomial function of bead height.

DNA molecule through the magnetic bead is approximately 8 pN at MD = 2 mm and 0.1 pN at MD = 12 mm although the force varies due to different magnet content of a bead.

3. Starting from MD = 1 mm, continue to increase MD to 15 mm with the speed = 0.1 mm/s.
4. Extract the position of the bead by comparing the profile of the bead with the calibration data (Look-Up Table, LUT) as described below. Check whether the full extension of a DNA molecule deduced from the position measurements at MD = 1 mm and 15 mm is in a reasonable range up to measurement error and deviation due to variation of

attachment point of DNA on the bead. If it is too short or too long, discard the molecule and look for another one.

3.4.1.5 Force calibration

1. In this step, we basically record the movement of (the diffraction pattern of) the magnetic bead attached to the DNA molecule at various values of MD, that is, different levels of force applied to the DNA molecule. Adjust the recording duration to be longer for weaker forces, as the relaxation time of fluctuation of the DNA molecule increases with decreased force.
2. For example, begin by setting the initial MD to 11 mm and record a bead movie for 30 min. Then, decrease the MD to 9 mm and record for 25 min. Continue recording with MD = 7, 5, 3, and 1 mm for 20, 15, 10, and 5 min, respectively.
3. Once all the recordings are completed, analyze the obtained data using a MATLAB code to obtain the lateral fluctuations of the bead and hence the vertical force applied to it by $F = \frac{k_B T < L >}{< \Delta x >^2}$.

3.4.1.6 Force-extension behavior of DNA

1. By measuring the vertical position of the bead at various values of force, we obtain the force-extension curve of DNA.
2. By fitting the curve to the worm-like chain model $\left(\text{e.g., } F = \frac{k_B T}{L_p}\left[\frac{L}{L_0} - \frac{1}{4} + \frac{1}{4}\left(1 - \frac{L}{L_0}\right)^{-2}\right]\right)$, we calculate the persistence length (L_p) and the contour length (L_0) of DNA. (Bouchiat et al., 1999)
3. Check whether the derived contour length and persistence length are in a reasonable range for the tether molecule. If not, search another molecule and repeat the quality check (see Note 10).

4. Analysis

4.1 Particle tracking

The analysis process starts with finding the position of the bead of interest in the images. For this purpose, a preprocessing step such as background subtraction is required. After recording and improving calibration and experimental bead images, we find and track the center of the bead and use the location and shape of the bead in a ROI to determine the length of the tether attached to it and the force applied to the tether.

4.1.1 Background subtraction

At first, we should remove any undesirable features in the image such as interference fringes, shade, and image of dirts present along the beam path. To achieve this, before recording actual images of the sample, we first capture some background images. Background images can be obtained by recording some blank regions with no beads or by imaging some middle of buffer solution far away from the sample surface so that any objects thereon are totally defocused. We create a background image by averaging between 50 and 100 individual frames. The intensity of the sample image is then adjusted to match that of this averaged background image, allowing for effective background removal. (see Fig. 2A).

4.1.2 Center finding

Once the background is removed, the next step is to track the center position (x, y) of the sample bead. There are several algorithms to find the center of a bead, such as 'flood fill' followed by averaging of the positions of filled pixels and least-square fit based on the convolution of a sample profile and LUT profiles. Here, we describe the radial symmetry method to find the center of a circular particle.(Parthasarathy, 2012) If the image of a bead is perfectly radially symmetric, the gradient of intensity calculated from each pixel will intersect at the center of the circular bead. This idea can be applied to read bead images, which may not be perfectly symmetric due to irregularity of the bead itself and background noise. With some algebra, we can calculate a point that minimizes the sum of distances between the point and radial lines created from each pixel and directed by the gradient on it. This point is eventually identified as the center of the bead (Parthasarathy, 2012). (see Fig. 3) We first use the center of ROI (region of interest) as an approximate center of the bead (seed) and run the radial symmetry method several times (for our case, 10 iterations) to find the center of the bead for better accuracy.

4.2 Profile analysis

4.2.1 Construction of the profile look-up table (LUT)

From the calculated center location, we can calculate the intensity profile of the bead. We generate the radial profile of the bead by averaging the intensity values of points at equal distance from the center of the bead. By accumulating or stacking the profile along z-axis or successive z positions, we can make a profile look-up table (LUT). More than 30 frames of image are recorded at every 20 nanometers, and the average intensity profile is

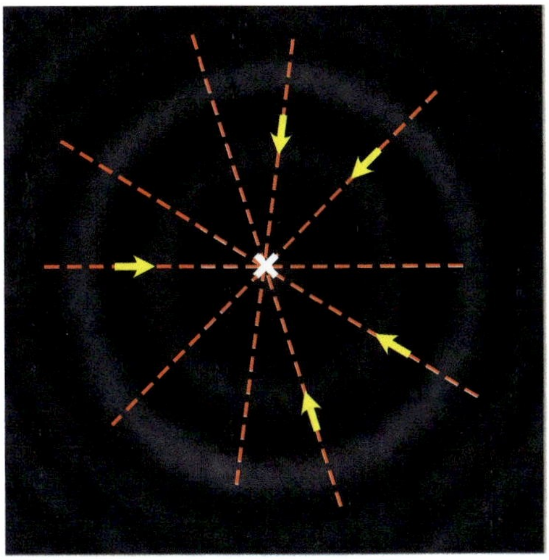

Fig. 3 Illustration of a center finding algorithm by the radial symmetry method.

assigned to the respective z-axis position. The (vertical) range of the recorded profile should be set sufficiently long to ensure that the position of a bead during measurement does not exceed the tracking range of the profile, typically at least twice the length of the tethered DNA. (see Fig. 2B).

4.2.2 Determination of tether length with LUT

The profile LUT created above is compared with the radial intensity profile of the bead bound to a DNA tether under investigation to determine the z position of the bead and hence the extension of the DNA tether. The comparison is made based on the 'profile discrepancy measure' defined as follows. First, we get the square of the difference of intensities at a radial position (in pixel or bin) of the profile of the bead under investigation and at the same position of the profiles in LUT acquired at different objective heights (z value). Next, we calculate the sum of the square over all the radial positions of profile, which we call 'profile discrepancy measure'. This is the sum of squared estimate of errors or residual sum of squares in profile analysis. Then, we plot the profile discrepancy measure with respect to the z value of profiles in LUT. (see Note 11) Last, we find the z value that gives the least profile discrepancy measure, the profile with which returns the intensity profile most similar to the one produced by the bead under investigation at the specific moment. It should be warned that some bead patterns may not be well analyzed, especially when they are produced from beads in a fully focused position. Therefore, we recommend using a range that does not include such bead patterns. (see Fig. 2C) The position obtained here needs to be converted into tether length to examine the change of length of DNA tether. For this purpose, an additional measurement for the reference position is required. The reference position can be obtained by a force smaller than 0.05 pN, which makes a bead sink to the surface of the chamber.

4.3 Force calibration

The variance of the lateral position (x, y) of a fluctuating bead is inversely proportional to the applied force, which depends on the magnet-chamber distance. To perform the force calibration through the power spectral density (PSD) method, it is necessary to have all three-dimensional position information of the bead; x-y position for calculating variance of fluctuation and z-position for tether length. Using the previously acquired 3-D location with several offset positions of sample bead, we can calibrate force exerted on the sample (Yu et al., 2014).

4.4 Drift correction

To correct the error of z position of a bead due to vertical drift of the chamber, it is highly desirable to track an additional bead (reference bead) which is tightly bound on the surface of the chamber and thus drifts together with the target bead and to record its image simultaneously on the same screen. Polystyrene beads non-specifically bound onto the bottom of the chamber can serve as fiducial markers. (Kuijpers, Laar, van, Janissen, & Dekker, 2022; Ostrofet, Papini, & Dulin, 2018) By calculating the position of the reference bead, we can correct the effect of vertical drift.

5. Example

Here, we provide an example of simultaneous observation of DNA extension and FRET efficiency by our integrated setup using a DNA tether molecule containing a TG17 repeat sequence. The sequence of TG17 is given in Table 1.

The TG repeat sequence in a DNA tether can form Z-DNA under a sufficient level of negative supercoiling. In addition, there is a tension threshold for forming Z-DNA, which competes against other modes of

Table 1 Sequences of core (TG17 repeats) and connector oligonucleotides.

Name	Sequence
Core-1	5′-Phos-GAT CCT GGT T**CA CAC ACA CAC ACA CAC ACA CAC A**/iCy5/**CA CAC ACA CA**A TCG AC-3′
Core-2	5′-Phos-TGC CGT CGA T**TG TGT GTG TGT GTG TGT GTG TGT G**/iCy3/**TG TGT GTG TG**A ACC AG-3′
Connector-1	5′-Phos-GGC CAG CGT CTG GCG ACG GCA GCG A-3
Connector-2	5′-Phos-GGC ATC GCT GCC GTC GCC AGA CGC T-3′

The underlined nucleotides are the overhangs of the hybridized duplex. Bold characters show core parts. The overhangs of Core-2 and Connector-2 are complementary and non-palindromic.

DNA conformation, writhes, and DNA bubble, which can also relax torsional stress. Fig. 4A and B show the images of a field of view of an area of the sample chamber where a bead-bound DNA tether molecule is located, taken by two cameras, one for bright-field image for MT and the other for smFRET, respectively. Fig. 4C shows an exemplary time traces of DNA extension and donor and acceptor fluorescence signals, hence FRET efficiency, measured simultaneously when the force applied to the bead changes in time by moving the magnets vertically. This DNA molecule had been underwound by 15 turns. At measurement time $t = 33$ s, the tension exerted to a TG17 molecule was decreased by raising the magnets from 7.8 to 8.2 mm. When the force applied changed from 0.95 pN to 0.81 pN, the extension of the DNA molecule was reduced by about 100 nm as shown in the bottom of Fig. 4C. Simultaneously, we measured the FRET efficiency, which exhibited frequent interconversion between two states, middle ($E_{FRET} = 0.5$) for B-form and low ($E_{FRET} = 0.2$) for Z-form, and observed that the molecule tended to visit the higher FRET state more frequently and longer, indicating the equilibrium shifted towards the B-form state when the force was reduced. As demonstrated here, we can track both global change of tether length by MT (and its bright-field imaging) and local conformational change of the TG17 core sequence by tracking FRET efficiency, simultaneously. This result shows the behavior of a DNA tether molecule expected from a simple mechanical picture of supercoiled DNA: under weaker tension, the supercoiled DNA prefers forming plectoneme structures rather than forming Z-DNA.

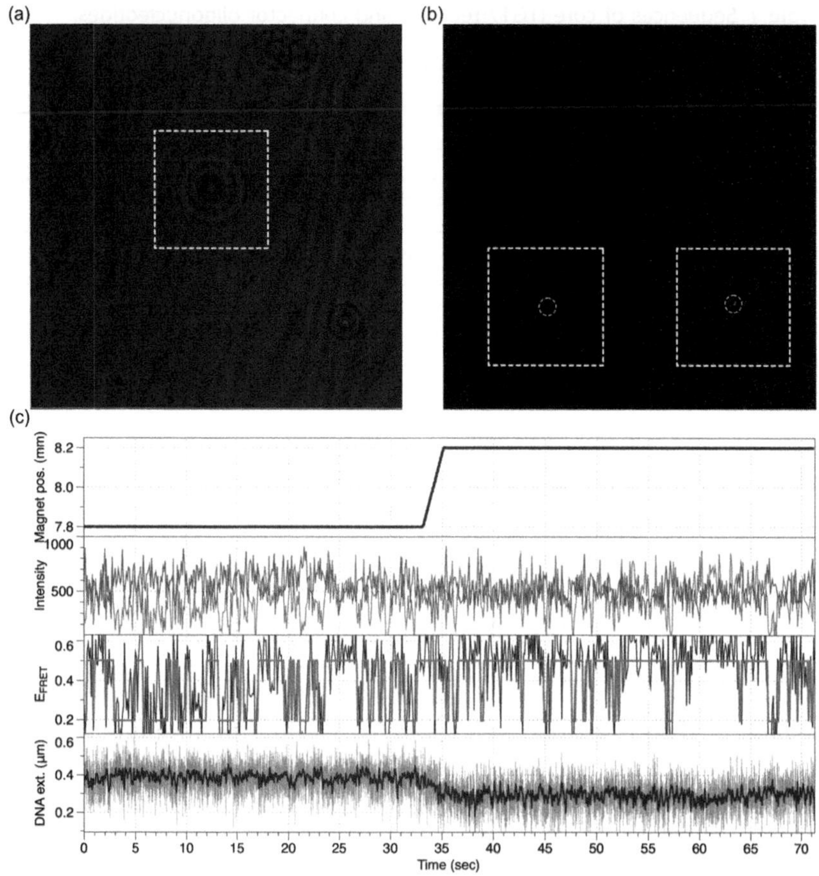

Fig. 4 Images of a field of view of the sample surface harboring a single bead-bound DNA tether molecule and real-time traces of fluorescent signals and DNA extension during the torsion-driven B-Z transition in TG17 repeats. (A) Bright-field image acquired during magnetic tweezers experiment and (B) fluorescence images taken in donor and acceptor channels are obtained simultaneously. Dashed squares in (A) and (B) capture the same area in the view field with (B) upside down and white dashed circles in (B) show the fluorescent signals from a pair of donor and acceptor dyes tagged in the DNA tether molecule. (C) Exemplary time traces obtained during simultaneous MT and smFRET experiments. (Top) Distance of magnets to the sample chamber. (Second) Traces of donor (green) and acceptor (red) fluorescence intensities. (Third) FRET efficiency calculated from the intensities shown above. Blue and red lines represent experimental FRET traces and its two-state approximation with variational Bayesian expectation maximization (Bronson, Fei, Hofman, Gonzalez, & Wiggins, 2009), respectively. (Bottom) DNA extension derived from LUT analysis. Gray and magenta lines represent raw and smoothened data by moving average, respectively.

6. Note

1. To avoid aggregation of beads, Tween 20 can be added to the solution to achieve a final concentration of 0.05% (v/v).
2. For assays with the hybrid MT-FRET technique, glucose oxidase supplemented with catalase can be also used as oxygen scavenger. However, if an experiment is sensitive to pH change or nuclease activity, we strongly recommend using the oxygen scavenger system based on rPCO and PCA.
3. The compositions and concentrations of imaging buffer vary widely depending on the type of experiment. Typically, an imaging buffer is based on 1X PBS or Tris-Cl containing some concentrations of cations such as Na^+ and Mg^{2+}. Used in fluorescence imaging, it must contain an oxygen scavenging cocktail such as PCA-rPCO or glucose-glucose oxidase-catalase. It also often contains Trolox to suppress triplet formation.
4. For efficient ligation of a longer fragment in reactions involving two fragments of unequal length, the molar ratio of the shorter fragment is adjusted to be three times that of the longer fragment. This means, for every single insert, you should add three of each type of linker, and nine each of the DNA core and connector.
5. In the context of our experiments, we found that 3-Glycidyloxypropyltrimethoxysilane (GPTMS) can be effectively substituted by 3-Aminopropyltriethoxysilane (APTES) and vice versa. This interchangeability is based on their comparable chemical functionalities and reactivity under the conditions employed in our study.
6. Both steps (1 and 2 in 3.3.2.2) require the passivation process for the chamber surface. This prevents non-specific binding of biomolecules and results in finding 'good' DNA samples more easily.
7. A pull-down process helps reducing unwanted fluorescent noise which is likely to be emitted by dye-labeled DNA molecules without beads non-specifically bound to the chamber. If there is too much fluorescence background detected, more runs of the pull-down process might be helpful.
8. To dislodge any DNA molecules that have non-specifically adhered to the bottom surface of the chamber, one can gently tap it while washing with 1X PBS.
9. A nick in double-stranded DNA refers to a localized disruption within the molecule's backbone, characterized by the cleavage of the phosphodiester bond that links two adjacent nucleotides in one of the strands.

10. The value of persistence length may vary depending on the attached point of the DNA tether on the magnetic bead. This can be corrected through geometric consideration (Shon, Rah, & Yoon, 2019).
11. As the profile LUT is constructed based on the profiles with the discrete z values, one may think that the resolution of z-position determination is set by the step size of LUT. To achieve better precision in z-position determination, we use a polynomial fit to determine the z value that provides the minimum profile discrepancy measure (see Fig. 3C).

Acknowledgments

This work was supported by RS-2023-00221182 and IBS-R023-D1.

References

Bouchiat, C., Wang, M. D., Allemand, J.-F., Strick, T., Block, S. M., & Croquette, V. (1999). Estimating the persistence length of a worm-like chain molecule from force-extension measurements. *Biophysical Journal, 76*(1), 409–413. https://doi.org/10.1016/s0006-3495(99)77207-3.

Bronson, J. E., Fei, J., Hofman, J. M., Gonzalez, R. L., & Wiggins, C. H. (2009). Learning rates and states from biophysical time series: A Bayesian approach to model selection and single-molecule FRET data. *Biophysical Journal, 97*(12), 3196–3205. https://doi.org/10.1016/j.bpj.2009.09.031.

Gosse, C., & Croquette, V. (2002). Magnetic tweezers: Micromanipulation and force measurement at the molecular level. *Biophysical Journal, 82*(6), 3314–3329. https://doi.org/10.1016/s0006-3495(02)75672-5.

Joo, C., & Ha, T. (2012). Single-molecule FRET with total internal reflection microscopy. pdb.top072058 *Cold Spring Harbor Protocols, 2012*(12), https://doi.org/10.1101/pdb.top072058.

Kim, S. H., Jung, H. J., Lee, I.-B., Lee, N.-K., & Hong, S.-C. (2021). Sequence-dependent cost for Z-form shapes the torsion-driven B–Z transition via close interplay of Z-DNA and DNA bubble. *Nucleic Acids Research, 49*(7), gkab153. https://doi.org/10.1093/nar/gkab153.

Kuijpers, L., Laar, T., van, Janissen, R., & Dekker, N. H. (2022). Characterizing single-molecule dynamics of viral RNA-dependent RNA polymerases with multiplexed magnetic tweezers. *STAR Protocols, 3*(3), 101606. https://doi.org/10.1016/j.xpro.2022.101606.

Lee, M., Kim, S. H., & Hong, S.-C. (2010). Minute negative superhelicity is sufficient to induce the B-Z transition in the presence of low tension. *Proceedings of the National Academy of Sciences, 107*(11), 4985–4990. https://doi.org/10.1073/pnas.0911528107.

Ostrofet, E., Papini, F. S., & Dulin, D. (2018). Correction-free force calibration for magnetic tweezers experiments. *Scientific Reports, 8*(1), 15920. https://doi.org/10.1038/s41598-018-34360-4.

Parthasarathy, R. (2012). Rapid, accurate particle tracking by calculation of radial symmetry centers. *Nature Methods, 9*(7), 724–726. https://doi.org/10.1038/nmeth.2071.

Shon, M. J., Rah, S.-H., & Yoon, T.-Y. (2019). Submicrometer elasticity of double-stranded DNA revealed by precision force-extension measurements with magnetic tweezers. *Science Advances, 5*(6), eaav1697. https://doi.org/10.1126/sciadv.aav1697.

Smith, S., Finzi, L., & Bustamante, C. (1992). Direct mechanical measurements of the elasticity of single DNA molecules by using magnetic beads. *Science (New York, N. Y.), 258*(5085), 1122–1126. https://doi.org/10.1126/science.1439819.

Yu, Z., Dulin, D., Cnossen, J., Köber, M., Oene, M. M., van, Ordu, O., ... ekker, N. H. (2014). A force calibration standard for magnetic tweezers. *Review of Scientific Instruments, 85*(12), 123114. https://doi.org/10.1063/1.4904148.

CHAPTER SEVEN

Horizontal magnetic tweezers and its applications in single molecule micromanipulation experiments

Santosh Gaire[a,b,*], Rajendra Bhandari[a,b], Ian Pegg[a,b], and Abhijit Sarkar[a,b]

[a]Department of Physics, The Catholic University of America, Washington, DC, United States
[b]Vitreous State Laboratory, The Catholic University of America, Washington, DC, United States
*Corresponding author. e-mail address: gaire@cua.edu

Contents

Abstract

Magnetic tweezers (MTs) have become indispensable tools for gaining mechanistic insights into the behavior of DNA-processing enzymes and acquiring detailed, high-resolution data on the mechanical properties of DNA. Currently, MTs have two distinct designs: vertical and horizontal (or transverse) configurations. While the vertical design and its applications have been extensively documented, there is a noticeable gap in comprehensive information pertaining to the design details, experimental procedures, and types of studies conducted with horizontal MTs. This article aims to address this gap by providing a concise overview of

Methods in Enzymology, Volume 694
ISSN 0076-6879, https://doi.org/10.1016/bs.mie.2023.12.015

the fundamental principles underlying transverse MTs. It will explore the multifaceted applications of this technique as an exceptional instrument for scrutinizing DNA and its interactions with DNA-binding proteins at the single-molecule level.

1. Introduction

In recent years, a number of force-based micromanipulation technologies have been developed to study the micromechanics of DNA and DNA–protein interactions at the single-molecule level. These include optical tweezers (OTs), magnetic tweezers (MTs), and atomic force microscopy (Ashkin, Dziedzic, Bjorkholm, & Chu, 2016; Binnig, Quate, & Gerber, 1986; Dulin et al., 2015; Neuman & Nagy, 2008). MTs is a versatile and powerful tool that provides a non-invasive and precise method to exert controlled mechanical forces and torques to stretch and twist individual DNA (and RNA and protein) molecules or DNA complexed with DNA-binding proteins (Abels, Moreno-Herrero, Van Der Heijden, Dekker, & Dekker, 2005; De Vlaminck & Dekker, 2012; Lipfert, Kerssemakers, Jager, & Dekker, 2010; Lipfert, Wiggin, Kerssemakers, Pedaci, & Dekker, 2011; Popa et al., 2016; Ribeck & Saleh, 2008; Sarkar & Rybenkov, 2016; Smith, Finzi, & Bustamante, 1992). In turn, such experiments allow researchers to investigate the mechanical properties and interactions of DNA and DNA–protein complexes at the single molecule level.

MTs are commonly found in two designs: vertical MTs and horizontal MTs (HMTs) (alternatively referred to as transverse MTs). In this article, we will focus on HMTs as exemplified by a system we have developed and iteratively optimized. We aim to provide comprehensive information into the current design and components, summarize some of the results of micromanipulation studies using the MT, and outline future plans to refine the design further.

2. Instrument configuration

The first HMTs was reported by Smith et al. (1992). Their approach involved a horizontal configuration, which they used to analyze the elastic characteristics of individual DNA molecules in a torsionally relaxed state. Subsequent progress was achieved by Yan et al. (Chen et al., 2011), whose setup facilitated forces exceeding 200 pN on DNA tethers, with a position resolution of approximately 30 nm. Further enhancements to this design were outlined by Hongxia et al. (Fu, Chen, Marko, & Yan, 2010) and

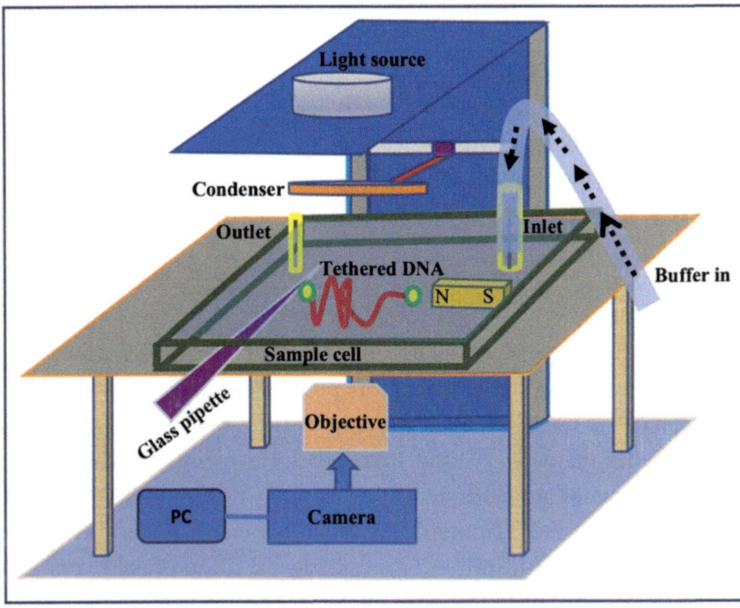

Fig. 1 A schematic of horizontal magnetic tweezers. A light source is positioned atop to illuminate the sample cell. An end-functionalized DNA molecule is positioned between two superparamagnetic beads. One bead is affixed to the surface of a functionalized glass pipette, while the other is linked to the opposite end of the DNA strand. A force on the tethered bead is generated by a permanent magnet in the direction of the field gradient. This tweezers setup allows for controlled manipulation of the stretching force experienced by the DNA molecule, as the magnets can be adjusted in position. The real-time movement of the bead is monitored using a digital camera.

Xinghua et al. (Fu et al., 2010), who utilized it to investigate distinct overstretched DNA states at varying temperatures. Building upon this foundation, we developed a relatively easy-to-assemble design capable of exerting forces ranging from 0.5 pN to over 100 pN (Fabian et al., 2018, 2021; Gaire, Fabian, Pegg, & Sarkar, 2022; McAndrew et al., 2016). A schematic of the current design with individual components annotated is shown in Fig. 1.

3. Method and materials

3.1 Working principles of horizontal magnetic tweezers (HMT)

This section provides concise descriptions of the procedures that must be performed before the experimental phase. We also list the materials needed and vendor and product details.

3.2 Functionalization of micro-pipette

3.2.1 Required materials

- 6-in. Borosilicate glass tubes
- micro-pipette puller
- Silane polyethylene glycol (PEG) biotin
- Ethanol
- DI water

3.2.2 Procedure

In our design, glass pipettes anchor one end of the DNA using magnetic beads—see Fig. 1. The functionalization process commences with the surface modification of the micro-pipette. We employ 6-in. borosilicate glass tubes with a 1.00 mm outer diameter (World Precision Instruments, Sarasota, FL, USA.) These micro-pipettes are fabricated using a micro-pipette puller (Sutter Instrument Co. Model P-97 Flaming/Brown Micropipette Puller), with the following settings: heat = 701, pull = 125, velocity = 110, and time = 140, resulting in high-quality, reproducible micro-pipettes. Subsequently, the tips of the micro-pipettes are immersed in a solution containing 10 mg of silane PEG Biotin (PG2-BSNL-2k, NANOCS New York, NY, USA) dissolved in 200 μL of a 90% ethanol (E7023, Sigma Aldrich, St. Louis, MO, USA) and water mixture for 1 h at room temperature. The surface-functionalized micro-pipettes are then stored in a dry environment at 4 °C for future use.

3.3 DNA end-functionalization

3.3.1 Required materials

- Lambda DNA (λ-DNA)
- Oligo 1
- Oligo 2
- T4 DNA Ligase
- 1X Tris-EDTA buffer solution
- DI water
- PE wash buffer
- QX1 solubilization buffer
- QIAEX II Suspension
- PBS (1X) buffer, pH 7.4

3.3.2 Procedure

The second step of the process is the end-functionalization of linear λ-DNA (N3011S, New England Biolabs, Ipswich, MA, USA) with 12 bases, single-stranded overhangs at its 3′- and 5′-end that are complementary to each other. The 3′-end overhang of the linear λ-DNA is ligated to an Oligo 2 (Integrated DNA Technologies, Coralville, IA, USA), which has a base sequence aggtcgccgcccBattggBattccBttccaBgttaB and the 5′-end overhang is ligated to an Oligo 1 (Integrated DNA Technologies, Coralville, IA, USA), which has a base sequence gggcggcgacctBattggBattccBttccaBgttaB. In both the sequences, B stands for Biotin. End-functionalization results in λ-DNA that can form high-strength bonds with super-paramagnetic beads coated with streptavidin—see Fig. 2A schematics.

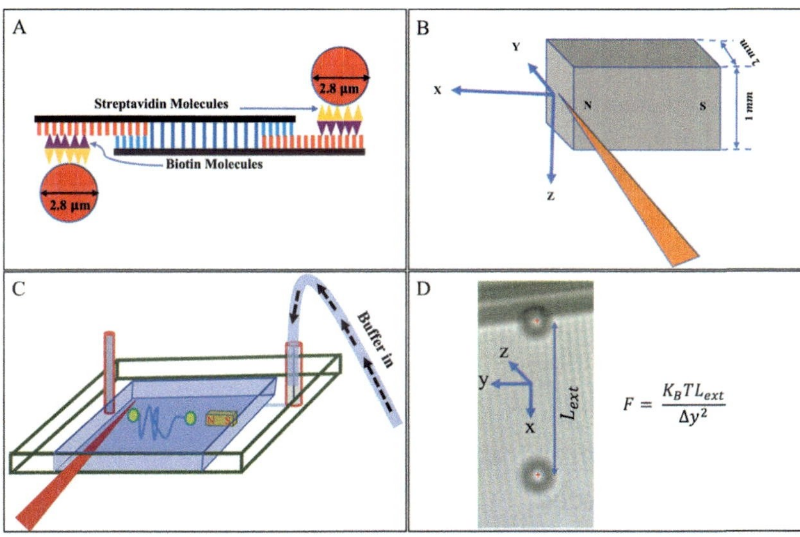

Fig. 2 (A) DNA end functionalization: In this step, the DNA molecules undergo end functionalization. Biotin molecules are attached to both ends of the DNA strand with allows the DNA to form multiple biotin-streptavidin linkages with streptavidin-coated magnetic beads. (B) Pipette positioning: This stage involves the precise positioning of the pipette tip at a known, reference position. This is crucial for accurately measuring the distance between the pipette and the magnet. (C) DNA-bead loaded sample cell: The sample cell is structured with glass coverslips, spacers, and a bar magnet. A surface-functionalized pipette is shown inserted into the sample cell. This design facilitates easy buffer exchange through two syringe pumps, which connect to inlet and outlet ports. (D) Experiment screenshot: This image from a typical experiment shows two beads simultaneously in focus. The distance between the centroids of these beads, adjusted for the bead diameter, provides the measurement of DNA extension.

3.4 Preparation of DNA-bead mixture

3.4.1 Materials needed

- 2.8 μm diameter superparamagnetic beads
- Phosphate-buffered saline (PBS)
- DI water
- Working buffer

3.4.2 Procedure

This part of the protocol starts with the bead cleaning process. We use 2.8 μm diameter super-paramagnetic beads (Dyna beads™ M-280, ThermoFisher Scientific) with its outer surface coated with streptavidin. Beads are vortexed and washed twice with 1× PBS buffer followed by a wash with 150 mM NaCl (working buffer). We use a permanent magnet to pull down the beads to remove the buffer. Once the beads are separated from the buffer, we pipette out the buffer taking care not to draw in any magnetic beads. We then add 20 μL of working buffer to the Eppendorf tube containing the cleaned magnetic beads and vortex for 30 s. The suspended beads are then pipetted to the Eppendorf tube that contains end-functionalized DNA. The streptavidin-coated beads then attach to the biotinylated ends of the λ-DNA forming DNA-bead complexes. Now our DNA is ready for experiment use; later we refer this as the DNA-bead mixture.

3.5 Sample cell preparation

3.5.1 Materials needed

- A custom 3D-printed spacer
- A neodymium bar magnet
- Coverslip 1 and coverslip 2
- All Purpose 100% Adhesive Sealant

3.5.2 Procedure

The sample cell is assembled by combining two coverslips of different dimensions, referred to as coverslip 1 (72198-20, Electron Microscopy Science, Hatfield, PA, USA) and coverslip 2 (72198-10, Electron Microscopy Science). Additionally, a custom 3D-printed spacer (Watershed XC 11122, Proto Labs Inc., Maple Plain, MN, USA) and a neodymium bar magnet measuring 3 mm × 2 mm × 1 mm (M0301, SUPERMAGNETMAN, Pelham, AL, USA) are employed in the construction. Coverslip 1 and coverslip 2 are affixed to the top and bottom of the spacer respectively, forming the upper and lower surfaces of the sample cell. The sides of the spacer constitute three of the four boundaries of the sample cell, leaving one side open to allow for the insertion of a pipette. The spacer is equipped with inlet and outlet ports for buffer exchange purposes. The bar magnet is securely attached to the underside of coverslip 2. To assemble these components, a clear room-temperature vulcanizing silicone sealant (All Purpose 100% Adhesive Sealant, DAP, Baltimore, MD, USA) is utilized. The overall dimensions of the sample cell measure 60 mm × 40 mm × 1 mm; however, the enclosed region measures 6 mm × 4 mm × 1 mm, resulting in a sample cell volume of approximately 100 μL.

3.6 DNA stretching experiments

The experiments start inserting the DNA-loaded sample cell and the surface-functionalized glass pipette into a HMTs setup—see Fig. 2C. The sample cell is affixed to a sample cell holder connected to a motorized stage system (X-MCB1, AB103B, XJOY3, Zaber) capable of facilitating movement in three spatial axes. Simultaneously, the functionalized glass pipette is clamped within a specially crafted, two-piece aluminum holder, which is held in a hydraulic, three-axis micromanipulator (MX630L S3432, Siskiyou Corporation).

At this point we introduce the pipette tip inside the sample cell and bring it into focus. The subsequent steps involve positioning the pipette tip precisely at a predetermined location, often referred to as the "origin" situated at the geometric center of the rectangular bar magnet—see Fig. 2B. The precise positioning of the tips is achieved as the micromanipulator is controlled by a custom LabVIEW program. This positioning is crucial as it enables us to determine the precise distance between the magnet and the glass pipette.

The DNA pulling experiment can now begin. Initially, we must locate a tethered bead, ensuring that both the tethered bead and the bead attached to the pipette are equally in focus in the same focus plane simultaneously—see Fig. 2D. Next, we proceed to confirm the presence of a single DNA molecule linking the pair of beads. To achieve this, we gently move the magnet toward the tethered bead at a constant speed of 10 μm/s while monitoring for the DNA overstretching transition. This transition is quite distinctive, characterized by a 70% increase in the DNA's contour length, and serves as confirmation that we have at least one DNA molecule tethered to the suspended bead. Once a single tether is verified, we reduce the force significantly by moving the magnet away from the tethered bead and start the video recording using a customized LabVIEW program. Typically, the data recording process commences with an initial separation of ~2 mm between the magnet and the glass pipette. From this starting point, a controlled movement of the magnet toward the tethered bead ensues, with periodic pauses at various locations. This deliberate approach continues until the overstretching transition is observed.

After completing the data recording process, the recorded movies are analyzed using a custom MATLAB code. This code employs a particle tracking algorithm to identify and track the centroids of the two tethered beads in each frame, extracting their coordinates. This information is crucial for calculating the tether's extension, L_{ext}, as well as quantifying the transverse fluctuations of the suspended bead, $\langle \Delta y^2 \rangle$. The applied force acting on the DNA is subsequently determined using various methods.

3.7 Force calibration

In principle, the force applied on the tethered bead can be computed using following relation:

$$\vec{F} = \frac{1}{2} \vec{\nabla} (\vec{m} . \vec{B})$$

The method involves using the saturated bead magnetization $(\vec{m})$ and applied magnetic field $(\vec{B})$ (Lipfert, Hao, & Dekker, 2009). Utility of this equation is limited with variations in material quantity within a batch of magnetic beads. Alternatively, the Equipartition theorem is employed, utilizing transverse (to the force) thermal fluctuations of the magnetic bead's centroid to calculate the force acting on the DNA:

$$\vec{F} = \frac{k_B T L_{ext}}{\langle \Delta y^2 \rangle}$$

Here, k_B as the Boltzmann constant, T as temperature, L_{ext} as the average molecular extension, and $\langle \Delta y^2 \rangle$ as the transverse fluctuation of the magnetic bead. Although straightforward to use, the computed force is sensitive to measurement biases in $\langle \Delta y^2 \rangle$. This limitation, especially prevalent at high forces, requires the removal of non-thermal noise sources from recorded motion. A robust solution involves time-domain filtering with the Allan variance (AV) method and frequency-domain filtering with the power spectral density (PSD) method (as detailed in Janssen et al., 2012; Lansdorp & Saleh, 2012; Te Velthuis, Kerssemakers, Lipfert, & Dekker, 2010).

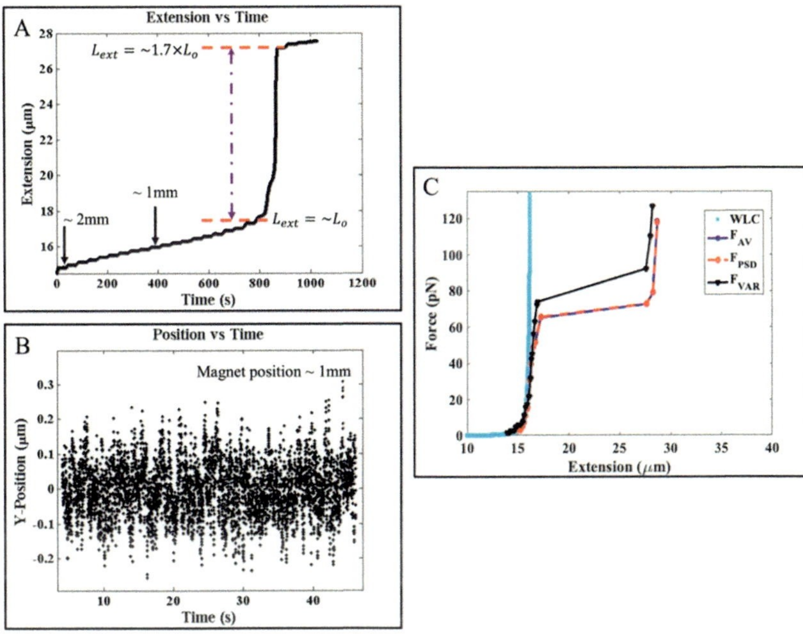

Fig. 3 illustrates the results from a typical single molecule micromanipulation experiment. In part (A), we show the tether extension versus time starting with a 2 mm separation between the magnet and tethered bead, where the DNA was approximately 14 μm long. The magnet was gradually moved toward the bead with pauses at different locations. The DNA extended until reaching the overstretching transition. This occurred within the narrow force range centered around 65 pN characteristic of this transition with an extension 1.7 times its contour length as expected. In part (B), we present a plot from the same experiment of the tethered bead's lateral position along the *y*-direction at a specific magnet position (around 1 mm). This information was later used to calculate the force exerted on the tethered bead. Part (C) displays a force versus extension plot. In order to calibrate the instrument three distinct methodologies were used to determine forces: the Allan variance method (blue curve), the power spectral density (PSD) method (red curve), and the Equipartition theorem (black curve). These values were then compared against the theoretical worm-like chain (WLC) model (cyan color). Additionally, calibration was performed by assessing the data against the overstretching transition.

In Fig. 3A, we show an extension versus time graph derived from a typical DNA pulling experiment. The data recording was initiated with the magnet positioned ~2 mm away from the tethered bead. Subsequently, data were recorded with the magnet at various fixed positions, each maintained for an extended duration. In Fig. 3B, we display the position distribution of the tethered bead when the magnet was held ~1000 μm from the bead. Force calculations were performed using three distinct methods: the Equipartition theorem, the AV method (Lansdorp & Saleh, 2012), and the PSD method (Lansdorp & Saleh, 2012). Fig. 3C shows force vs extension measurements for a complete DNA-pulling experiment using these methods and plotted against the worm-like chain (WLC) model of DNA (Marko & Siggia, 1995). This approach reduces force measurement biases and generates higher accuracy experimental data.

Our force-extension data correlates well with the WLC model up to ~40 pN, covering low-force non-linear entropic and high-force linear Hookean elastic responses. Beyond this in between 40 and 60 pN, a deviation occurs, possibly due to changes in DNA's elastic modulus (k_o) and factors like strand breaks. At ~65 pN, tethered DNA undergoes the well-known overstretching transition, with a 70% contour length increase. This reversible and highly non-linear elastic regime reproduces known results (Strick, Allemand, Croquette, & Bensimon, 2000), providing a robust calibration for our measurements.

3.8 Study of DNA–protein interactions using HMT

So far we have used our design to study DNA–protein interactions under a range of mechanical stresses. Specifically, we have focused on studies using native and hyperacetylated histones. DNA–protein interactions are governed by a variety of factors including the amino acid sequence (primary structure), the presence of alpha helices or beta sheets (secondary structure), and the overall three-dimensional shape of the protein (tertiary structure), and others characterizing the cellular or in vitro environment including temperature, ionic strength, protein concentration, and others. In vivo, these interactions are vital for maintaining genetic integrity as well as the proper functioning of cells.

Our tweezers can take advantage of the horizontal geometry of the tethered DNA to enable differential detection of the centroids of the beads. The custom-designed and built sample cell allows buffer exchange and the on-demand addition of DNA-binding proteins. Using our apparatus and the associated biochemical protocols, we studied the mechanical stability of our in situ-assembled DNA-core particle arrays to mechanical disturbances in a force range relevant to the action of genetic machinery inside cells. In this section, we will provide a brief overview of the protein experiments; readers are also referred to Fabian et al. (2021) and Gaire et al. (2022) for more detailed information.

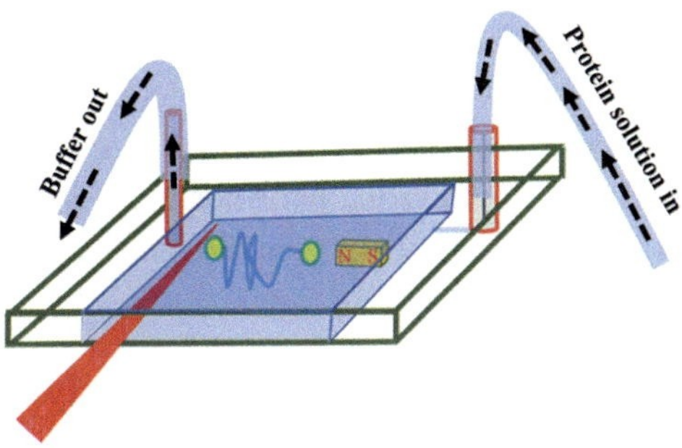

Fig. 4 Schematics of the buffer exchange system. Balanced exchange is facilitated by employing calibrated syringe pumps (not shown,) one at each end. To prevent buffer overflow, the rates of the buffer solution entering and exiting the system are maintained at a constant level, considering the variations in diameters between the inlet and outlet ports.

Experiments start like the DNA pulling assays discussed above to ensure the presence of single DNA tether. Next, we introduce proteins—histones in our case—into the sample cell using a buffer exchange method—see Fig. 4. Briefly, we start by making the core histones and NAP1 solutions (protein solution) in the buffer of choice (in our case, 150 mM NaCl) and adjusting the final concentrations of core histones to ~0.05 mg/mL and NAP1 to ~0.1 mg/mL. While introducing proteins into the sample cell, we normally maintain the magnet ~1600 μm from the tethered bead, which corresponds to a force of less than 4 pN on the tether. The custom designed buffer exchange system is used to extract buffer from the outlet while simultaneously pumping proteins in through inlet in a balanced manner—see schematic in Fig. 4. Balanced exchange is achieved through the use of calibrated syringe pumps, one at each end. In order to avoid buffer overflow, the rates of the buffer solution entering and leaving the system are kept constant taking account of the differences in diameters of the inlet and outlet ports.

We now activate the video recorder in LabVIEW with the appropriate settings. As soon as the visible compaction of DNA molecules is evident (magnetic beads at each end of the DNA approach each other, which can be seen clearly on the monitor), the two syringe pumps are turned off simultaneously. The magnet is then moved even farther away to ~1800 μm to allow more histones to interact with and bind to the DNA.

We now increase the pulling force by moving the magnet toward the tethered bead. The magnet is typically moved at a speed of 0.5–1 μm/s until we see the suspended bead separating from the tether. A typical experiment takes around two hours to complete. To further analyze the findings, we process our recorded videos using a custom developed offline particle tracking algorithm (Tyson, McAndrew, Tuma, Pegg, & Sarkar, 2015). All studies of protein–DNA interactions use the same methodology, regardless of histone types.

3.9 DNA-histone interactions and data analysis

The data files obtained from the DNA–protein experiments are processed using an offline particle tracking algorithm. During the data analysis, we consider the following: single-molecule DNA tethers undergo pre-extension at low forces. Subsequently, histones are introduced and allowed to bind to the tether, compacting it. The force is then systematically increased, and the extension of the tether is measured as a function of the applied force. (We plot the extension data against time, i.e., video frame, or force, as necessary.) Beyond the critical force, denoted as f^* (the minimal force required to disrupt the nucleosome arrays), octamers start to unbind from the tether, releasing DNA that was previously wrapped around them. Our method can detect individual unbinding events, characterized by a distinctive step-like change in DNA extension, with each step aligning with (integer multiples of) the expected length of DNA released when a single nucleosome is mechanically disassembled. Furthermore, we observe multistep events corresponding to the simultaneous unbinding of two or more histone octamers.

Fig. 5 presents the force-versus-extension plot derived from a representative protein experiment. The experimental data re juxtaposed with the WLC model (depicted in cyan) and the data from a preceding DNA pulling experiment conducted before the introduction of proteins into the sample cell (represented by the red line). As elucidated in the preceding paragraph, while introducing the proteins into the sample cell the DNA tether to held at a low force, ~1 pN, where the DNA extension is approximately 15 μm (indicated as point A in Fig. 5).

DNA compaction was seen within a few minutes, resulting in a decrease in DNA extension from point A to point B. Throughout this stage, the position of the magnet remained constant. Subsequently, the magnet was incrementally brought continuously toward the tethered bead at a low speed, ensuring a low force loading rate of approximately 0.008 pN/s. In the interval between points B and C, the DNA extended up to its contour length, approximately 16.4 μm. Notably, the force-versus-extension plot during this period deviated from the pattern observed in the DNA

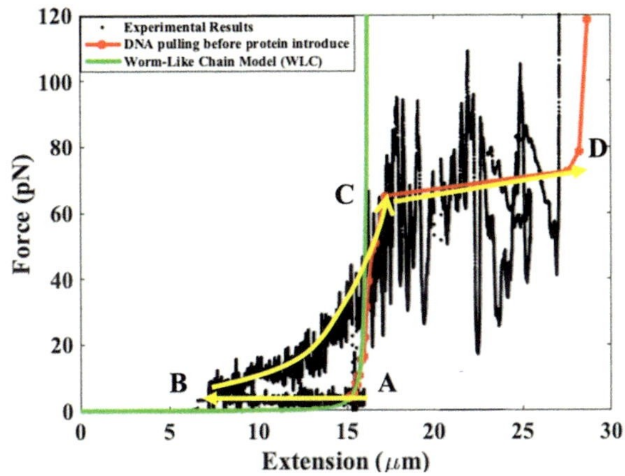

Fig. 5 depicts DNA force-versus-extension data derived from a protein experiment. The green curve represents the theoretical WLC model prediction. The baseline, depicted by the red line, illustrates force-versus-extension data for the same tether before protein introduction. Force-versus-extension data after histones are introduced are in black. Yellow lines serve as guides, aiding in the interpretation of the experiment. Notably, compaction of the tether occurs between points A and B, as indicated by the yellow lines. Subsequently, with increasing force, DNA extension progresses to point C, where the force reaches approximately 65 pN. A critical observation is the occurrence of the overstretching transition between points C and D. Within this interval, the DNA exhibits pronounced instability, leading to substantial fluctuations in the tethered bead position and, consequently, noticeable force fluctuations in the region between points C and D. This instability is a signature of the significant structural changes in the DNA tether during the overstretching transition.

pulling experiment, a phenomenon attributed to the increased force required to dissociate histone octamers from the DNA. Theoretically, at around 40 pN, it is anticipated that most nucleosomes would dissociate from the DNA tether. Beyond this threshold, the behavior of the DNA resembles its pre-protein introduction state.

Upon reaching a force of approximately 65 pN, the DNA undergoes an overstretching transition. This reconfirms that the selected tether for our study indeed was a single molecule. With the successful validation step, we proceed to the subsequent phase, focusing on analyzing the compaction and decompaction regions. A custom-developed step-finding algorithm is employed to identify potential stepwise extensions in the DNA tether. These step-like extensions carry crucial information about the behavior of DNA-bound histones, which is important for understanding the mechanical properties of chromatin.

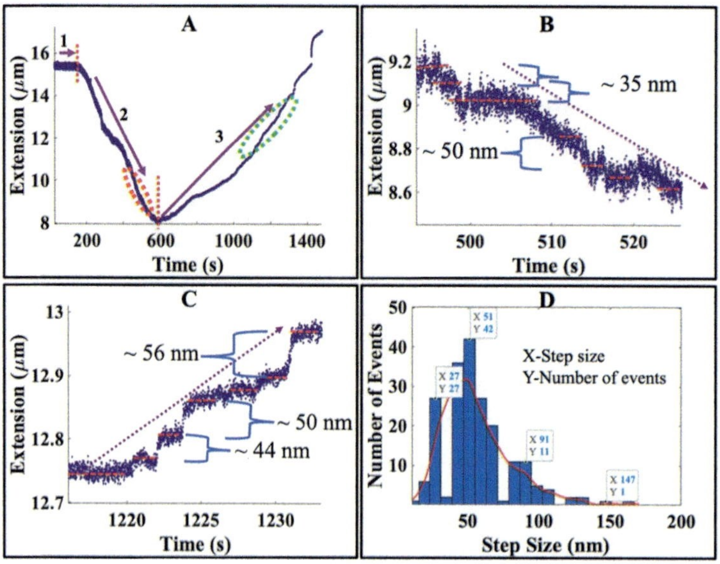

Fig. 6 Our data analysis workflow for protein experiments. The extension versus time graph in (A) depicts a typical DNA–protein experiment with three distinct phases. In the first part—the pre-compaction stage—proteins are loaded into the reaction chamber, but the interaction between DNA and proteins has not yet commenced (spanning arrow 1). The second phase is the compaction stage, where DNA and histones interact, forming nucleosomes and reducing DNA extension from ~15.5 nm to ~8 nm (arrow 2). The third region is the decompaction stage, as indicated by arrow 3, where force is gradually increased by moving the magnet toward the tethered bead. (B) Zooming into the DNA compaction region within the red dotted circle from panel A, the red dotted lines assist in identifying steps and pauses in DNA compaction. In contrast, the black dotted arrow indicates the overall compaction of the DNA. (C) These data focus on the decompaction stage, magnified from the region marked by the green dotted circle in panel A. (D) The step size histogram for the overall experiment, featured in panel D, highlights that most step sizes are concentrated around ~50 nm. This value corresponds to the length of DNA wrapped around a nucleosome.

The extension versus time graph in Fig. 6A is annotated to depict three distinct regimes in the experimental data. The initial phase, denoted by arrow 1, represents the pre-compaction stage of the experiment, where the force is maintained below 4 pN by keeping the magnet position constant. This position is held for a duration until DNA compaction is observed. Following arrow 2 in Fig. 6A, DNA undergoes compaction from approximately 15.5 μm to about 8 μm while subjecting to a constant force. Once DNA is fully compacted, a waiting period ensues to facilitate maximal nucleosome formation. Subsequently, the force is increased by moving the magnet continuously toward the tethered bead at a constant and low speed to minimize the loading rate as indicated by arrow 3 in Fig. 6A.

Nucleosomes represent complexes formed by DNA and histone proteins, wherein approximately 145 base pairs (equivalent to ~50 nm) of DNA wrap around the octameric complex of histones in one-and-three-quarter turns. Evidence supporting nucleosome formation is presented in Fig. 6B and C. In these two panels, we have zoomed in on selected areas from Fig. 6A within red and green dotted circles. Upon closer examination of small regions within these two areas, we have identified step-like events on both the.

compaction and decompaction regimes. Most strikingly, most of these events exhibit a consistent step length of approximately 50 nm.

To identify and measure the step lengths, we employed a custom-developed step-finding algorithm (as outlined in Tyson et al., 2015). The histogram presented in Fig. 6B serves as additional evidence for nucleosome formation, with most steps centered around a length of approximately 50 nm. Notably, additional peaks are observed at around 100 and 150 nm, indicating simultaneous rupturing of two and three nucleosomes.

It's important to note that this description provides a simplified overview of our data analysis procedures. For a comprehensive discussion, please refer to Fabian et al. (2021) and Gaire et al. (2023). Furthermore, our analytical approach is the same regardless of the specific type of histones used in our experiments.

3.10 Conclusion and future perspectives

In this article, we explored the design and applications of HMTs. We briefly traced the evolution of this technique from its introduction by Smith et al. (1992). Notably, a HMTs is characterized by its simplicity, affordability, minimal data noise, and high measurement accuracy and precision. However, challenges persist in achieving high throughput and implementing studies to manipulate the torsional state of DNA molecules. Recent strides in computational hardware and camera technology have made highly parallel MT setups possible. Moreover, video-based microscopy enables extension measurements with resolutions approaching those of OTs. Future research will focus on streamlined HMTs that maintain exceptional spatial resolution, high throughput, system integration to apply twist, and designs that integrate force application with parallel single-molecule fluorescence detection. The continuous advancement of single-molecule techniques promises to yield detailed mechanistic insights into DNA's mechanical properties and the functional mechanisms of DNA-bound proteins. We believe HMTs will be an important tool for achieving this goal.

Acknowledgments

The authors gratefully acknowledge the assistance of the following individuals: Christopher McAndrew, Christopher Tyson, Roberto Fabian, and Pamela L Tuma.

Author contributions

A Sarkar and I Pegg provided project supervision. Santosh Gaire, Rajendra Bhandari, and A Sarkar authored the manuscript. All authors have reviewed and approved the final version of the manuscript for publication.

Financial and competing interests disclosure

A Sarkar and S Gaire acknowledge financial support from the Catholic University of America and Vitreous State Laboratory. The authors declare no other affiliations or financial associations with organizations or entities having a financial interest or conflict related to the subject matter or materials discussed in the manuscript, except those already disclosed.

References

Abels, J. A., Moreno-Herrero, F., Van Der Heijden, T., Dekker, C., & Dekker, N. H. (2005). *Single-molecule measurements of the persistence length of double-stranded RNA, 88*, Elsevier,4.

Ashkin, A., Dziedzic, J. M., Bjorkholm, J. E., & Chu, S. (2016). Observation of a single-beam gradient force optical trap for dielectric particles. *Optical Angular Momentum, 11*(5), 196–198. https://doi.org/10.1364/ol.11.000288.

Binnig, G., Quate, C. F., & Gerber, C. H. (1986). Atomic force microscope. *Physical Review Letters, 56*(9), https://doi.org/10.1201/9781420075250.

Chen, H., Fu, H., Zhu, X., Cong, P., Nakamura, F., & Yan, J. (2011). Improved high-force magnetic tweezers for stretching and refolding of proteins and short DNA. *Biophysical Journal, 100*(2), 517–523. https://doi.org/10.1016/j.bpj.2010.12.3700.

Dulin, D., Cui, T. J., Cnossen, J., Docter, M. W., Lipfert, J., & Dekker, N. H. (2015). High spatiotemporal-resolution magnetic tweezers: Calibration and applications for DNA dynamics. *Biophysical Journal, 109*(10), 2113–2125. https://doi.org/10.1016/j.bpj.2015.10.018.

De Vlaminck, I., & Dekker, C. (2012). Recent advances in magnetic tweezers. *Annual Review of Biophysics, 41*(1), 453–472. https://doi.org/10.1146/annurev-biophys-122311-100544.

Fabian, R., Gaire, S., Tyson, C., Adhikari, R., Pegg, I., & Sarkar, A. (2021). A horizontal magnetic tweezers for studying single DNA molecules and DNA-binding proteins. *Molecules (Basel, Switzerland), 26*(16), 4781. https://doi.org/10.3390/molecules26164781.

Fabian, R., Tyson, C., Striz, A., Tuma, P. L., Pegg, I. L., & Sarkar, A. (2018). Design of novel magnetic tweezers and its use for studying DNA-compacting proteins. *Biophysical Journal, 114*(3), 91a. https://doi.org/10.1016/j.bpj.2017.11.537.

Fu, H., Chen, H., Marko, J. F., & Yan, J. (2010). Two distinct overstretched DNA states. *Nucleic Acids Research, 38*(16), 5594–5600. https://doi.org/10.1093/nar/gkq309.

Gaire, S., Fabian, R. L., Adhikari, R., Tuma, P. L., Pegg, I. L., & Sarkar, A. (2023). Micromechanical study of hyperacetylated nucleosomes using single molecule transverse magnetic tweezers. *International Journal of Molecular Sciences, 24*(7), https://doi.org/10.3390/ijms24076188.

Gaire, S., Fabian, R., Pegg, I., & Sarkar, A. (2022). Magnetic tweezers: Development and use in single-molecule research. *Biotechniques, 72*(2), 65–72. https://doi.org/10.2144/btn-2021-0104.

Janssen, X. J. A., Lipfert, J., Jager, T., Daudey, R., Beekman, J., & Dekker, N. H. (2012). Electromagnetic torque tweezers: A versatile approach for measurement of single-molecule twist and torque. *Nano Letters, 12*(7), 3634–3639. https://doi.org/10.1021/nl301330h.

Lansdorp, B. M., & Saleh, O. A. (2012). Power spectrum and Allan variance methods for calibrating single-molecule video-tracking instruments. *The Review of Scientific Instruments, 83*(2), https://doi.org/10.1063/1.3687431.

Lipfert, J., Hao, X., & Dekker, N. H. (2009). Quantitative modeling and optimization of magnetic tweezers. *Biophysical Journal, 96*(12), 5040–5049. https://doi.org/10.1016/j.bpj.2009.03.055.

Lipfert, J., Kerssemakers, J. W. J., Jager, T., & Dekker, N. H. (2010). Magnetic torque tweezers: Measuring torsional stiffness in DNA and RecA-DNA filaments. *Nature Methods,* 7(12), 977–980. https://doi.org/10.1038/nmeth.1520.

Lipfert, J., Wiggin, M., Kerssemakers, J. W. J., Pedaci, F., & Dekker, N. H. (2011). Freely orbiting magnetic tweezers to directly monitor changes in the twist of nucleic acids. *Nature Communications, 2*(1), https://doi.org/10.1038/ncomms1450.

Marko, J. F., & Siggia, E. D. (1995). Stretching DNA. *Macromolecules, 28*(26), 8759–8770. https://doi.org/10.1021/ma00130a008.

McAndrew, C. P., et al. (2016). Simple horizontal magnetic tweezers for micro-manipulation of single DNA molecules and DNA–protein complexes. *Biotechniques, 60*(1), 21–27. https://doi.org/10.2144/000114369.

Neuman, K. C., & Nagy, A. (2008). Single-molecule force spectroscopy: Optical tweezers, magnetic tweezers and atomic force microscopy. *Nature Methods, 5*(6), 491–505. https://doi.org/10.1038/nmeth.1218.

Popa, I., et al. (2016). A HaloTag anchored ruler for week-long studies of protein dynamics. *Journal of the American Chemical Society, 138*(33), 10546–10553. https://doi.org/10.1021/jacs.6b05429.

Ribeck, N., & Saleh, O. A. (2008). Multiplexed single-molecule measurements with magnetic tweezers. *The Review of Scientific Instruments, 79*(9), https://doi.org/10.1063/1.2981687.

Sarkar, R., & Rybenkov, V. V. (2016). A guide to magnetic tweezers and their applications. *Frontiers in Physics, 4*. https://doi.org/10.3389/fphy.2016.00048.

Smith, S. B., Finzi, L., & Bustamante, C. (1992). Direct mechanical measurements of the elasticity of single DNA molecules by using magnetic beads. *Science (80-.), 258*(5085), 1122–1126. https://doi.org/10.1126/science.1439819.

Strick, T., Allemand, J. F., Croquette, V., & Bensimon, D. (2000). Twisting and stretching single DNA molecules. *Progress in Biophysics and Molecular Biology, 74*(1–2), 115–140. https://doi.org/10.1016/S0079-6107(00)00018-3.

Tyson, C., McAndrew, C., Tuma, P. L., Pegg, I., & Sarkar, A. (2015). Automated non-parametric method for detection of step-like features in biological data sets. *Cytometry Part A, 87*(5), 393–404. https://doi.org/10.1002/cyto.a.22631.

Te Velthuis, A. J. W., Kerssemakers, J. W. J., Lipfert, J., & Dekker, N. H. (2010). Quantitative guidelines for force calibration through spectral analysis of magnetic tweezers data. *Biophysical Journal, 99*(4), 1292–1302. https://doi.org/10.1016/j.bpj.2010.06.008.

CHAPTER EIGHT

Magnetic tweezers characterization of the entropic elasticity of intrinsically disordered proteins and peptoids

Hoang P. Truong[a] **and Omar A. Saleh**[a,b,c,*]
[a]Materials Department, University of California, Santa Barbara, CA, United States
[b]Biomolecular Sciences and Engineering Program, University of California, Santa Barbara, CA, United States
[c]Physics Department, University of California, Santa Barbara, CA, United States
*Corresponding author. e-mail address: saleh@ucsb.edu

Contents

Abstract

Understanding the conformational behavior of biopolymers is essential to unlocking knowledge of their biophysical mechanisms and functional roles. Single-molecule force spectroscopy can provide a unique perspective on this by exploiting entropic elasticity to uncover key biopolymer structural parameters. A particularly powerful approach involves the use of magnetic tweezers, which can easily generate lower stretching forces (0.1–20 pN). For forces at the low end of this range, the elastic response of biopolymers is sensitive to excluded volume effects, and they can be described by Pincus blob elasticity model that allow robust extraction of the Flory

Methods in Enzymology, Volume 694
ISSN 0076-6879, https://doi.org/10.1016/bs.mie.2023.12.011

polymer scaling exponent. Here, we detail protocols for the use of magnetic tweezers for force-extension measurements of intrinsically disordered proteins and peptoids. We also discuss procedures for fitting low-force elastic curves to the predictions of polymer physics models to extract key conformational parameters.

1. Introduction

Biopolymers are the building blocks of all living systems, from assembling structural networks and storing information to acting as molecular machines. In many of their functional roles, biopolymers undergo significant conformational changes, involving a nuanced and complex interplay of a vast range of intra- and inter-molecular interactions. Detailed knowledge of the conformations of biopolymers in any environment is critical to elucidating their functional roles, as well as aiding their rational de novo design (DeGrado & Korendovych, 2020; Kosuri & Church, 2014). Different experimental methods have been explored to provide detailed characterization of biopolymer structures and conformations, such as NMR, crystallography, X-ray scattering, and fluorescence spectroscopy (Kosol, Contreras-Martos, Cedeño, & Tompa, 2013; Millar, 1996; Petsko & Ringe, 1984; Schuler, Soranno, Hofmann, & Nettels, 2016).

Single-molecule force spectroscopy has emerged as a powerful tool for the investigation of biopolymer structures and dynamics (De Vlaminck & Dekker, 2012; Neuman & Nagy, 2008). As opposed to bulk characterization methods, single-molecule stretching approach allows direct observation and manipulation of individual polymer chains at high resolution. While there are various experimental approaches that differ in their details, all such techniques feature the immobilization and attachment of a polymer chain to a probe sensitive to an external field and capable of applying tension. Under an applied force, the configurational freedom of the polymer resists its extension, giving rise to the entropic elastic response (Rubinstein & Colby, 2003). Prior single-molecule force spectroscopy studies have successfully exploited entropic elasticity measurements of proteins and nucleic acids to quantify their conformations (Bustamante, Smith, Liphardt, & Smith, 2000; Hughes & Dougan, 2016; Solanki, Neupane, & Woodside, 2014).

Measurements with atomic force microscopy or optical tweezers are typically performed at very high forces (2–1000 pN), where the polymer is strongly stretched to near its contour length and has minimal configurational freedom. However, the entropic elasticity at low forces (below 2 pN), when

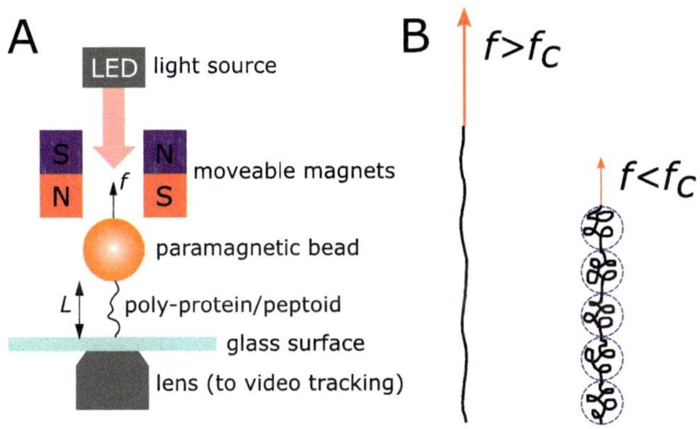

Fig. 1 Schematic of a standard magnetic tweezers experiment and entropic elasticity force regimes. (A) A polyprotein or polypeptoid construct is tethered between a glass surface and a paramagnetic bead. A pair of magnets generate a magnetic field gradient, which attracts the bead and applies tension on the polymer. (B) The low-force capability (0.1–20 pN) of magnetic tweezers allows access to distinct entropic elastic regimes: intermediate-force asymptotic stretching (at forces above the crossover force f_c) and low-force Pincus blobs (at forces below f_c).

the polymer has much more configurational dynamics, are often overlooked in such experiments. Pincus pointed out that at these forces, where a polymer contains significant loops and bends, it can be characterized as a chain of blobs aligned in the direction of force. Within each blob, the polymer configuration follows the Flory scaling law (Pincus, 1976). Thus, force-extension measurements at low forces allow direct estimate of the Flory exponent, which is a simple yet effective way to describe and compare the conformations of biopolymers (Marsh & Forman-Kay, 2010).

Of the common single-molecule force spectroscopy techniques, magnetic tweezers provides the most reliable access to low stretching forces (Neuman & Nagy, 2008). In a standard magnetic tweezers experiment, a polymer is immobilized to a glass surface on one end and attached to a paramagnetic bead on the other end (Fig. 1A). In the low-to-intermediate force range (0.1–20 pN) enabled by the technique, the entropic elastic response of a flexible polymer exhibits distinct regimes that are controlled by different structural length scales, shown in Fig. 1B (Saleh, 2015). Direct analysis of these elastic regimes is achieved by measuring the force-extension curve, which is obtained by measuring the extension of the polymer in successive stable magnet positions (fixed forces) (Saleh, 2015).

Magnetic tweezers have been found widely useful in the study of the structures and dynamics of biopolymers. Various articles have discussed the application and methodology of magnetic tweezers in the investigation of nucleic acids and polysaccharides (Bae et al., 2014; Berezney & Saleh, 2017; Haber & Wirtz, 2000; Kriegel, Ermann, & Lipfert, 2017; Ruiz-Gutierrez et al., 2022; Shon, Rah, & Yoon, 2019; Vilfan, Lipfert, Koster, Lemay, & Dekker, 2009). While the protocols described below are applicable to those biopolymers, we here specifically focus on intrinsically disordered proteins (IDPs) and polypeptoids as they present unique challenges to magnetic tweezers experiments that have not been widely discussed.

Magnetic tweezers offer a novel approach to the investigation of the conformational behaviors of IDPs through the study of their low-force entropic elasticity. Despite not folding into a stable structure, IDPs nonetheless carry out many biological functions and are involved in several disease pathways (Oldfield & Dunker, 2014; Uversky, 2011; Van Der Lee et al., 2014). Due to the lack of a well-defined three-dimensional structure, studies of IDP conformation have eluded conventional characterization techniques that are typically used for folded proteins. Alternative methods, notably small-angle X-ray scattering (SAXS) and single-molecule Förster resonance energy transfer (smFRET), have been explored to quantify IDP conformational ensembles (Bernadó & Svergun, 2012; Brucale, Schuler, & Samorì, 2014). Generally, these methods involve measurements of chain extent (e.g., radius), which are then used to estimate the Flory exponent through approximation of the polymer scaling relationship. However, several discrepancies in experimental results from SAXS and smFRET have been reported, which complicate the goal of forming a unified view on IDP conformation (Ruff & Holehouse, 2017; Yoo et al., 2012). Thus, there is a need to explore alternate means to characterize IDP conformational ensembles, such as magnetic tweezers entropic elasticity measurements.

In addition, there has been significant interest in developing fundamental understanding of biopolymer sequence-conformation relationship, as this knowledge is valuable to the elucidation of sequence-dependent biophysical processes and the rational design of sophisticated bioinspired materials (Chang et al., 2017; Numata, 2020; Perry & Sing, 2020; Sing, 2020). Indeed, many studies leveraged synthetic capabilities of peptides and peptoids for the experimental investigation of such effects (Murnen, Khokhlov, Khalatur, Segalman, & Zuckermann, 2012; Murnen, Rosales, Dobrynin, Zuckermann, & Segalman, 2013; Rosales, Murnen, Kline, Zuckermann, & Segalman, 2012; Shi, Adhikari, Asthagiri, & Marciel, 2023). Single-molecule entropic elasticity measurements can provide detailed, high-resolution observations of their conformations and responses to different solution environments. We discuss here how magnetic tweezers can be leveraged to probe conformational behaviors of these synthetic biopolymers. While the protocols below have only been exercised for the study of polypeptoids, they will undoubtedly be applicable for future studies on synthetic polypeptides (Truong, Mengel, Yu, Segalman, & Saleh, 2023).

In this article, we first describe strategies to form long polymer chains from short IDPs and peptoids sequences through different click chemistry routes. We then provide protocols on surface chemistry and tethering for different polymer types. Finally, we discuss magnetic tweezers measurements of force-extension behavior, including quantitative details of methodology and data interpretation strategies that reveal information the conformations of biopolymers.

2. Design, synthesis, and attachment of biopolymers for magnetic tweezers experiments

2.1 Construction of long polymer chains

Very short polymers (contour length <200 nm, or ~500 protein residues) tend to be problematic for magnetic tweezers experiments due to systematic errors and nonspecific bead-surface adhesion (Neuman & Nagy, 2008). This presents a major challenge to studying IDPs and intrinsically disordered regions, whose sequences are typically 30–500 residues in length, much shorter than the effective limit of magnetic tweezers (Van Der Lee et al., 2014). Further, current solid-phase synthetic capabilities for peptides and peptoids are limited to lengths below 70-mers (Behrendt, White, & Offer, 2016; Zuckermann, Kerr, Kent, & Moos, 1992).

Long chain constructs can be formed from shorter chains through continuous end-to-end linkage of these biopolymers using a specific attachment chemistry (Figs. 2A and 2C). The synthesis or expression of these biopolymers must result in terminal functional groups that can enable such reactions. Generally, the polymerized biopolymers should reach at least 100 kDa in molecular weight (MW), corresponding to a ~900-residue protein with a contour length of ~350 nm, for successful magnetic tweezers experiments. In addition, the long biopolymer chains require terminal labeling with functional groups that facilitate attachments to glass surface and paramagnetic bead. The surface immobilization reaction can vary depending on the available functional groups on the

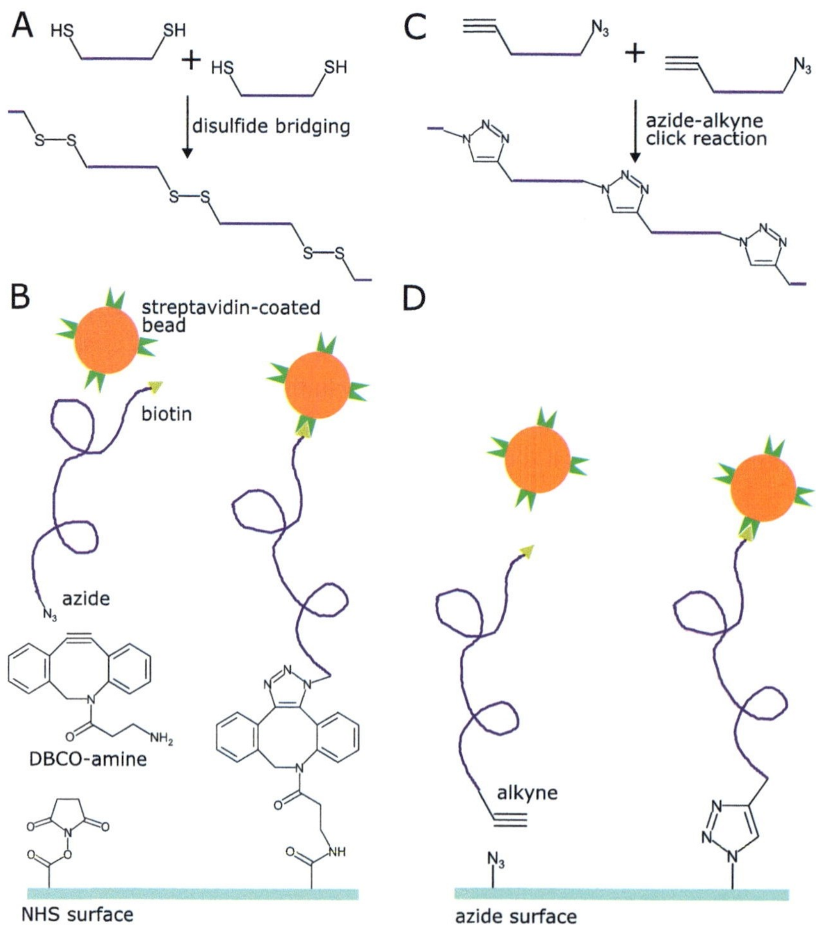

Fig. 2 Long polymer synthesis and tethering schemes for IDPs and polypeptoids. Figures are not drawn to scale. (A) Native IDP sequences can be modified to contain terminal cysteines. Long polyprotein constructs are formed through end-to-end disulfide bridging. (B) Tethering of polyproteins with terminal azide and biotin labeling. Enabled by NHS-amine click chemistry, an NHS-functionalized coverslip is treated with DBCO-amine, transforming the surface functionalization to DBCO. The azide label of polyproteins can then bind to the DBCO through a copper-free click reaction. A streptavidin-coated paramagnetic bead can attach to the biotin end of the bound polyprotein. (C) Synthetic peptoids with terminal azide and alkyne can use end-to-end click reactions to form long polymer chains. (D) Tethering of polypeptoids with terminal alkyne and biotin. The alkyne label can bind to an azide-functionalized surface using the same click chemistry. Streptavidin-biotin affinity allows attachment of a paramagnetic bead.

biopolymer (Figs. 2B and 2D). For bead attachment, the biopolymer typically is labeled with biotin, allowing high-affinity bonding to streptavidin-coated beads. We discuss below two chemistry strategies for the construction of long chains dependent on the polymer type.

2.1.1 Polyprotein construct by disulfide bridging

Recombinant expression in *Escherichia coli* (*E. coli*) is a simple method to generate larger proteins compared to sequence-controlled solid-phase synthesis; however, it is also more challenging to introduce unnatural side chains that are capable of click chemistry, such as maleimide, alkene, or azide (Francis & Page, 2010). Therefore, the most straightforward path is to include terminal thiol-containing cysteines, allowing disulfide bridge formation between individual proteins (Fig. 2A). However, disulfide bonds are rather unstable, especially in the presence of reducing agents. In addition, this linkage scheme can cause circularization, which reduces the yield of long polyproteins.

The protocol below assumes the protein sample, with terminal cysteines, has undergone proper cleavage and purification. The desired number of units per polyprotein chain is dependent on the MW of the protein (for example, a 10 kDa protein needs at least 10 units to form polyproteins above 100 kDa); therefore, the stoichiometry of labeling reagents must be adjusted accordingly. The amount of reagents should be approximately $10 * n_{\text{units per polyprotein}}$ times less than the concentration of protein in solution. For proteins with high sequence charge, salt can be used to screen electrostatic repulsion, allowing protein units to achieve the close contacts needed for disulfide bridging.

2.1.1.1 Materials

1. Protein of interest in solution (~100 μM)
2. Dithiothreitol (DTT)
3. 50 mM 2-(N-morpholino)ethanesulfonic acid (MES) buffer pH 7
4. Maleimide-biotin (Sigma-Aldrich, B1267)
5. Azido-PEG3-maleimide kit (BroadPharm, BP-22468)
6. Sodium chloride (NaCl)
7. Amicon Ultra-0.5 Centrifugal Filter Units (Millipore), one with a molecular weight cut-off (MWCO) of 100 kDa and another at below the MW of the protein
8. SDS-PAGE setup

2.1.1.2 Protocol

1. Form free thiols by reducing the protein with DTT.
2. Exchange protein buffer solution with 50 mM MES pH 7 using a centrifugal unit with a MWCO below the MW of the protein to fully remove DTT.
3. Add the appropriate amount of functional labels (maleimide-biotin and azido-PEG3-maleimide) to the protein sample. Add 500 mM NaCl to the protein sample (optional). The protein mixture is allowed one week for polymerization.
4. Perform SDS-PAGE to characterize reaction. Because of large polyprotein sizes, a 3% stacking acrylamide gel is suggested (Morgan, Avinery, Rahamim, Beck, & Saleh, 2020).
5. Clean protein mixture with a 100 kDa centrifugal filter unit to remove short polyproteins and unreacted labeling molecules. The desired polyproteins are at least 100 kDa in MW and have azide and biotin labels on each terminus.

2.1.2 Long polypeptoid constructs through end-to-end azide-alkyne click chemistry

Automated solid-phase synthesis of peptides and peptoids, despite yielding shorter unit sequences, allows introduction of various functional groups that facilitate click chemistry, which results in specific reactions that form stable bonds. In addition, having two different functional groups on each terminus imposes sequence unidirectionality to the long polymer constructs. However, such chemistry usually leads to the formation of linkages that are not orthogonal to amide bonds in these synthetic biopolymers, which can cause unwanted physical effects that confound experimental results (Bird, Lemmel, Yu, & Zhou, 2021). To confirm that the linkage chemistry does not influence their physical properties, control experiments are highly suggested. For example, the lengths of the unit polymers can be significantly varied to change the relative amount of linkages within the chain; any effects (or lack thereof) can be observed by comparing elastic behaviors between different constructs.

Particularly, copper(I)-catalyzed azide-alkyne cycloaddition has been explored as a robust click chemistry to form long polymer chains (Luo, Yang, Kopečková, & Kopeček, 2011; Murnen et al., 2012). In addition, the resulting 1,4-triazoles have been found to be compatible to amide bonds, owing to the similarities in polarity, planarity, and length (Angell & Burgess, 2007; Holub & Kirshenbaum, 2010). We discuss below the protocol to form long constructs of polypeptoids by exploiting the azide-alkyne click chemistry (Fig. 2C). This protocol assumes the unit polymers contain terminal azide and alkyne functional groups from the synthesis protocol.

2.1.2.1 Materials

1. Peptoids of interest, lyophilized
2. Copper(II) sulfate ($CuSO_4$)
3. Tris((1-benzyl-4-triazolyl)methyl)amine (TBTA)
4. Ascorbic acid
5. Triethylammonium acetate buffer pH 7
6. Amicon Ultra-0.5 Centrifugal Filter Units 100 kDa MWCO (Millipore, UFC5100)

2.1.2.2 Protocol

1. Prepare Cu(II)-TBTA complex stock solution by mixing equal volumes of $CuSO_4$ in water and TBTA in acetonitrile in 1:1 molar ratio.
2. In a pressure-tight vial or round-bottom flask with a stir bar, dissolve azide-alkyne polymers in water to a final reaction concentration of 1–5 mM.
3. For peptoids containing negative charges, add triethylammonium acetate buffer to make 0.2 M final concentration.
4. Add 5 molar equivalents (mol eq) of Cu(II)-TBTA complex.
5. Add 5 mol eq of ascorbic acid.
6. Seal the vial with a rubber septum. Bubble the reaction solution with nitrogen gas for at least 2 min.
7. Place the reaction container in a heat bath (bead bath) at 50 °C and stir for 24 h.
8. Collect the reaction solution and spin through 100 kDa MWCO centrifugal filter unit. Add water and spin several times until catalysts and low-MW unit polymers are sufficiently removed.

2.1.2.3 Tips

1. For best results, prior to reaction, peptoid units should be purified using analytical techniques.
2. Upon mixing all reagents and catalysts, if precipitation is observed, gently heat the reaction solution with a heat gun and vortex. Alternate between heat and vortex mixing until the solution becomes clear. Be careful to not cause the reaction solution to boil.

3. At the end of the reaction, dark precipitants (likely metallic Cu) might be present. Let the solids sediment, collect the supernatant, and proceed with the filtering step.
4. The concentrations of long polypeptoid constructs can be measured by UV-Vis spectroscopy.
5. Due to the large polydispersity, characterization using analytical techniques, such as liquid chromatography and mass spectrometry, can be challenging. As with the polyproteins described above, the presence of high-MW polymer chains can be observed with SDS-PAGE.

For magnetic tweezers experiments, the long polypeptoids can be labeled with a terminal biotin using the same click chemistry. Depending on the functionalization of the glass surface, either the azide or alkyne group on the polymer can remain. For example, if an azide-functionalized glass surface is used later, the azide ends of the long polymer chains can be capped with biotin, leaving the alkyne for surface attachment.

2.1.2.4 Materials

1. Long polypeptoid constructs with azide and alkyne terminal groups
2. Biotin-alkyne (Lumiprobe, C37B0)
3. CuSO4
4. Ascorbic acid
5. Amicon Ultra-0.5 Centrifugal Filter Units 100 kDa MWCO (Millipore, UFC5100)

2.1.2.5 Protocol

1. In an airtight vial or microcentrifuge tube, combine the polypeptoids (~0.2–2 μM final), biotin-alkyne (pre-dissolved in acetonitrile, 0.5 mM final), $CuSO_4$ (0.5 mM final), and ascorbic acid (0.5 mM final) in an aqueous solution.
2. Tightly close the reaction vial, wrap with parafilm if necessary, and agitate on a shaker or rotator for at least 1 h to overnight at room temperature.
3. Spin reaction solution through 100 kDa centrifugal filter unit. Add water and spin again until excess biotin has been sufficiently removed.

2.1.3 Tips

1. The amount of biotin-alkyne in the sample can be estimated from the dilution factor based on the spin time of the centrifugal filter unit. Spins should be repeated until the excess biotin concentration is <1 nM to prevent interference with surface attachment later.

2.2 Surface and bead attachment

Different attachment chemistry strategies can be applied depending on the polymer labeling and surface functionalization (Figs. 2B and 2D). Additionally, some protocols might be incompatible for certain biopolymers. For polyproteins with disulfide bridges, an attachment chemistry requiring a reducing agent, such as ascorbic acid, is unsuitable. For synthetic biopolymers, the linkage chemistry can be exploited again for surface immobilization.

The biopolymers are typically attached to paramagnetic beads through streptavidin-biotin bonding; another bead attachment chemistry would be adequate, as long as the long polymer constructs contain the appropriate labeling for such reaction. Specifically, for low-force elasticity measurements, $1\mu m$-diameter beads are used because they can consistent apply 0.1–10 pN of force on the polymers. Bead-polymer attachment is fast and straightforward.

Beyond attachment chemistry, additional consideration must be given to nonspecific adhesion, which reduces data throughput and measurement precision. Commercial functionalized surfaces are often coated with a hydrophilic thin brush layer or a polymer matrix to improve passivation. However, additional steps can be taken, if needed, to further prevent nonspecific adhesion. Bovine serum albumin (BSA) blocks hydrophobic areas on the surface, allowing biopolymers to correctly bind to attachment sites (Jeyachandran, Mielczarski, Mielczarski, & Rai, 2010). Surfactants, such as Tween-20, shield biopolymers from nonspecific interactions (Pan, Xia, Qin, Cao, & Wang, 2015). We include below several strategies to prevent nonspecific adhesion.

2.2.1 Flow cell construction

Here, we describe a primitive flow cell design for magnetic tweezers experiments. A flow channel is formed by sandwiching a parafilm cut-out between two glass coverslips. The bottom coverslip is functionalized with attachment chemistry, and it can either be purchased from a supplier or prepared in the laboratory. In addition, because single-molecule measurements are sensitive to external noise, inert reference beads are immobilized on the surface for noise subtraction.

2.2.1.1 Materials

1. Functionalized bottom glass coverslip ($25 * 60 * 0.1$ mm)
2. Top glass coverslip ($22 * 22 * 0.1$ mm)
3. Parafilm with channel cutout ($25 * 60$ mm in dimension, $6 * 30$ mm channel at the center, can be prepared with a laser cutter or a Cricut machine)

4. Polystyrene (PS) beads (~2.8 μm, suspended in 200-proof ethanol at 0.1 mg/mL, Bangs Laboratories, PS05002)
5. 2 M sodium hydroxide (NaOH)
6. Rain-X Original Glass Water Repellant

2.2.1.2 Protocol

1. Wash the top glass coverslip with NaOH, then thoroughly rinse with water. Use strong nitrogen flow to dry the coverslip.
2. Plasma clean the top glass coverslip for at least 5 min.
3. Apply Rain-X on the top side of the top glass coverslip and wipe off any excess.
4. Spread 50 μL of the PS bead suspension on the top side of the bottom glass coverslip. Allow the ethanol to fully evaporate in an enclosed container.
5. Place the bottom glass coverslip on a dust-free heat plate and lightly melt the PS beads at 100 °C for 1 min and 45 s
6. Assemble the coverslip. Melt the parafilm using a soldering iron at ~720°F (382 °C)

2.2.1.3 Tips

1. Functionalized glass coverslips must be stored in a vacuum desiccator in the dark to preserve reactivity.
2. Coverslip and parafilm sizes can be adjusted based on experimental requirements. Channel width must be below 10 mm to ensure laminar flow.
3. The purpose of Rain-X is to create a hydrophobic top surface, forcing solutions to enter the flow channel instead of spreading on top; Sigmacote can be used in place of Rain-X. Rain-X should fully cover the top surface of the top glass coverslip, but it must not be present on any surfaces inside the flow channel.
4. PS beads must be suspended in 200-proof ethanol to prevent residues upon evaporation.
5. Bead melting time can be adjusted to fit specific laboratory conditions. Ideally, beads should be fixed firmly onto the glass surface but still retain a spherical shape for optical image tracking.
6. Generally, experiments must be done in a clean, dust-free environment. Gloves must be worn during flow cell construction.

2.2.2 Polyprotein attachment

Surface immobilization of polyproteins with a terminal azide labeling can be done through the biorthogonal click reaction with dibenzocyclooctyne

(DBCO), which does not require Cu catalyst in a reducing environment. Because commercial functionalized coverslips typically do not offer DBCO coating, one additional step before polymer incubation is necessary to convert an available surface chemistry to DBCO (Fig. 2B). Alternatively, DBCO-functionalized coverslips can be self-prepared using several published protocols (Dadfar, Sekula-Neuner, Bog, Trouillet, & Hirtz, 2018; Zimmermann, Nicolaus, Neuert, & Blank, 2010).

2.2.2.1 Materials

1. Long polyproteins with azide and biotin terminal labels (in 10 mM MES buffer pH 7, 50 mM NaCl, 0.1% Tween-20)
2. N-hydroxysuccinimide ester (NHS) functionalized coverslip (PolyAn 3D-NHS, 104 00 407)
3. DBCO-amine (dissolved in DMSO at 10 mg/mL, BroadPharm, BP-22066)
4. 1X phosphate-buffered saline (PBS) pH 8.4
5. 10 mM tris buffer pH 7.4, 0.1% Tween-20
6. 10 mM tris buffer pH 7.4, 1% BSA, 0.1% Tween-20
7. 10 mM MES buffer pH 7, 50 mM NaCl, 0.1% Tween-20

2.2.2.2 Protocol

1. Construct a flow cell from an NHS-functionalized coverslip following the protocol in Section 2.2.1.
2. Mix equal volumes of DBCO-amine and PBS pH 8.4 and flow 40 μL of that solution into channel. Incubate at room temperature for 1 h in a humidified chamber.
3. Rinse the channel thoroughly with 400 μL of tris buffer and let incubate for 15 min at room temperature to neutralize any unreacted NHS sites.
4. Rinse the channel thoroughly with 400 μL of MES buffer.
5. Flow in polyprotein sample (final concentration in channel of 10–100 nM). Incubate at 4 °C for 12–24 h in a humidified chamber.
6. Rinse the channel thoroughly with 800 μL of MES buffer.
7. Flow 200 μL of BSA-containing buffer and let incubate for 1 h at room temperature.
8. Rinse the channel thoroughly with 800 μL of MES buffer.

2.2.2.3 Tips

1. Aggregation of polyprotein samples reduces binding efficiency; de-aggregation is necessary if this is an issue.

2. Because NHS is sensitive to hydrolysis, prevent contact with moisture prior to DBCO incubation.
3. Avoid any amine-containing buffers and chemicals prior to DBCO incubation.
4. The reaction of NHS to amine is pH-sensitive and most efficient at pH 8.3–8.5.
5. Avoid azide-containing buffers, which can reduce reactivity of DBCO.

2.2.3 Attachment of long polypeptoids

2.2.3.1 Materials

1. Long polypeptoid constructs with alkyne and biotin terminal labels
2. Azide functionalized coverslip (PolyAn 2D-Azide, 104 00 626)
3. CuSO4
4. Ascorbic acid
5. 1X PBS pH 7.4, 0.1% Tween-20
6. 10 mM tris buffer pH 7.4, 0.1% Tween-20
7. 10 mM tris buffer pH 7.4, 1% BSA, 0.1% Tween-20

2.2.3.2 Protocol

1. Construct a flow cell from an azide-functionalized coverslip following the protocol in Section 2.2.1.
2. Rinse the flow cell with 400 μL of PBS buffer without BSA.
3. In a microcentrifuge tube, mix the long polymer constructs (10–100 nM final), $CuSO_4$ (2.5 mM final), ascorbic acid (2.5 mM final), and PBS without BSA to a final volume of 20 μL.
4. Flow the long polymer solution into the channel. Incubate at room temperature for 2 h in a humidified chamber.
5. Rinse the channel thoroughly with 800 μL of tris buffer.
6. Flow in 200 μL of the BSA-containing buffer. Let incubate at room temperature for 1 h in a humidified chamber.
7. Rinse the channel thoroughly with 800 μL of tris buffer.

2.2.4 Bead attachment

The size of the attached paramagnetic bead controls the magnitude of forces applied on the polymer chain. While a bead with a 1 μm diameter can apply a force range of 0.01–20 pN, a higher range (1–100 pN) can be achieved with a 2.8 μm-diameter bead. For measurements of sub-picoNewton elasticity, the former is thus preferred. Of the commercially available paramagnetic bead options, we strongly recommend Invitrogen Dynabeads

MyOne Streptavidin C1 (1 μm diameter), as it offers the smallest polydispersity (<5%), ensuring the lowest variation in forces between polymer tethers. In addition, the surface chemistry of these beads is carboxylic acid, providing hydrophilicity and preventing nonspecific adhesion.

2.2.4.1 Materials

1. Dynabeads MyOne Streptavidin C1 (Invitrogen, 65001)
2. 10 mM tris buffer pH 7.4, 1% BSA, 0.1% Tween-20
3. Experimental buffer of choice

2.2.4.2 Protocol

1. Wash 2 μL beads at least 2 times, each with tris BSA buffer.
2. Resuspend beads in 40 μL tris BSA buffer for a final concentration of 0.5 mg/mL.
3. Incubate 10 μL of the bead solution in the flow cell for 10 min.
4. Rinse beads off the channel thoroughly with the experimental buffer of choice (at least 1 mL).

2.2.4.3 Tips

1. If the carboxylic acid surface of C1 beads is incompatible with the experiment, the T1 beads with tosylactivated, hydrophobic surface can be an alternative.
2. Beads must be added from the outlet opposite from the buffer inlet to allow complete removal of unbound beads.
3. Bead incubation can be observed on the microscope to make sure that beads have significantly covered the glass surface; however, the magnets must not be present during incubation.
4. If the beads are left in the channel for an extended amount of time, they can nonspecifically adhere to the surface.

3. Force-extension measurements using magnetic tweezers

Single-molecule elastic measurements enabled by magnetic tweezers provides novel insights to biopolymer conformations. Specifically, interpreting force-extension data using polymer physics theories on force-extension measurements can reveal key parameters of biopolymer physical properties (Saleh, 2015). To generate force-extension curves, displacements of tethered beads are measured at decreasing discrete forces. We present below strategies for bead selection, force-extension measurements, and data analysis. We do not discuss instrumental aspects of magnetic tweezers, such as microscopic setup and image processing for three-dimensional bead tracking, as these topics have been elaborated in previous articles (Choi, Kim, Shon, & Yoon, 2022; Gosse & Croquette, 2002; Ribeck & Saleh, 2008; Seol & Neuman, 2011).

3.1 Bead selection

After successful biopolymer and bead attachments, tethers must be examined prior to force-extension measurements to prevent factors that complicate experimental results. Only tethers that consist of a single biopolymer without significant nonspecific adhesion can be considered. These criteria can be achieved through proper experimental setup, as discussed in Section 2.2. First, surface passivation with BSA and Tween-20 can prevent nonspecific adhesion. Second, the biopolymer concentration during flow cell incubation must be low enough to allow enough spacing between attachments and reduce the probability of more than one chain binding to the same bead. Besides preparative methods, we discuss ways to screen tethers and select a field-of-view.

Generally, tethers are identified by their diffraction patterns, which indicate that the beads are at a higher z-position under force compared to those that are on the glass surface due to the extension of the polymeric tether. To be able to easily detect such tethers qualitatively, the applied force should be at least 3 pN to create significant differences in height between attached and nonspecifically stuck beads. In addition, for accurate tracking, there should be enough distance between beads so that diffraction rings do not overlap.

3.1.1 Protocol

1. Under a force >3 pN, search for a field-of-view that contains biopolymer tethers and at least 3 reference beads.
2. Vary the applied forces between 0.1 and 10 pN, observe any change in z-position through the diffraction patterns of tether beads. If the tethers show significant response with changing forces, the beads should be selected. If the beads do not show any changes in height under force variation, they should be ignored.
3. To check for double tethers (two polymers bound to one bead), at a force of 2 pN, rotate the magnets at least 100 times to apply torque on the beads. If a bead is a double tether, its attached polymer chains will be braided, and the bead will collapse to the surface.

3.2 Force-extension measurements

As the applied tension on a polymer is sensitive to the paramagnetic bead size, the non-negligible polydispersity of commercial beads requires accurate force calibration of individual tethers. Through the Langevin equation, the motion of a tethered bead under an applied force can be modeled as a harmonic oscillator subject to stochastic thermal fluctuations (Berg-Sørensen & Flyvbjerg, 2004). From the time trajectory of a bead under a constant tension, this analysis allows direct estimation of force for individual tethers. This method bypasses the need for a calibration curve $f(Z_{mag})$, where the force is determined as a function of the magnet position Z_{mag}. The power spectral density and Allan variance analyses of bead time trajectory for force calibration have been previously described (Lansdorp & Saleh, 2012). Within the Python programming language, the calculation of force from experimental data is simplified by the use of the open package Tweezepy (Morgan & Saleh, 2021). From a corrected time trajectory, Tweezepy applies maximum likelihood estimation to yield fitting parameters, notably the tether spring constant κ, through either power spectral density or Allan variance analysis (Morgan & Saleh, 2021).

To obtain a force-extension curve, the extension of a polymer is measured at decreasing steps of discrete forces. By moving the magnets away in equal increments, an exponential ladder of forces can be achieved (Fig. 3A). In addition, due to the timescale of bead fluctuations, as modeled

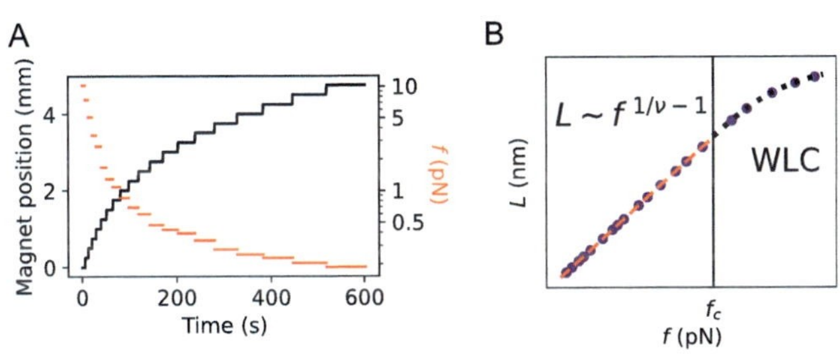

Fig. 3 Magnetic tweezers force-extension measurement. (A) Magnet position (black) and corresponding forces (red) to generate a force-extension curve. Starting from the closest position to the flow cell (0 mm), magnets are moved away in successive equal steps. For accurate calibration of low forces, the tracking time at each step is increased by a factor of 1.15–1.175 from the previous step. (B) A typical force-extension curve of a flexible disordered polyprotein chain. Two distinct elastic regimes can be observed. At higher forces, the asymptotic regime can be fitted by the WLC mode (black dotted line). At low forces, the force-extension follows a power law as predicted by the Pincus blob model (red dashed line). The crossover force f_c marks the transition between the two elastic regimes.

by the Langevin equation, the length of time spent at each force must be longer at lower forces for accurate force estimation. An "acquisition scaling" factor can be applied to increase the trace length from the previous trace multiplicatively. The table below provides general guidance on the parameters for a single force-extension curve; depending on the application and experimental setup, these values may be tuned through trial and error.

Parameters	Value
number of forces	20–31
acquisition scaling	1.15–1.175
magnets step size (mm)	−0.25
total acquisition time	10–20 min

3.2.1 Protocol

1. Before starting a force-extension measurement, move the magnets to the closest position to the flow cell to apply the highest force.
2. Measure the force-extension response of biopolymer tethers following the parameters above, along with any necessary modifications.
3. To measure the same tethers in a different solution condition, move the magnets to apply a force of 2 pN. Wash flow channel with at least 200 μL of solution of interest.
4. In the case of small displacements during channel washing, readjust the position of the field-of-view.
5. Check the tethers for signs of nonspecific adhesion. If no such interactions occur, proceed with force-extension measurements in the new solution condition.

In single-molecule force microscopy literature, force-extension curves are usually visualized as force as a function of extension. However, in the protocol that we presented, changes in the extension are observed at constant, calibrated forces. Therefore, we plot force-extension curves as extension as a function of force.

3.3 Analysis of force-extension results

Single-molecule elasticity, quantified by force-extension curves, provides insights to the conformations of biopolymers on a wide range of length scales. By fitting equations derived from polymer physics models to the force-extension curve, different parameters of chain conformation, such as persistence length l_p and Flory exponent ν, can be obtained (Fig. 3B). Further, comparing elastic responses in changing solution environments can reveal information on intramolecular interactions, such as electrostatic repulsion and hydrophobic attraction.

Applying force to a biopolymer chain divides it into statistical segments of size given by the "tensile screening length", $\xi = k_B T/f$ (Pincus, 1976). Within this length scale, the chain conformation fluctuates with thermal energy; beyond it, the chain aligns in the direction of the force. For flexible biopolymers (l_p in the order of 1 nm), the typical experimental force range (f = 0.1–20 pN) allows access to two distinct force regimes: an intermediate-force asymptote and a low-force Pincus blobs regime. At sub-picoNewton forces, the polymer sees significant loops and bends; the chain conformation within a tensile blob follows the Flory scaling law, $R \sim N^\nu$, where R is extent of the blob and N is the number of statistical monomers within the blob. As the applied force increases, ξ becomes shorter, and long-range monomer-monomer interactions become disallowed as the polymer chain is stretched to near its contour length, L_c. The crossover between these two regimes occurs when the tensile screening length is equal to the statistical monomer Kuhn length, l_K, corresponding to a crossover force of $f_c \sim k_B T/l_K$. We discuss below strategies to apply the appropriate polymer physics models to these two force regimes and extract conformational parameters.

3.3.1 Persistence length: WLC model at intermediate force

The force-extension relationship in the asymptotic stretching regime is typically described by the freely joined chain or worm-like chain (WLC) models (James & Guth, 1943; Kuhn & Grün, 1942; Marko & Siggia, 1995). Particularly, the latter has been found useful in describing the entropic elasticity of semiflexible biopolymers, such as double-stranded nucleic acid and hyaluronic acid (Berezney & Saleh, 2017; Bouchiat et al., 1999). However, for flexible biopolymers with fixed bond angles and bond lengths, the WLC model has been shown to be inexact (Schwarzl, Liese, Brünig, Laudisio, & Netz, 2020). Nevertheless, the WLC model has been used for IDPs and polypeptides, and it is a good starting point to describe the elastic response at intermediate-to-high force of many biopolymers (Singh & Lapidus, 2008; Solanki et al., 2014). If the exact contour length of a polyprotein or polypeptoid tether is known, a modified freely rotating chain model has been demonstrated to be more accurate than the WLC model (Schwarzl et al., 2020).

The WLC model describes the asymptotic behavior through the approximate force-extension relationship, $L \approx L_c(1 - \sqrt{k_B T / 4 f l_p})$ (Marko & Siggia, 1995). The WLC fitting thus consists of two independent parameters, L_c and l_p. For a more robust fitting, a more precise numerical formula should be used, such as one derived by Bouchiat (Bouchiat et al., 1999). From force-extension curves, only points above the crossover force, $f_c = k_B T / l_K = k_B T / 2 l_p$, should be included in this fitting; however, this requires a priori knowledge of the l_p value of the biopolymer of interest. Initial estimation of the cutoff force can be calculated from the predicted persistence length; for flexible biopolymers such as IDPs and polypeptides, l_p is typically close to 1 nm, corresponding to $f_c = 2\text{pN}$ (Buscaglia, Lapidus, Eaton, & Hofrichter, 2006; Chin, Toptygin, Elam, Schrank, & Hilser, 2016; Murnen et al., 2013; Rosales et al., 2012; Singh & Lapidus, 2008). Based on goodness-of-fit determination, such as a residuals plot, the cutoff force can be adjusted as needed. Generally, we have found that the fitting result is not significantly sensitive to the cutoff choice if it is reasonably close to the actual crossover force.

Comparison of fitting results on an individual biopolymer chain measured in different solution conditions provides insights to its inter- and intramolecular interactions. For example, measurements in salt (such as sodium chloride, NaCl) reveal electrostatic interactions, while denaturants (such as guanidinium hydrochloride, GuHCl) elicit responses driven by hydrophobicity. However, there are several considerations when performing this analysis. Solutions with high solute concentrations have higher refractive indices compared to buffers at low ionic strengths, which can alter measurements of absolute extensions; thus bead tracking measurements (which are sensitive to the solution refractive index) need to be corrected for any refractive index changes. In addition, data stochasticity can lead the fitting of individual force-extension curves in different solution conditions to output small variations in L_c, despite being from the same polymer chain. To factor for this, a global WLC fitting of all curves can be performed in which there are $x + 1$ independent parameters: one globally-fitted contour length L_c and x persistence length variables, one for each of the x different solution conditions.

3.3.2 Flory exponent: Pincus blob model at low force

The conformational behavior of a polymer chain at low force can be described as a chain of Pincus blobs of size ξ aligned in the direction of the applied tension (Pincus, 1976; Rubinstein & Colby, 2003). Within these blobs, the polymer configuration is a random walk with a Flory exponent ν. In this regime, the force-extension relationship follows the scaling model $L \sim f^{\gamma}$, with an exponent that is directly related to the Flory scaling, $\gamma = 1/\nu - 1$. This exponent can be extracted from a force-extension curve by performing a power-law fitting of data points below the crossover force; whichever cutoff force choice is used for WLC fitting can be reapplied for this regime. As a precaution, this fitting is sensitive to outliers. In force-extension curves, they are typically present in the lowest forces; as biopolymers are most susceptible to nonspecific adhesion at such forces, extension measurements can be significantly affected. Therefore, for accurate estimate of exponents, the data quality must be examined, and points that show evidence of surface interactions should be discarded.

4. Application to the study of a disordered neurofilament subunit protein tail region

Within the axon of myelinated neurons, neurofilaments are a critical component of the cytoskeletal network that maintains cellular mechanical stability (Fuchs & Cleveland, 1998). Neurofilaments are composed of low-, medium-, and high-molecular-weight subunit proteins, all of which features an α-helical N-terminus and unstructured C-terminal tail. Upon self-assembly of these subunit proteins, the folded domains form the rigid core of a bottlebrush, with the disordered tails protruding from the core. "Handshake" interactions, or crossbridging, between these sidearms control inter-filament interactions, as well as the organization of larger neurofilament hydrogel networks (Beck, Deek, Jones, & Safinya, 2010; Beck, Deek, & Safinya, 2012). Previous studies have linked aberrations in side arm interactions to motor-neuron diseases, such as Parkinson's and amyotropic lateral sclerosis (Liu et al., 2004; Xu, Cork, Griffin, & Cleveland, 1993). Therefore, developing a thorough understanding of the conformational behavior of the disordered neurofilament tail region, especially using orthogonal experimental techniques, is of particular interest. Magnetic tweezers have been demonstrated to be able to provide a unique perspective on this disordered protein region through entropic elasticity measurements (Truong, Morgan et al., 2023).

4.1 Neurofilament tail polyprotein construct and magnetic tweezers experiment

A 163-residue construct from the neurofilament low-molecular-weight subunit protein C-terminal tail domain (NFLt) was synthesized through recombinant protein expression in E. coli with the appropriate encoding gene. The sequence was modified to included terminal cysteines. Following the protocols described in Section 2.1.1, long polyprotein constructs of NFLt with terminal azide and biotin labels were generated. For magnetic tweezers experiments, the polyproteins were immobilized (Section 2.2.2) in a flow cell (Section 2.2.1), and paramagnetic beads were attached (Section 2.2.4). Force-extension curves of this polyprotein constructs in different solution conditions were obtained through magnetic tweezers measurements, as discussed in Section 3.

4.2 NFLt conformational characterization through analysis of force-extension curves

The force-extension curve of NFLt in the low-to-intermediate force range shows two distinct elastic regimes: Pincus blobs and asymptotic stretching. The crossover force between these two regimes is approximately 2 pN, when the tensile screening length is equal to the Kuhn length of the polyprotein (~2 nm). At forces above 2 pN, the polyprotein's elastic behavior is well-described by the WLC model, revealing a persistence length l_p of 1 nm. This value is within the range of reported persistence lengths of various unstructured IDPs and polypeptides, suggesting that the polyprotein tether is an individual chain and not an oligomerized aggregate. As mentioned in Section 3.3.1, the WLC model is not optimized to describe protein elasticity due to its ignorance of specific peptide bond lengths and bond angles. Nevertheless, unless the exact persistence length is required, the simplicity of the WLC model allows decently accurate estimation of polyprotein flexibility.

The NFLt construct is a disordered protein region with significant charge content (net charge per residue: −0.23). From an intuitive prediction based on its sequence, the NFLt would assume a self-avoiding random walk conformation ($\nu = 0.6$), similar to a random-coil unfolded polypeptide, due to intramolecular electrostatic repulsion. However, as shown in Fig. 4, in low ionic strength buffer solution, the configuration of NFLt is surprisingly near that of an ideal chain, with a Flory exponent $\nu = 0.53 \pm 0.01$ extracted from Pincus blob analysis of the force-extension curve below 2 pN (Section 3.3.2). This deviation from self-avoiding chain behavior suggests that there are attractive forces counterbalancing electrostatic repulsion, bringing the net intramolecular interaction to nearly zero.

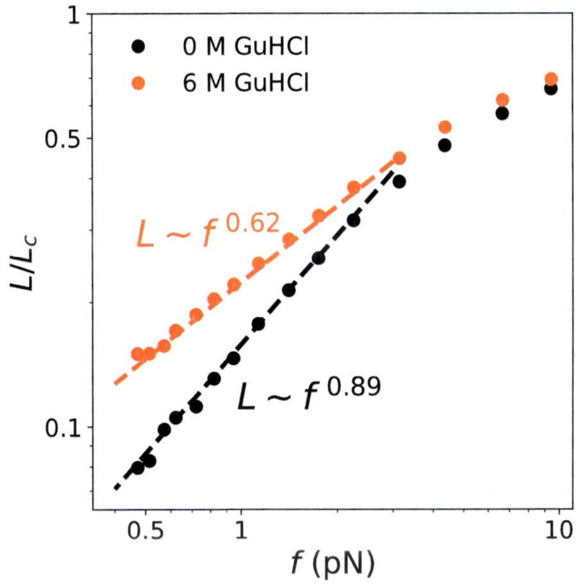

Fig. 4 Force-extension curves of a single NFLt polyprotein in 0 M GuHCl (10 mM MES buffer pH 7, black) and 6 M GuHCl (red). The extensions are normalized by the contour length L_c from WLC fitting. At 0 M GuHCl, the force-extension power law is 0.89 ± 0.01, corresponding to a Flory exponent $\nu = 0.53 \pm 0.01$. At 6 M GuHCl, this power law is instead 0.62 ± 0.01, indicating a Flory exponent $\nu = 0.62 \pm 0.02$. *Figure is adopted from Truong, H. P., Morgan, I. L., Koren, G., Avinery, R., Beck, R., & Saleh, O. A. (2023). Pincus blob elasticity in an intrinsically disordered protein. The European Physical Journal E, 46(100). https://doi.org/10.1140/epje/s10189-023-00360-0.*

The nature of such attractive interactions can further be probed by measuring the polyprotein's elastic response in denaturant. On the same polyprotein tether in strong denaturant (6 M GuHCl), NFLt expands and behaves as a self-avoiding random walk, indicating that these attractions are removed with increasing solvent quality (Fig. 4). From this observation, it can be hypothesized that these interactions are likely disordered hydrophobic globules. Indeed, another magnetic tweezers study on the NFLt found glassy dynamics in a similar construct that were attributed to such interactions (Morgan et al., 2020). Generally, the study on NFLt demonstrates the value of magnetic tweezers in providing a unique approach to the study of IDP and polypeptide conformation: important structural parameters can be extracted from entropic elasticity measurements through simple polymer physics models.

5. Conclusions

In this article, we have discussed strategies to apply magnetic tweezers to the investigation of the conformations and intramolecular interactions of biopolymers, particularly for IDPs and peptoids. To overcome the short chain size of natural protein sequences and synthetic biopolymers, we have described an end-to-end attachment chemistry that can be used to form long polymers suitable for stretching experiments. From force-extension curves, fitting of WLC and Pincus blob models reveals biopolymer physics parameters, such as persistence length and Flory exponent, that describe chain conformation. In addition, comparing measurements in different solution conditions provides insights to their intramolecular and polymer-solvent interactions. Overall, single-molecule entropic elasticity measurements offer a unique perspective in the study of the polymeric aspects of biomolecular behavior, and we suggest this approach supplies a useful complement to scattering and fluorescence microscopy techniques.

References

Angell, Y. L., & Burgess, K. (2007). Peptidomimetics via copper-catalyzed azide–alkyne cycloadditions. *Chemical Society Reviews, 36*(10), 1674–1689. https://doi.org/10.1039/b701444a.

Bae, W., Kim, K., Min, D., Ryu, J. K., Hyeon, C., & Yoon, T. Y. (2014). Programmed folding of DNA origami structures through single-molecule force control. *Nature Communications, 5*(1), 8. https://doi.org/10.1038/ncomms6654.

Beck, R., Deek, J., Jones, J. B., & Safinya, C. R. (2010). Gel-expanded to gel-condensed transition in neurofilament networks revealed by direct forcemeasurements. *Nature Materials, 9*(1), 40–46. https://doi.org/10.1038/nmat2566.

Beck, R., Deek, J., & Safinya, C. R. (2012). Structures and interactions in "bottlebrush" neurofilaments: The role of charged disordered proteins in forming hydrogel networks. *Biochemical Society Transactions, 40*(5), 1027–1031. https://doi.org/10.1042/BST20120101.

Behrendt, R., White, P., & Offer, J. (2016). Advances in Fmoc solid-phase peptide synthesis. *Journal of Peptide Science, 22*(1), 4–27. https://doi.org/10.1002/psc.2836.

Berezney, J. P., & Saleh, O. A. (2017). Electrostatic effects on the conformation and elasticity of hyaluronic acid, a moderately flexible polyelectrolyte. *Macromolecules, 50*(3), 1085–1089. https://doi.org/10.1021/acs.macromol.6b02166.

Berg-Sørensen, K., & Flyvbjerg, H. (2004). Power spectrum analysis for optical tweezers. *Review of Scientific Instruments, 75*(3), 594–612. https://doi.org/10.1063/1.1645654.

Bernadó, P., & Svergun, D. I. (2012). Structural analysis of intrinsically disordered proteins by small-angle X-ray scattering. *Molecular Biosystems, 8*(1), 151–167. https://doi.org/10.1039/c1mb05275f.

Bird, R. E., Lemmel, S. A., Yu, X., & Zhou, Q. A. (2021). Bioorthogonal chemistry and its applications. *Bioconjugate Chemistry, 32*(12), 2457–2479. https://doi.org/10.1021/acs.bioconjchem.1c00461.

Bouchiat, C., Wang, M. D., Allemand, J. F., Strick, T., Block, S. M., & Croquette, V. (1999). Estimating the persistence length of a worm-like chain molecule from force-extension measurements. *Biophysical Journal, 76*(1 I), 409–413. https://doi.org/10.1016/S0006-3495(99)77207-3.

Brucale, M., Schuler, B., & Samorì, B. (2014). Single-molecule studies of intrinsically disordered proteins. *Chemical Reviews, 114*(6), 3281–3317. https://doi.org/10.1021/cr400297g.

Buscaglia, M., Lapidus, L. J., Eaton, W. A., & Hofrichter, J. (2006). Effects of denaturants on the dynamics of loop formation in polypeptides. *Biophysical Journal, 91*(1), 276–288. https://doi.org/10.1529/biophysj.105.071167.

Bustamante, C., Smith, S. B., Liphardt, J., & Smith, D. (2000). Single-molecule studies of DNA mechanics. *Current Opinion in Structural Biology, 10*, 279–285. ⟨http://apps.who.int/medicinedocs/en/d/Js6169e/5.4.html⟩.

Chang, L. W., Lytle, T. K., Radhakrishna, M., Madinya, J. J., Vélez, J., Sing, C. E., & Perry, S. L. (2017). Sequence and entropy-based control of complex coacervates. *Nature Communications, 8*(1), 1–7. https://doi.org/10.1038/s41467-017-01249-1.

Chin, A. F., Toptygin, D., Elam, W. A., Schrank, T. P., & Hilser, V. J. (2016). Phosphorylation increases persistence length and end-to-end distance of a segment of Tau protein. *Biophysical Journal, 110*(2), 362–371. https://doi.org/10.1016/j.bpj.2015.12.013.

Choi, H., Kim, H. G., Shon, M. J., & Yoon, T. (2022). High-resolution magnetic tweezers. *Annual Review of Biochemistry, 91*, 33–59.

Dadfar, S. M. M., Sekula-Neuner, S., Bog, U., Trouillet, V., & Hirtz, M. (2018). Site-specific surface functionalization via microchannel cantilever spotting (μCS): Comparison between Azide–alkyne and thiol–alkyne click chemistry reactions. *Small (Weinheim an der Bergstrasse, Germany), 14*(21), https://doi.org/10.1002/smll.201800131.

De Vlaminck, I., & Dekker, C. (2012). Recent advances in magnetic tweezers. *Annual Review of Biophysics, 41*(1), 453–472. https://doi.org/10.1146/annurev-biophys-122311-100544.

DeGrado, W. F., & Korendovych, I. V. (2020). De novo protein design, a retrospective. *Quarterly Reviews of Biophysics, 53*(e3), 1–33. https://doi.org/10.1017/S0033583519000131.

Francis, D. M., & Page, R. (2010). Strategies to optimize protein expression in *E. coli*. *Current Protocols in Protein Science, 61*(1), 29. https://doi.org/10.1002/0471140864.ps0524s61.

Fuchs, E., & Cleveland, D. W. (1998). A structural scaffolding of intermediate filaments in health and disease. *Science (New York, N.Y.), 279*(5350), 514–519. https://doi.org/10.1126/science.279.5350.514.

Gosse, C., & Croquette, V. (2002). Magnetic tweezers: Micromanipulation and force measurement at the molecular level. *Biophysical Journal, 82*(6), 3314–3329. https://doi.org/10.1016/S0006-3495(02)75672-5.

Haber, C., & Wirtz, D. (2000). Magnetic tweezers for DNA micromanipulation. *Review of Scientific Instruments, 71*(12), 4561–4570. https://doi.org/10.1063/1.1326056.

Holub, J. M., & Kirshenbaum, K. (2010). Tricks with clicks: Modification of peptidomimetic oligomers via copper-catalyzed azide-alkyne [3 + 2] cycloaddition. *Chemical Society Reviews, 39*(4), 1325–1337. https://doi.org/10.1039/b901977b.

Hughes, M. L., & Dougan, L. (2016). The physics of pulling polyproteins: A review of single molecule force spectroscopy using the AFM to study protein unfolding. *Reports on Progress in Physics, 79*(7), https://doi.org/10.1088/0034-4885/79/7/076601.

James, H. M., & Guth, E. (1943). Theory of the elastic properties of rubber. *The Journal of Chemical Physics, 11*(10), 455–481. https://doi.org/10.1063/1.1723785.

Jeyachandran, Y. L., Mielczarski, J. A., Mielczarski, E., & Rai, B. (2010). Efficiency of blocking of non-specific interaction of different proteins by BSA adsorbed on hydrophobic and hydrophilic surfaces. *Journal of Colloid and Interface Science, 341*(1), 136–142. https://doi.org/10.1016/j.jcis.2009.09.007.

Kosol, S., Contreras-Martos, S., Cedeño, C., & Tompa, P. (2013). Structural characterization of intrinsically disordered proteins by NMR spectroscopy. *Molecules (Basel, Switzerland), 18*(9), 10802–10828. https://doi.org/10.3390/molecules180910802.

Kosuri, S., & Church, G. M. (2014). Large-scale de novo DNA synthesis: Technologies and applications. *Nature Methods, 11*(5), 499–507. https://doi.org/10.1038/nmeth.2918.

Kriegel, F., Ermann, N., & Lipfert, J. (2017). Probing the mechanical properties, conformational changes, and interactions of nucleic acids with magnetic tweezers. *Journal of Structural Biology, 197*(1), 26–36. https://doi.org/10.1016/j.jsb.2016.06.022.

Kuhn, W., & Grün, F. (1942). Beziehungen zwischen elastischen Konstanten und Dehnungsdoppelbrechung hochelastischer Stoffe. *Kolloid-Zeitschrift, 101*(3), 248–271. https://doi.org/10.1007/BF01793684.

Lansdorp, B. M., & Saleh, O. A. (2012). Power spectrum and Allan variance methods for calibrating single-molecule video-tracking instruments. *Review of Scientific Instruments, 83*(2), 0–10. https://doi.org/10.1063/1.3687431.

Liu, Q., Xie, F., Siedlak, S. L., Nunomura, A., Honda, K., Moreira, P. I., ... Perry, G. (2004). Neurofilament proteins in neurodegenerative diseases. *Cellular and Molecular Life Sciences, 61*(24), 3057–3075. https://doi.org/10.1007/s00018-004-4268-8.

Luo, K., Yang, J., Kopečková, P., & Kopeček, J. (2011). Biodegradable multiblock poly[N-(2-hydroxypropyl)methacrylamide] via reversible addition-fragmentation chain transfer polymerization and click chemistry. *Macromolecules, 44*(8), 2481–2488. https://doi.org/10.1021/ma102574e.

Marko, J. F., & Siggia, E. D. (1995). Stretching DNA. *Macromolecules, 28*(26), 8759–8770. https://doi.org/10.1021/ma00130a008.

Marsh, J. A., & Forman-Kay, J. D. (2010). Sequence determinants of compaction in intrinsically disordered proteins. *Biophysical Journal, 98*(10), 2383–2390. https://doi.org/10.1016/j.bpj.2010.02.006.

Millar, D. P. (1996). Fluorescence studies of DNA and RNA structure and dynamics. *Current Opinion in Structural Biology, 6*(3), 322–326. https://doi.org/10.1016/S0959-440X(96)80050-9.

Morgan, I. L., Avinery, R., Rahamim, G., Beck, R., & Saleh, O. A. (2020). Glassy dynamics and memory effects in an intrinsically disordered protein construct. *Physical Review Letters, 125*(5), 58001. https://doi.org/10.1103/PhysRevLett.125.058001.

Morgan, I. L., & Saleh, O. A. (2021). Tweezepy: A Python package for calibrating forces in single-molecule video-tracking experiments. *PLoS One, 16*(12 December), 1–18. https://doi.org/10.1371/journal.pone.0262028.

Murnen, H. K., Khokhlov, A. R., Khalatur, P. G., Segalman, R. A., & Zuckermann, R. N. (2012). Impact of hydrophobic sequence patterning on the coil-to-globule transition of protein-like polymers. *Macromolecules, 45*(12), 5229–5236. https://doi.org/10.1021/ma300707t.

Murnen, H. K., Rosales, A. M., Dobrynin, A. V., Zuckermann, R. N., & Segalman, R. A. (2013). Persistence length of polyelectrolytes with precisely located charges. *Soft Matter, 9*(1), 90–98. https://doi.org/10.1039/c2sm26849c.

Neuman, K. C., & Nagy, A. (2008). Single-molecule force spectroscopy: Optical tweezers, magnetic tweezers and atomic force microscopy. *Nature Methods, 5*(6), 491–505. https://doi.org/10.1038/nmeth.1218.

Numata, K. (2020). How to define and study structural proteins as biopolymer materials. *Polymer Journal, 52*(9), 1043–1056. https://doi.org/10.1038/s41428-020-0362-5.

Oldfield, C. J., & Dunker, A. K. (2014). Intrinsically disordered proteins and intrinsically disordered protein regions. *Annual Review of Biochemistry, 83*(1), 553–584. https://doi.org/10.1146/annurev-biochem-072711-164947.

Pan, H., Xia, Y., Qin, M., Cao, Y., & Wang, W. (2015). A simple procedure to improve the surface passivation for single molecule fluorescence studies. *Physical Biology, 12*(4), 0–7. https://doi.org/10.1088/1478-3975/12/4/045006.

Perry, S. L., & Sing, C. E. (2020). 100th anniversary of macromolecular science viewpoint: Opportunities in the physics of sequence-defined polymers. *ACS Macro Letters, 9*(2), 216–225. https://doi.org/10.1021/acsmacrolett.0c00002.

Petsko, G. A., & Ringe, D. (1984). Fluctuations in protein structure from X-ray diffraction. *Annual Review of Biophysics and Bioengineering, 13*, 331–371.

Pincus, P. (1976). Excluded volume effects and stretched polymer chains. *Macromolecules, 9*(3), 386–388. https://doi.org/10.1021/ma60051a002.

Ribeck, N., & Saleh, O. A. (2008). Multiplexed single-molecule measurements with magnetic tweezers. *Review of Scientific Instruments, 79*(9), 0–6. https://doi.org/10.1063/1.2981687.

Rosales, A. M., Murnen, H. K., Kline, S. R., Zuckermann, R. N., & Segalman, R. A. (2012). Determination of the persistence length of helical and non-helical polypeptoids in solution. *Soft Matter, 8*(13), 3673–3680. https://doi.org/10.1039/c2sm07092h.

Rubinstein, M., & Colby, R. H. (2003). *Polymer physics*. Oxford University Press,.

Ruff, K. M., & Holehouse, A. S. (2017). SAXS versus FRET: A matter of heterogeneity? *Biophysical Journal, 113*(5), 971–973. https://doi.org/10.1016/j.bpj.2017.07.024.

Ruiz-Gutierrez, N., Rieu, M., Ouellet, J., Allemand, J. F., Croquette, V., & Le Hir, H. (2022). Novel approaches to study helicases using magnetic tweezers. *Methods in Enzymology, 673*, 359–403. https://doi.org/10.1016/bs.mie.2022.03.035.

Saleh, O. A. (2015). Single polymer mechanics across the force regimes. *Journal of Chemical Physics, 142*(19), https://doi.org/10.1063/1.4921348.

Schuler, B., Soranno, A., Hofmann, H., & Nettels, D. (2016). Single-molecule FRET spectroscopy and the polymer physics of unfolded and intrinsically disordered proteins. *Annual Review of Biophysics, 45*(1), 207–231. https://doi.org/10.1146/annurev-biophys-062215-010915.

Schwarzl, R., Liese, S., Brünig, F. N., Laudisio, F., & Netz, R. R. (2020). Force response of polypeptide chains from water-explicit MD simulations. *Macromolecules*. https://doi.org/10.1021/acs.macromol.0c00138.

Seol, Y., & Neuman, K. C. (2011). Magnetic tweezers for single-molecule manipulation. *Methods in Molecular Biology, 783*, 265–293. https://doi.org/10.1007/978-1-61779-282-3_15.

Shi, W. H., Adhikari, R. S., Asthagiri, D. N., & Marciel, A. B. (2023). Influence of charge block length on conformation and solution behavior of polyampholytes. *ACS Macro Letters, 12*(2), 195–200. https://doi.org/10.1021/acsmacrolett.2c00665.

Shon, M. J., Rah, S. H., & Yoon, T. Y. (2019). Submicrometer elasticity of double-stranded DNA revealed by precision force-extension measurements with magnetic tweezers. *Science Advances, 5*(6), https://doi.org/10.1126/sciadv.aav1697.

Sing, C. E. (2020). Micro- to macro-phase separation transition in sequence-defined coacervates. *Journal of Chemical Physics, 152*(2), https://doi.org/10.1063/1.5140756.

Singh, V. R., & Lapidus, L. J. (2008). The intrinsic stiffness of polyglutamine peptides. *Journal of Physical Chemistry B, 112*(42), 13172–13176. https://doi.org/10.1021/jp805636p.

Solanki, A., Neupane, K., & Woodside, M. T. (2014). Single-molecule force spectroscopy of rapidly fluctuating, marginally stable structures in the intrinsically disordered protein α-synuclein. *Physical Review Letters, 112*(15), 1–6. https://doi.org/10.1103/PhysRevLett.112.158103.

Truong, H. P., Mengel, S., Yu, B., Segalman, R. A., & Saleh, O. A. (2023). Single-molecule polypeptoid stretching reveals effects of charge sequence on chain conformation. *Macromolecules, 56*, 8558–8564.

Truong, H. P., Morgan, I. L., Koren, G., Avinery, R., Beck, R., & Saleh, O. A. (2023). Pincus blob elasticity in an intrinsically disordered protein. *The European Physical Journal E, 46*, 100. https://doi.org/10.1140/epje/s10189-023-00360-0.

Uversky, V. N. (2011). Intrinsically disordered proteins from A to Z. *International Journal of Biochemistry and Cell Biology, 43*(8), 1090–1103. https://doi.org/10.1016/j.biocel.2011.04.001.

Van Der Lee, R., Buljan, M., Lang, B., Weatheritt, R. J., Daughdrill, G. W., Dunker, A. K., ... Babu, M. M. (2014). Classification of intrinsically disordered regions and proteins. *Chemical Reviews, 114*(13), 6589–6631. https://doi.org/10.1021/cr400525m.

Vilfan, I. D., Lipfert, J., Koster, D. A., Lemay, S. G., & Dekker, N. H. (2009). Magnetic tweezers for single-molecule experiments. *In Handbook of Single-Molecule Biophysics*, 371–396. https://doi.org/10.1007/978-0-387-76497-9.

Xu, Z., Cork, L. C., Griffin, J. W., & Cleveland, D. W. (1993). Increased expression of neurofilament subunit NF-L produces morphological alterations that resemble the pathology of human motor neuron disease. *Cell, 73*(1), 23–33. https://doi.org/10.1016/0092-8674(93)90157-L.

Yoo, T. Y., Meisburger, S. P., Hinshaw, J., Pollack, L., Haran, G., Sosnick, T. R., & Plaxco, K. (2012). Small-angle X-ray scattering and single-molecule FRET spectroscopy produce highly divergent views of the low-denaturant unfolded state. *Journal of Molecular Biology, 418*(3–4), 226–236. https://doi.org/10.1016/j.jmb.2012.01.016.

Zimmermann, J. L., Nicolaus, T., Neuert, G., & Blank, K. (2010). Thiol-based, site-specific and covalent immobilization of biomolecules for single-molecule experiments. *Nature Protocols, 5*(6), 975–985. https://doi.org/10.1038/nprot.2010.49.

Zuckermann, R. N., Kerr, J. M., Kent, S. B. H., & Moos, W. H. (1992). Efficient method for the preparation of peptoids [Oligo(N-substituted glycines)] by submonomer solid-phase synthesis. *Journal of American Chemical Society, 114*, 10646–10647.

CHAPTER NINE

Exploring the free energy landscape of proteins using magnetic tweezers

Hao Sun[a], Shimin Le[b], Zilong Guo[a,*], and Hu Chen[a,b,*]
[a]Center of Biomedical Physics, Wenzhou Institute, University of Chinese Academy of Sciences, Wenzhou, P.R. China
[b]Department of Physics, Research Institute for Biomimetics and Soft Matter, Fujian Provincial Key Lab for Soft Functional Materials Research, Xiamen University, Xiamen, P.R. China
*Corresponding authors. e-mail address: guozilong@wiucas.edu.cn; chenhu@xmu.edu.cn

Contents

Abstract

Proteins fold to their native states by searching through the free energy landscapes. As single-domain proteins are the basic building block of multiple-domain proteins or protein complexes composed of subunits, the free energy landscapes of single-domain proteins are of critical importance to understand the folding and unfolding processes of proteins. To explore the free energy landscapes of proteins over large conformational space, the stability of native structure is perturbed by biochemical or mechanical means, and the conformational transition process is measured. In single molecular manipulation experiments, stretching force is applied to proteins, and the folding and unfolding transitions are recorded by the extension time course. Due to the broad force range and long-time stability of magnetic tweezers, the free energy

Methods in Enzymology, Volume 694
ISSN 0076-6879, https://doi.org/10.1016/bs.mie.2023.12.008

landscape over large conformational space can be obtained. In this article, we describe the magnetic tweezers instrument design, protein construct design and preparation, fluid chamber preparation, common-used measuring protocols including force-ramp and force-jump measurements, and data analysis methods to construct the free energy landscape. Single-domain cold shock protein is introduced as an example to build its free energy landscape by magnetic tweezers measurements.

1. Introduction

Most proteins have stable three-dimensional structures which are determined by their amino acid sequences. Protein folding refers to the process of conformational transition from a random coil conformation to the well-defined stable native structure (Dill & MacCallum, 2012; Finkelstein, 2018). On the other hand, the inverse process called protein unfolding is involved in the protein degradation process (Aubin-Tam, Olivares, Sauer, Baker, & Lang, 2011; Majumder & Baumeister, 2020; Matouschek, 2003). The stability and dynamic properties of proteins are crucial to their specific functions. Misfolding or disfunction in degradation of proteins are related to serious diseases such as Alzheimer and diabetes (Chiti & Dobson, 2006; Knowles, Vendruscolo, & Dobson, 2014).

Protein folding problem involves several aspects: (1) How to predict the structure from amino acid sequence? (2) How to design protein to have required structure and function? (3) What is the mechanism of protein folding and other conformational transitions? Artificial intelligence (AI) has been widely used to solve the first and second questions and obtained tremendous improvements (Baek et al., 2021; Jumper et al., 2021). But one main shortcoming of AI is that it is difficult to get the rational understanding of the molecular mechanism. Therefore, more experiment and simulation works are focusing on solving the mechanism problem of protein folding.

To enable diverse biological functions, proteins are often composed of multiple domains, either linked together within a single polypeptide chain or assembled from separate subunits into a protein complex. The folding of individual domains and their assembly into multi-domain structures can occur through sequential or coupled processes. Understanding this complex folding and assembly pathway requires first mapping the underlying free energy landscapes of individual domains (Dill & MacCallum, 2012; Stigler, Ziegler, Gieseke, Gebhardt, & Rief, 2011). In this article, we focus on elucidating the folding free energy landscape of single protein domain.

Traditional biochemistry methods use denaturant or temperature control to study the folding and unfolding process of proteins (Gursky & Atkinson, 1996; Jha & Marqusee, 2014). Sometimes, observation of the non-linear dependence of the transition rate vs denaturant concentration indicates the existence of intermediate states (Huang, Craggs, Christodoulou, & Jackson, 2007; Oliveberg & Wolynes, 2005; Parker, Spencer, & Clarke, 1995). However, predicting precise free energy landscapes remains challenging because the interactions between denaturants and proteins are complex, depending on the solvent exposure of each amino acid. Additionally, signals from traditional bulk biochemical assays represent the average behavior of heterogeneous protein conformations, obscuring transient intermediate states that populate lowly.

Comparing to bulk biochemistry study, single-molecule techniques detect signal from a single molecule to avoid the averaging problem. Additional benefit from single molecule manipulation technique is that force is used to regulate the protein stability and transition rates (Guo et al., 2020; Yao et al., 2016). The work done by stretching force is well known from classical mechanics, which makes the construction of a quantitative free energy landscape possible.

Since the development of single molecule manipulation techniques more than twenty years ago, the mechanical responses of proteins are a research focus (Hughes & Dougan, 2016; Neuman & Nagy, 2008). Atomic force microscopy (AFM) based force spectroscopy technique was firstly used to pull proteins (Rief, Gautel, Oesterhelt, Fernandez, & Gaub, 1997). Typical saw-tooth force-extension curves were obtained from the stretching experiment working on proteins with multiple protein domain repeats. Usually the pulling speed is fast, and the unfolding forces are about tens of pN to hundreds of pN. With the development of force-clamp AFM, the folding processes were recorded at constant forces (Fernandez & Li, 2004). But the results might not be very reliable due to the low force resolution and mechanical drift of AFM.

Another popular single molecular manipulation technique, optical tweezers, is also used to stretch protein (Kellermayer, Smith, Granzier, & Bustamante, 1997). In optical tweezers, usually the two microspheres must be separated by about one micron to avoid the interference between the two tethered microspheres. Therefore, DNA handles have to be used to link the protein of interest to the microspheres. This makes the sample preparation complicated. Another limitation of optical tweezers comes from the heating effect of focused high-power laser which needs special attention (Peterman, Gittes, & Schmidt, 2003; Seol, Carpenter, & Perkins, 2006).

In protein stretching system, extension and force are two conjugate variables. Force is the controlled external condition to the protein, and its conjugate variable extension is a natural selection of the reaction coordinate to build the free energy landscape (Sun et al., 2021). To explore the free energy landscape over large extension range, it is required to apply force from several pN to hundreds of pN, which can be realized by magnetic tweezers (Chen, Fu, et al., 2011). Another advantage of magnetic tweezers is that a single-domain protein can be stretched directly without the need for DNA handles in optical tweezers or multiple protein repeats in AFM (Guo et al., 2021; Ma et al., 2022). Mechanobiology is an emerging research area which focuses on how mechanical signal are sensed by cells and transduced into cells to modulate the behavior of cells. Magnetic tweezers have been widely used to study the response of mechano-sensing proteins or mechano-transduction systems of cells (Le et al., 2017; Le, Yu, & Yan, 2019; Le, Yu, & Yan, 2021).

In the following sections, we discuss several key considerations for studying protein dynamics and free energy landscape construction using magnetic tweezers. In Section 2, we describe magnetic tweezer designs that enable stable force control and long-duration measurements to characterize protein dynamics. Section 3 covers protein construct design, including N- and C-terminal modifications for tethering proteins between coverslip surface and beads. Detailed protocols for molecular cloning, protein expression, and purification are also provided. Section 4 gives experimental protocols for performing magnetic tweezers measurements. Section 5 discusses data analysis methods for deriving dynamic parameters and free energy landscapes from magnetic tweezers data. Finally, we summarize the method and give some perspectives.

2. Instrumentation

Different from AFM and optical tweezers, the force control system and the bead tracking system in magnetic tweezers are decoupled, which makes magnetic tweezers have the capability to resist external perturbation. Additionally, the applied stretching force is intrinsically constant without any feedback control (Guo & Chen, 2021). Magnetic tweezers can be classified into two types based on the force orientation: vertical magnetic tweezers apply force perpendicular to the microscope focal plane, while transverse magnetic tweezers pull biomolecules parallel to the focal plane

(Yan, Skoko, & Marko, 2004). To directly pull protein without long DNA handles, vertical magnetic tweezers are preferred. The vertical magnetic tweezers were invented by Strick and coworkers to study the supercoiling properties of double stranded DNA (Strick, Allemand, Bensimon, Bensimon, & Croquette, 1996). We made some revisions to their original design to make it more suitable to directly pull protein in a large force range (Chen, Zhu, et al., 2011). Firstly, we use two permanent magnetic rods attached to each other without gap with the opposite direction above the sample to apply vertical stretching force upward. The zero-gapped magnets have larger magnetic field with greater gradient, which increases the largest force available. Because there is no gap between the two anti-parallel magnets, bright field illumination must be replaced with back-scattered illumination through the same light path of the fluorescence illumination of an inverted microscope, except replacing the dichromic mirror with a beam splitter with 50% transmission and 50% reflection. 100× oil immersion objective with numerical aperture of 1.3 was used to image the bead. CMOS cameras with hundreds to thousands frame per second are available to capture the image of beads with time resolution of milli-second.

The beads can be imaged in focus over a range of several microns using backscattered illumination. As the microscope focus is adjusted, effectively changing the vertical position of the beads, the ring pattern of each bead changes accordingly (Gosse & Croquette, 2002). Thus, by analyzing the ring patterns in images, the vertical position of a bead can be determined. For detecting the nanometer-scale extension steps that occur during folding and unfolding transitions of single-domain proteins, high spatial resolution is required. The ring pattern analysis method provides sub-nanometer spatial resolution, sufficient to resolve the extension steps of a few nanometers up to tens of nanometers upon protein conformational changes (Guo et al., 2021; Yao, Goult, et al., 2014).

As force is well controlled in magnetic tweezers, the tether usually remains connected between paramagnetic bead and glass surface when protein is unfolded. Most unfolded proteins can refold to their native state when force is decreased to lower than its critical force or zero. Then in magnetic tweezers experiment, the same protein can be stretched multiple times in force-increasing and force-decreasing cycles. The experiment on the same protein molecule can last for several hours to days (Chen et al., 2015; Popa et al., 2016). During this kind of long-time measurement, magnetic tweezers setup and the sample will drift due to mechanical vibration or temperature fluctuation. Though the protein extension is obtained by the relative vertical position of the tethered paramagnetic bead and the fixed reference bead, the focus drift will still induce extension changes of several nanometers. To further minimize these drift effects, we implement active feedback control of the reference bead position by adjusting microscope focus through a piezo objective actuator. This keeps the reference bead at the same position, allowing more accurate measurements of protein extensions.

Force is determined by the properties of permanent magnets and paramagnetic beads and the distance between them. In the magnetic tweezers experiment, force is controlled by the distance between the permanent magnets and paramagnetic beads. Before experiment, we can calibrate the force as a function of distance $F(d)$ by measuring the drift velocity of paramagnetic bead in 90–95% glycerol solution. Over force range from 1 pN to more than 100 pN, $F(d)$ can be fitted with double exponential function:

$$F(d) = C_1[\exp(-d/r_1) + C_2\exp(-d/r_2)] \quad (1)$$

where d is the distance between the bottom surfance of the permanent magnets and the sample, r_1, r_2, and C_2 are fitting parameters specific to the permanent magnets size and arrangement, C_1 is the fitting parameter which reflects the heterogeneity of paramagnetic beads (Chen, Zhu, et al., 2011; Popa et al., 2016).

In our MT setup, we use cylindrical magnets with a diameter of 3 mm, corresponding parameters r_1, r_2, and C_2 are shown in Fig. 4B. For each specific tethered paramagnetic bead, we can get the force calibration done with a force marker, such as the SpyTag/SpyCatcher unzipping transition at 30 pN or the DNA overstretching transition at 65 pN. If there is no force marker along the protein construct, thermal fluctuation of the bead can also be used to calibrate force.

3. Design and preparation of protein constructs

3.1 Protein construct design

Starting from the earliest realization of surface tethering proteins based on non-specific adhesion to the specific covalent linkage, biophysicists have developed a wide variety of protein tags to link the protein on glass surface or beads (Choi, Kim, Shon, & Yoon, 2022; De Vlaminck & Dekker, 2012). In general, researchers introduce conjugation sites at the intended

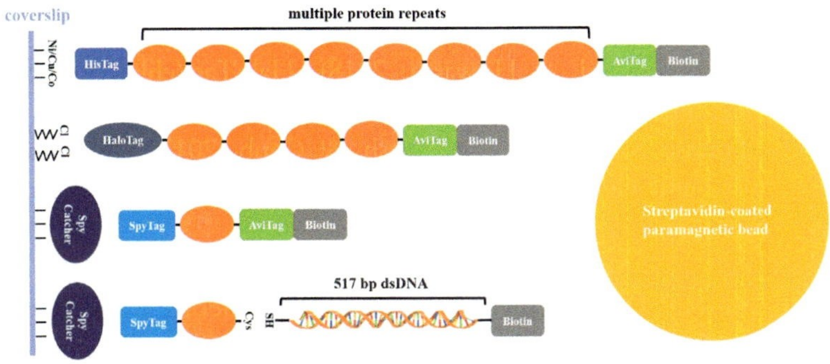

Fig. 1 Protein constructs for studying the mechanical characteristics of proteins in magnetic tweezers experiments. Four representative designs are shown with HisTag, Halotag, and SpyTag for surface attachment, and AviTag for bead binding. DNA handles can be inserted between an internal cysteine and the bead to prevent steric clashing or non-terminal force loading.

tethering positions, usually the N- or C-termini of the protein of interest, by using protein fusion techniques. In this section, we introduce various designs of protein constructs for magnetic tweezers experiments (Fig. 1).

Some protein tags form non-covalent bond with their binding partners, such as the most popular HisTag (6 ×His) which is widely employed for the purification of recombinant proteins. HisTag forms coordination bond with Ni (or Cu) ion-NTA. Proteins with HisTag were linked on glass surface coated with NTA-Ni ion in early magnetic tweezers experiments and successfully probed the mechanical properties of several force-sensing proteins (Filamin A, talin, and α-catenin) (Chen, Zhu, et al., 2011; del Rio et al., 2009; Yao, Qiu, et al., 2014). However, the fragility of this non-covalent HisTag linkage limits its application.

Another strong non-covalent tag is AviTag which can be ligated with a biotin molecule by BirA biotin ligase. Biotin binds tightly with streptavidin protein which is coated on the surface of commercial paramagnetic beads. Though biotin-streptavidin is almost the strongest non-covalent linkage, it was further improved by Gaub and Lipfert using monovalent streptavidin which forms stronger linkage with biotin than the commercially available tetravalent streptavidin (Gruber et al., 2020; Sedlak, Schendel, Gaub, & Bernardi, 2020).

Specific covalent linkages are the ideal choice for supporting long-duration measurements at large forces. There are lots of tags available which form covalent linkage through either autocatalytic reactions or reactions catalyzed by other enzymes. Examples include SpyTag/SpyCatcher (Zakeri et al., 2012),

DogTag/DogCatcher (Keeble et al., 2022), SnoopTag/SnoopCatcher (Buldun, Jean, Bedford, & Howarth, 2018), HaloTag/chloroalkane (Popa et al., 2013), Histidine/PS (Lei et al., 2022), SNAP/BG (Li et al., 2018), LPXTG (Sortase A) (Durner, Ott, Nash, & Gaub, 2017; Ott, Durner, & Gaub, 2018), and ybbRTag/CoA (Sfp) (Löf et al., 2019; Ott et al., 2018) where enzyme is inside the parentheses.

SpyCatcher/SpyTag complex is from *Streptococcus pyogenes* fibronectin-binding protein FbaB, which contains a spontaneous isopeptide bond between Lys31 and Asp117 (Zakeri et al., 2012). Reaction occurred in high yield simply upon mixing under diverse conditions of pH, temperature, and buffer. SpyTag could be fused at either terminus or internally of the protein of interest, and it readily forms a covalent bond with SpyCatcher coated on the substrate. By choosing a suitable pulling geometry, SpyTag/SpyCatcher complex can feature equilibrium zipping/unzipping transition signals with 3.5 nm in a narrow force range of ~30 pN (Guo et al., 2021). It will facilitate the applications of SpyTag/SpyCatcher as a fingerprint and force marker in single molecule manipulation experiments. Additionally, Howarth lab further optimized the reaction efficiency and binding intensity of SpyCatcher/SpyTag complex (Keeble et al., 2017), and proposed another two covalent self-assembled Catcher/Tag conjugation, SnoopCatcher/SnoopTag (Buldun et al., 2018), DogCatcher/DogTag (Keeble et al., 2022), which were also engineered from the *Streptococcus pneumoniae* adhesin and enabled isopeptide bonds formation without any additional reagents.

HaloTag is an engineered dehalogenase that could directly form a specific covalent bond with chloroalkane ligand without requiring any additional reagents or enzymes (England, Luo, & Cai, 2015; Los et al., 2008), this permits prolonged observation of the target protein for several days (Chen et al., 2015; Popa et al., 2016). HaloTag system has been extensively used as the attachment method for substrate immobilization (Rivas-Pardo et al., 2016; Tapia-Rojo, Eckels, & Fernández, 2019). However, the use of HaloTag has a non-negligible problem due to the fact that it occasionally unfolds at high force, which can provide tens of nanometers of extension change, and thus can introduce unnecessary signal that would complicate the experimental analysis.

DNA handle is used to reduce the effect of volume-exclusion problem when paramagnetic bead is too close to the sample. In the polymerase chain reaction amplification process, a variety of functional groups can be readily added to the ends of double-stranded DNA with chemically modified

DNA primers (Choi et al., 2019; Janissen et al., 2014; Shon, Rah, & Yoon, 2019). One end of the double-stranded DNA handle, typically modified with a thiol group, can be used to link the protein construct. Correspondingly, cysteine mutation with thiol group exposed on the protein surface is usually used to define the pulling point which is not limited to the N- or C-termini of the protein of interest. The other end of the DNA handle is modified with biotin or digoxigenin, which can be employed to attach the DNA handle to streptavidin-coated magnetic bead or anti-digoxigenin-coated glass surface (Le et al., 2019; Min, Arbing, Jefferson, & Bowie, 2016; Paik, Roskens, & Perkins, 2013). It is worth noting that overstretching of the linked DNA double helix also can serve as a finger print signal to identify the correct protein construct and a force sensor to calibrate the force for each tethered paramagnetic bead (Chen, Fu, et al., 2011).

3.2 Protocol to express and purify proteins

Below a detailed protocol to make protein construct (HisTag-AviTag-CSP-SpyTag) is given as an example.

1. Synthesize or amplify the gene of cold shock protein (CSP).
2. Insert the gene of CSP into a vector pET151 (ampicillin resistant) containing HisTag-AviTag and SpyTag through HiFi DNA Assembly kit (New England Biolabs) or traditional restriction enzyme cutting and ligation method.
3. Transform the pET151 plasmid with HisTag-AviTag-CSP-SpyTag into *Escherichia coli* DH5α in LB buffer with antibiotic ampicillin.
4. Plate the transformed *E. coli* cells on LB agar plates containing ampicillin. Screen the colonies with restriction enzyme digestion and sequencing.
5. Co-transform plasmids of pET151-HisTag-AviTag-CSP-SpyTag and pBirA (Biotin ligase expression plasmid, chloramphenicol resistant) into the *E. coli* strain BL21 (DE3).
6. Culture the selected transformants in LB medium (supplemented with chloramphenicol, ampicillin, and D-biotin) at 37 °C with 220 rpm shaking until the optical density (OD) reached 0.6.
7. 30 min later, induce the expression with 0.8 mM isopropyl β-d-1-thio-galactopyranoside for 12 h at 25 °C with continuous shaking of 180 rpm.
8. Harvest the *E. coli* cells by centrifugation at 5000 rpm and resuspend them in lysis buffer (50 mM Tris, 500 mM NaCl, 10% glycerol, 5 mM imidazole, 5 mM 2-mercaptoethanol, pH 8.0).

9. Sonicate the *E. coli* cells for 10 min to lyse the cells.
10. Centrifuge the lysate at 12,000 rpm for 40 min at 4 °C.
11. Collect the supernatant and purify the protein by Ni-NTA resin (Sangon Biotech).
12. Analyze the purified protein using SDS–PAGE to confirm the size is correct.
13. Aliquot the purified proteins and flash-freeze them in liquid nitrogen and store at −80 °C.

4. Magnetic tweezers measurement

4.1 Flow chamber preparation

The protein of interest will be anchored on a functionalized coverslip which is used to make a flow chamber (Fig. 2). Below is a protocol to build flow chambers with SpyCatcher-coated coverslips.

1. Clean coverslips (24 × 32 mm and 24 × 24 mm) by sonicating in 5% Decon 90 detergent solution for 10 min. Rinse thoroughly with DI water and dry at 80 °C in an oven.
2. Treat 24 × 32 mm coverslips with oxygen plasma for 15 min to activate the surface.
3. Immerse activated 24 × 32 mm coverslips in 1% APTES methanol solution for 1 h to aminate the surface for protein attachment.
4. Wash the salinized coverslips by methanol and DI water, and dried in oven at 80 °C for 40 min
5. Make flow chambers by sandwiching a piece of functionalized coverslip and another piece of coverslip (24 × 24 mm) with parafilm in between, as shown in Fig. 1.

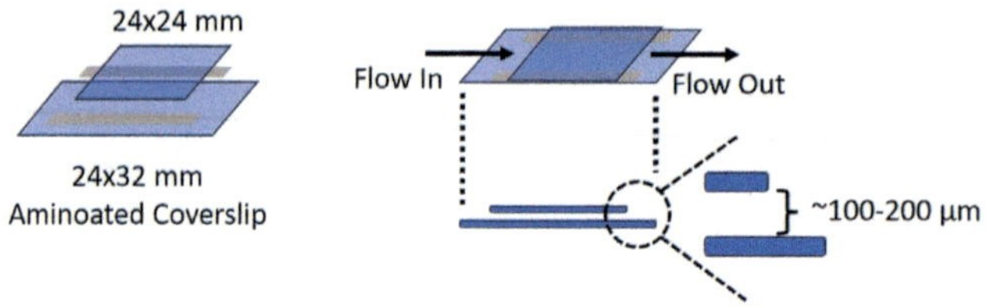

Fig. 2 Magnetic tweezer flow chamber. The chamber is constructed by sandwiching a salinized 24 × 32 mm coverslip and a 24 × 24 mm coverslip with parafilm spacer in between. This creates a channel with ~100–200 µm thickness through which solution can flow.

6. Polybead microspheres (cat. 17145, Polysciences) with diameter of 3.0 μm were flowed into chamber and incubated for 20 min to get stuck on the coverslip.
7. Rinse the chamber by 200 μL 1 ×PBS buffer to remove free beads.
8. Flush 1% Sulfo-SMCC (SE 247420, Thermo Science) in DI water into the chamber and incubated for 20 min
9. Rinse with 1 ×PBS buffer, and flow SpyCatcher proteins in 1 ×PBS into chamber to coat the functionalized surface for 1 h.
10. Flow blocking buffer (1% BSA in Tris Buffer pH 7.4 with 150 mM NaCl) into chamber and incubate overnight at 4 °C to passivate the surface.

4.2 Sample preparation for magnetic tweezers

The flow chambers can be stored for one month at 4 °C. Before magnetic tweezers experiment, the protein of interest and paramagnetic beads will be added to form the protein tethers (Fig. 3).

1. Flow the protein constructs with SpyTag at a concentration of about 1 nM in blocking buffer and incubate for 20 min. The concentration of protein constructs needs to be adjusted to have a higher chance of finding correct tethered beads.
2. Flush the chamber with blocking buffer to remove the free protein constructs.

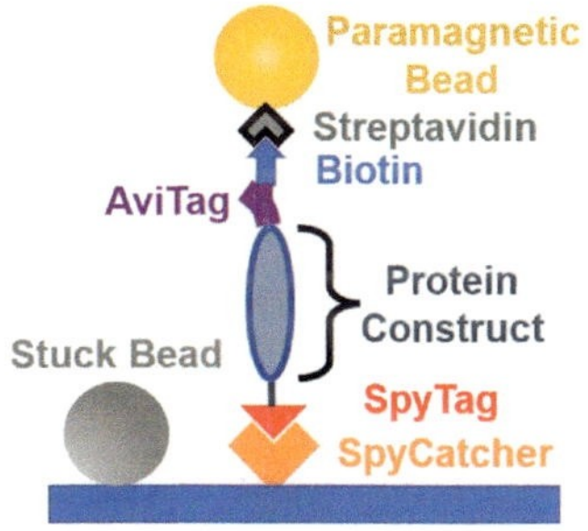

Fig. 3 Schematic of protein-mediated tethering of a paramagnetic bead to a functionalized coverslip. In this example, the protein construct has an N-terminal AviTag and C-terminal SpyTag. The AviTag forms a covalent bond with biotin, which is attached to the streptavidin-coated paramagnetic bead. The SpyTag forms an isopeptide bond with SpyCatcher immobilized on the coverslip via Sulfo-SMCC. Stuck polystyrene bead is used as reference to remove mechanical drift.

3. Flow in the chamber with streptavidin-coated paramagnetic beads Dynabead M270 (cat. 65305, Invitrogen) in blocking buffer and incubate for 5 min to form protein-tethered paramagnetic beads.
4. Flush the chamber with measuring buffer (5 mM L-ascorbic acid sodium in blocking buffer) to remove the free paramagnetic beads.
5. Put the flow chamber with protein-tethered paramagnetic beads on the sample stage of magnetic tweezers, then adjust the focus and move the sample stage to locate at least one stuck reference polystyrene bead and one fluctuating tethered paramagnetic bead in the same field of view.

Now we are ready for the magnetic tweezers experiment.

4.3 Procedures of magnetic tweezers experiment

To start a single molecule magnetic tweezers experiment, we must first identify paramagnetic beads that are tethered by a single protein of interest. When we increase the force, usually the protein will unfold with signal of extension jump (Hong et al., 2021; Ma et al., 2022). If the unfolding step size is as expected from the number of amino acids of the protein, and the bead moves up vertically without translocation in the focal plane during unfolding transition, then we have high confidence that the tether is by a single protein.

4.3.1 Force calibration

We also need to calibrate the force before or after the experiment to compensate the heterogeneity of each paramagnetic bead. To calibrate the force for the specific tethered bead, we just need to measure the force at a magnets-sample distance (Fig. 4). The transvers fluctuation of the bead can be used to measure the force by formula $F = k_B T\,(r + z)/\delta^2 y$ (Chen, Fu, et al., 2011), where k_B is the Boltzmann constant, T the absolute temperature, z the extension of the tether along the force direction, r the radius of the bead, and $\delta^2 y$ the variance of the bead fluctuation along a direction perpendicular to both the force and magnetic field directions (Fig. 5). This formula works when the tether does not interact with the glass surface. Please note that correction might be required when the frame rate of the camera is comparable than the corner frequency of the fluctuation (Guo & Chen, 2021; Wong & Halvorsen, 2006).

The force calibration can be easily done if there is an intrinsic force marker in the molecule construct. DNA overstretching transition at 65 pN is a highly reversible and cooperative transition at room temperature in physiological buffer for double-stranded DNA (Smith, Cui, & Bustamante, 1996).

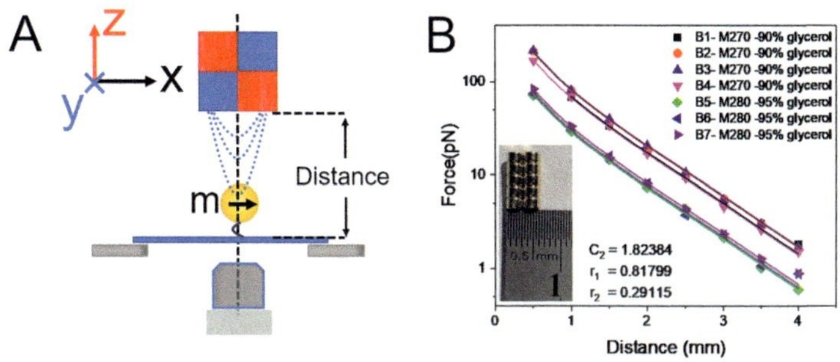

Fig. 4 Force calibration for magnetic tweezers. (A) Schematic defining coordinate axes—*x* is magnetic field and bead magnetization direction; *z* is the force direction; *y* is perpendicular to *x*–*z*. Magnet distance is controlled precisely. (B) Force at different magnet distances for different paramagnetic beads (M270 and M280, Invitrogen) calibrated by Stokes' law in 90 or 95% glycerol solution. Solid lines show the global fitting results using Eq. (1).

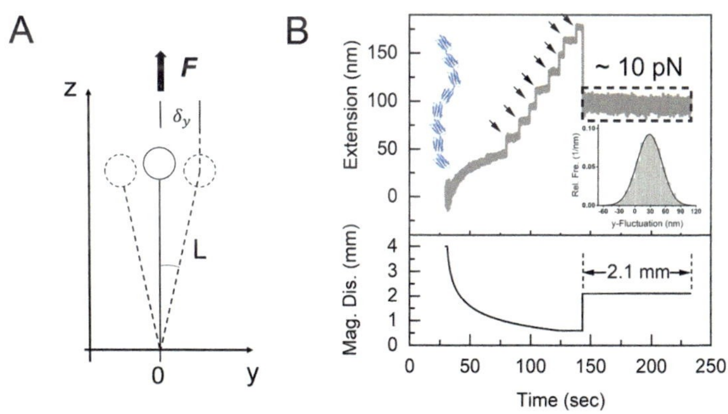

Fig. 5 Force calibration by fluctuation along *y*-direction at around 10 pN. (A) Schematic of bead fluctuation as a pendulum for force measurement. (B) As the double magnets approach to the sample, eight tandem repeats of GB1 unfold in a stepwise fashion. After eight unfolding events, the distance between magnets and coverslip is set to 2.1 mm to apply ~10 pN force. Extension is recorded for ~100 s to obtain the variance of the bead fluctuation along *y*-direction for force calibration at this position.

Therefore, the force function can be determined from the location of the magnets when overstretching transition is recorded (Fig. 6). Another force marker is the SpyTag/SpyCatcher when force is applied on N-termini of both SpyTag and SpyCatcher (Fig. 7). At critical force of 30 pN, part of the SpyTag unzips from the SpyTag/SpyCatcher with step size of about 3.5 nm

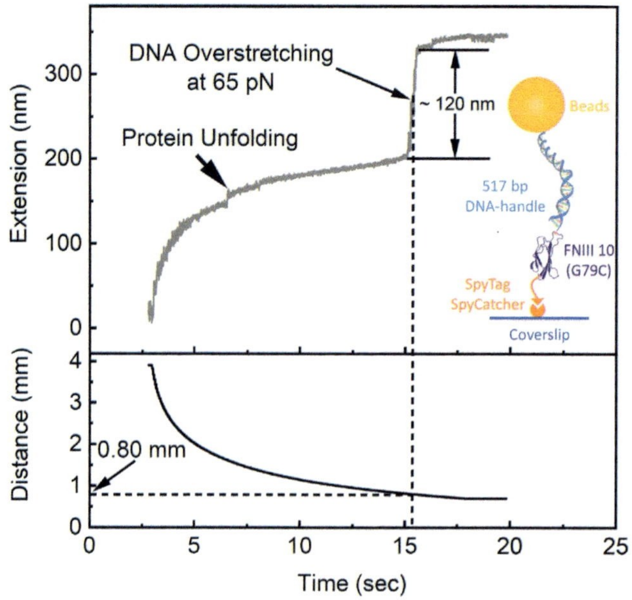

Fig. 6 Force calibration by DNA Overstretching. The inset shows FNIII10 domain tethered between surface and bead via a 517 bp DNA handle. Upon increasing force, the extension exhibits two distinct conformational transitions: protein unfolding followed by DNA overstretching at 65 pN which happens at a magnet distance of 0.80 mm.

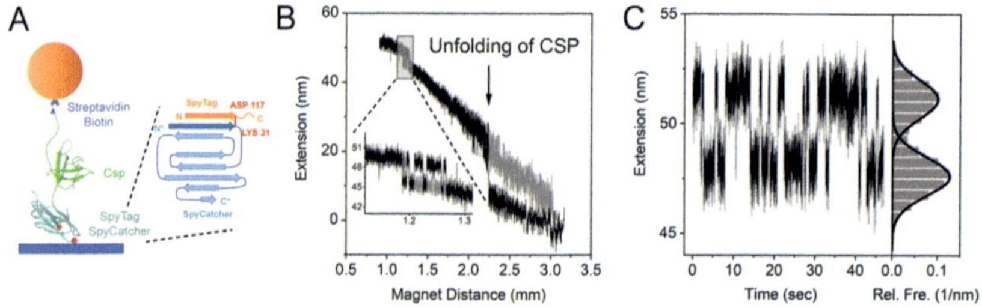

Fig. 7 Force calibration using SpyTag/SpyCatcher. (A) Sketch of a single CSP tethered to a glass surface through N-SpyTag and N-SpyCatcher. (B) Approaching magnets induced CSP unfolding at ~2.2 mm and SpyTag/SpyCatcher unzipping at ~1.2 mm. (C) At fixed magnet position ($d = 1.23$ mm), the extension time course shows a 44.2% unzipping probability. The force is calculated as 29.7 pN using the inverse of Eq. (2), giving $C_1 = 118.8$ in the force function (Eq. (1)).

(Guo et al., 2021). If the force is not exactly 30 pN, the force can be determined by the force-dependent probability of the native structure:

$$P(F) = \frac{1}{\exp[(F - f_c)\Delta x/k_B T] + 1} \tag{2}$$

where the critical force $f_c = 30$ pN, and $\Delta x = 3.5$ nm.

4.3.2 Constant loading rate measurement

With the force function $F(d)$, the stretching force can be controlled to increase with a constant loading rate. Constant loading rate measurement is usually the first measurement done for a protein tether, as it will give us general information about the mechanical stability of a protein in short time.

Using protein GB1 as an example, eight unfolding steps with regular step size were recorded for protein construct $GB1_8$ during the force-increasing process with constant loading rate of 1.0 pN/s (Fig. 8B). After that, the force was dropped to below 1 pN, driving unfolded GB1 protein domains fold to native state. Successful folding was confirmed by similar unfolding signals upon stretching with the same loading rate. Multiple force-ramp cycles on a protein tether produce the unfolding force distribution of GB1 at loading rate of 1.0 pN/s, from which the kinetic unfolding rates and related parameters can be obtained (Fig. 8C).

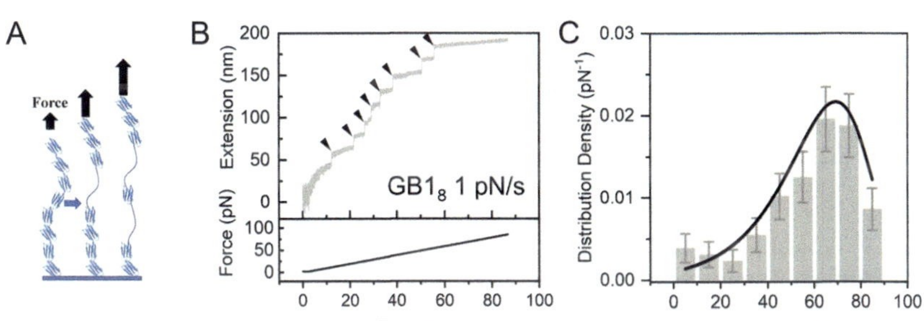

Fig. 8 Analysis of the unfolding force distribution from force-ramp experiment. (A) Schematic of the random unfolding process of $GB1_8$ with increasing force. (B) Example trace showing eight unfolding steps when force increases at a constant loading rate of 1 pN/s. (C) The unfolding force distribution is fitted with Eq. (3) to derive the dynamic parameters in Bell's model.

4.3.3 Force-jump and constant force equilibrium measurement

Magnetic tweezers could provide intrinsic constant force control with superior stability. The force-dependent conformational transitions can be obtained directly from constant force experiments, regardless of how slow the transitions are. For protein constructs of a single CSP domain or tandem repeats $GB1_8$, the unfolding and folding rates at constant forces can be obtained from typical force-jump measurements (Fig. 9A and B). First, protein domains remained folded at 1 pN. The force was then increased abruptly to 10 or 20 pN, and the corresponding extension time course was recorded with unfolding steps.

After eight GB1 domains were unfolded, the force was dropped to 4 pN, and the corresponding extension dropped suddenly with decreasing force due to the entropic collapse of the unfolded peptide, followed by a further shortening process caused by folding transition of GB1 (Fig. 9C). The successful folding to the native state of GB1 can be verified by a subsequent pulling test. Following similar procedures, force-dependent unfolding process over a large force range and force-dependent folding process were recorded.

Taking advantage of the stability of MT, it is worthwhile to measure the folding and unfolding dynamics of protein at equilibrium, which was found to exist at a narrow force range. At forces close to the critical force,

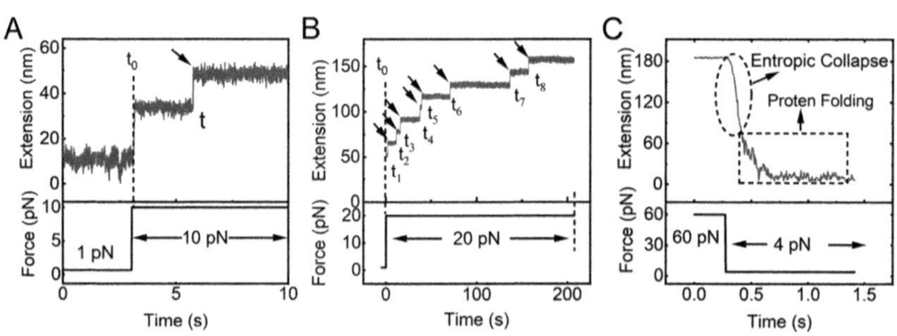

Fig. 9 Typical force-jump measurements. (A) Force-jump measurement on a single CSP domain. The force was first held at ~1 pN to allow CSP folding to the native state. At t_0, the force jumped to 10 pN. The time between t_0 and the unfolding event gives the CSP unfolding time. (B) Force-jump measurement on eight repeats of GB1 domain. The force on $GB1_8$ increased from ~1 to 20 pN at t_0. The intervals between t_0 and unfolding events t_i represent the unfolding times of individual GB1 domains. (c) After confirming that $GB1_8$ domains had unfolded, the force was reduced to 4 pN. The resulting time course shows initial entropic chain collapse (circled) followed by the folding process of GB1 domains (boxed).

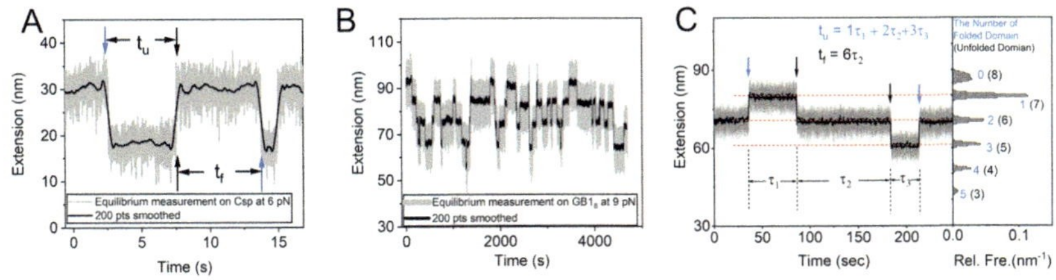

Fig. 10 Equilibrium folding and unfolding measurement. (A) Extension time course of a single CSP domain at 6 pN. The time interval between a folding event (blue arrow) and the subsequent unfolding event (black arrow) was recorded as the unfolding time t_u. Similarly, an interval time between an unfolding event and the next folding event gives the folding time t_f. (B) Equilibrium folding/unfolding dynamics of GB1 at 9 pN, starting with all domains unfolded. (C) Definition of pseudo unfolding/folding dwell times for equilibrium time course of multiple repeats. In this time window of 250 s, two unfolding events (blue arrows) and two folding events (black arrows) occur. The numbers of folded and unfolded GB1 domains are shown on the right panel. Pseudo unfolding dwell time for each unfolding event is calculated from the sum of the products of the time interval and the corresponding number of folded domains between adjacent unfolding events, while pseudo folding dwell time is obtained in a similar way.

the single CSP protein and eight tandem repeats of GB1 folded and unfolded randomly (Fig. 10). When the measuring duration is long enough, the probability of each state with certain number of unfolded domains will approach to the equilibrium binomial distribution.

5. Data analysis

5.1 Unfolding force distribution

Unfolding force distribution is obtained in constant loading rate experiments with the native state as the initial state. If the force-dependent unfolding rate follows Bell's model (Bell, 1978): $k_u(F) = k_0 \exp(Fx_u/k_B T)$, where k_0 is the unfolding rate at zero force, x_u denotes the unfolding distance which is defined as the distance between native state and the unfolding transition state, the unfolding force distribution is given by equation (Evans & Ritchie, 1997):

$$P_f(F) = \frac{k_0}{r} \exp\left[\frac{Fx_u}{k_B T} + \frac{k_0 k_B T}{x_u r}\left(1 - \exp\frac{Fx_u}{k_B T}\right)\right] \tag{3}$$

where r is the loading rate.

By fitting the unfolding force distribution, parameters in Bell's model k_0 and x_u can be determined (Fig. 8C). If the number of unfolding events is too limited to give a smooth distribution histogram, the unfolding distance can be estimated by $x_u = k_B T/\sigma$, where σ is the standard deviation of unfolding forces.

The above analysis is based on the assumptions of Bell's model. If the unfolding rates do not follow Bell's model, a more general method can give the force-dependent unfolding rate from the unfolding force distribution (Dudko, Hummer, & Szabo, 2008):

$$k_u(F) = \frac{\dot{F}(F)\,P(F)}{\int_F^\infty P(f)\,df} \tag{4}$$

where $\dot{F}(F)$ is the force loading rate when force is F, which is not constrained to remain constant. The unfolding rate can be obtained at forces with non-zero distribution.

5.2 Data analysis for constant force measurements

Though the unfolding force distribution can shift to low force with decreasing loading rate and shift to high force with increasing loading rate, it is still difficult to obtain the unfolding rates over large force range. Force-jump or constant-force measurements provide the most straightforward method to obtain the unfolding rates over a wide range of forces.

From the life time of the native state, the unfolding probability as a function of time can be obtained from the accumulative distribution of life time, which can be fitted with an exponential function to get the unfolding rate directly for two-state systems (Fig. 11A). To check whether a single exponential function fits the data points well, the survival probability of the native state as a function of time can be plotted in logarithm scale (Fig. 11B). Even if the transition step cannot be identified to get the lifetime, which is the case for the folding process at low forces, exponential function fitting of the average extension time course yields the transition rate (Fig. 11C).

Folding transitions and unfolding transitions can be observed at a constant force close to the critical force. Therefore, the folding and unfolding rates can be obtained from equilibrium measurements (Fig. 10A). If the protein construct has multiple repeats of the same protein domain, the observed folding/unfolding transition can be from anyone of unfolded/folded protein repeats (Fig. 10B). Hence, the dwell time between two

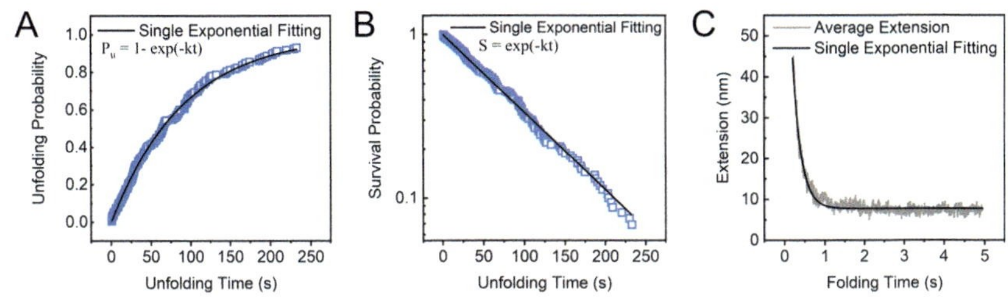

Fig. 11 Data analysis for determining unfolding and folding rates. (A) The unfolding probability is obtained from the cumulative distribution of unfolding times or pseudo-unfolding dwell times. Fitting the probability with a single exponential equation yields the unfolding rate constant. (B) To verify the unfolding probability follows single exponential kinetics, the survival probability should show a linear dependence on time when plotted on a logarithmic scale. Non-linear behavior may indicate the presence of an intermediate or misfolded state. (C) Multiple folding time courses at 4 pN ($n > 10$) are averaged and fitted to an exponential decay function to derive the folding rate constant.

adjacent transitions should be converted to pseudo dwell time to get the folding/unfolding rate (Fig. 10C) (Chen et al., 2015).

5.3 Free energy landscape construction

The stretching force and the extension are two conjugate variables in statistical physics. With stretching force as the control variable to disturb the conformation of proteins, extension along the force direction is the natural reaction coordinate to build the free energy landscape (Fig. 12).

The folding free energy ΔG_0, which is the free energy difference between the naive state and unfolded state at zero force, can be obtained from the critical force and unfolding step size $\Delta x(f)$ which is a function of force:

$$\Delta G_0 = \int_0^{f_c} \Delta x(f)\, df. \tag{5}$$

The folding process of proteins usually can only be observed at forces smaller than 10 pN. The extension of the unfolded state is much longer than the folding transition state, which makes the folding rate very sensitive to force and its force-dependent behavior mainly determined by the entropic elasticity of the unfolded polypeptide chain. Unfolded polypeptide chain can be modeled as a worm-like chain whose force-extension curve is well characterized. Therefore, the size of the folding transition state can be estimated.

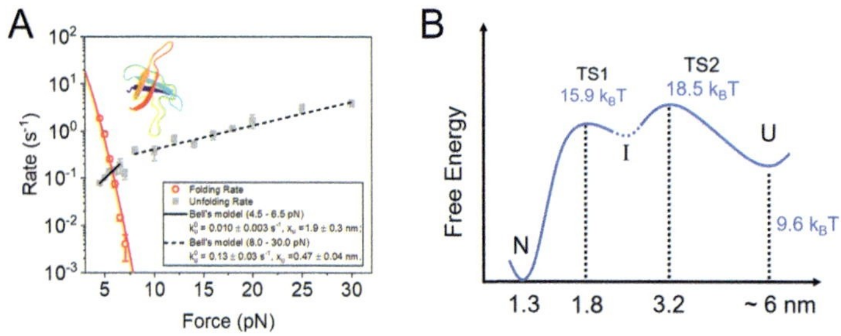

Fig. 12 Construction of the free energy landscape of CSP based on force-dependent folding and unfolding rates. (A) Force-dependent folding and unfolding rates of a single CSP domain (Fig. 7) at forces ranging from 4.5 to 30 pN were fitted using Arrhenius' law and Bell's model in certain force ranges, respectively. (B) Free energy landscape of CSP at zero force was plotted as a function of *N–C* distance. With free energy of *N* as a reference, the free energy of *U* is obtained by Eq. (5). Free energies of TS1 and TS2 were calculated from fitting parameters of Bell's mode k_0 over 8–30 pN and 4.5–6.5 pN, respectively using Eq. (6). An intermediate state located between TS1 and TS2 may exist but was not sufficiently stable to be recorded experimentally. Please note that the unfolding rates of CSP is greater than our published results with I27 domains in the protein construct (Hong et al., 2021) which needs further study.

The slope of the logarithm of the unfolding rate vs force gives the distance between the native state and the transition state (unfolding barrier along the free energy landscape) using the formula of Bell's model. The extrapolated zero force unfolding rate k_0 can be used to estimate the height of the free energy barrier by

$$k_0 = k^* \exp(-\Delta G^{\ddagger}/k_B T) \tag{6}$$

where k^* is the preexponential factor which is estimated in the order of magnitude 10^6 s^{-1}, and $\Delta G^{\ddagger}$ is the free energy barrier (Best & Hummer, 2006).

As the unfolding process can be recorded at large force range from several pN to tens of pN, the unfolding rate might deviate from Bell's model in large force range to show non-linear force dependence, i.e., the slope is force-dependent (Fig. 12). As a result, the dominant transition pathway or the transition state with the highest free energy barrier along the same pathway might change with force.

There are two types of deviations from Bell's model. For the case with large slope at low force and small slope at high force, the unfolding process need to cross two barriers along a single unfolding pathway (Hong et al., 2021; Ma et al., 2022). At small force, the transition state with long extension has the highest free energy. When force increases, this transition state is depressed and another transition state with short extension has the highest free energy to determine the unfolding rate. For the other case with small slope at low force and large slope at high force, the unfolding dynamics can be reproduced by a model with two parallel unfolding pathways (Jagannathan, Elms, Bustamante, & Marqusee, 2012). Sometimes, the elasticity of the transition state can be used to explain the non-linear behavior of force-dependent unfolding dynamics, even the non-monotonic force dependence with catch-bond behavior at low forces (Guo et al., 2020; Yuan et al., 2017).

6. Conclusions and perspectives

In summary, magnetic tweezers provide key advantages of large force range and long-term stability, which makes this technique suitable to study the protein folding and unfolding dynamics over wide force range. The measurement at low force reveals the free energy at long extensions with large conformational fluctuation, while the measurement at large force helps to build the free energy landscape at short extension with conformation very close to the native state. Therefore, magnetic tweezer measurements can effectively map protein free energy landscapes over a large conformational space. The broad force control and temporal stability are unique capabilities enabling comprehensive characterization of protein energy landscapes from unfolded state to folded state.

In the futher, proteins with two or more domains which interact with each other will be studied, not only the folding and unfolding transition, but also transition between different conformations which is usually related to its function. Though in most cases the stretching force is along the N- and C-termini direction, it can also be applied on other residues. Therefore, the free energy landscape along different extensions as the reaction coordinates can be constructed.

Acknowledgments

We thank Dr Haiyan Hong, Zhuwei Zhang, Ping Yu, and Zhenyong Xue from Chen lab for contributing the experimental protocols in this research. This work was supported by research grants from the National Natural Science Foundation of China (12174322 to H.C., 12204124 to Z.G.), 111 project (B16029), and Wenzhou Institute.

References

Aubin-Tam, M. E., Olivares, A. O., Sauer, R. T., Baker, T. A., & Lang, M. J. (2011). Single-molecule protein unfolding and translocation by an ATP-fueled proteolytic machine. *Cell, 145*, 257–267.

Baek, M., DiMaio, F., Anishchenko, I., Dauparas, J., Ovchinnikov, S., Lee, G. R., et al. (2021). Accurate prediction of protein structures and interactions using a three-track neural network. *Science (New York, N.Y.), 373*, 871–876.

Bell, G. I. (1978). Models for the specific adhesion of cells to cells. *Science (New York, N.Y.), 200*, 618–627.

Best, R. B., & Hummer, G. (2006). Diffusive model of protein folding dynamics with Kramers turnover in rate. *Physical Review Letters, 96*, 228104.

Buldun, C. M., Jean, J. X., Bedford, M. R., & Howarth, M. (2018). SnoopLigase catalyzes peptide-peptide locking and enables solid-phase conjugate isolation. *Journal of the American Chemical Society, 140*, 3008–3018.

Chen, H., Fu, H. X., Zhu, X. Y., Cong, P. W., Nakamura, F., & Yan, J. (2011). Improved high-force magnetic tweezers for stretching and refolding of proteins and short DNA. *Biophysical Journal, 100*, 517–523.

Chen, H., Yuan, G. H., Winardhi, R. S., Yao, M. X., Popa, I., Fernandez, J. M., et al. (2015). Dynamics of equilibrium folding and unfolding transitions of titin immunoglobulin domain under constant forces. *Journal of the American Chemical Society, 137*, 3540–3546.

Chen, H., Zhu, X. Y., Cong, P. W., Sheetz, M. P., Nakamura, F., & Yan, J. (2011). Differential mechanical stability of filamin A rod segments. *Biophysical Journal, 101*, 1231–1237.

Chiti, F., & Dobson, C. M. (2006). Protein misfolding, functional amyloid, and human disease. *Annual Review of Biochemistry, 75*, 333–366.

Choi, H. K., Kim, H. G., Shon, M. J., & Yoon, T. Y. (2022). High-resolution single-molecule magnetic tweezers. *Annual Review of Biochemistry, 91*, 33–59.

Choi, H. K., Min, D., Kang, H., Shon, M. J., Rah, S. H., Kim, H. C., et al. (2019). Watching helical membrane proteins fold reveals a common N-to-C-terminal folding pathway. *Science (New York, N.Y.), 366*, 1150–1156.

De Vlaminck, I., & Dekker, C. (2012). Recent advances in magnetic tweezers. *Annual Review of Biophysics, 41*, 453–472.

del Rio, A., Perez-Jimenez, R., Liu, R. C., Roca-Cusachs, P., Fernandez, J. M., & Sheetz, M. P. (2009). Stretching single talin rod molecules activates vinculin binding. *Science (New York, N.Y.), 323*, 638–641.

Dill, K. A., & MacCallum, J. L. (2012). The protein-folding problem, 50 years on. *Science (New York, N.Y.), 338*, 1042–1046.

Dudko, O. K., Hummer, G., & Szabo, A. (2008). Theory, analysis, and interpretation of single-molecule force spectroscopy experiments. *Proceedings of the National Academy of Sciences of the United States of America, 105*, 15755–15760.

Durner, E., Ott, W., Nash, M. A., & Gaub, H. E. (2017). Post-translational sortase-mediated attachment of high-strength force spectroscopy handles. *ACS Omega, 2*, 3064–3069.

England, C. G., Luo, H. M., & Cai, W. B. (2015). HaloTag technology: A versatile platform for biomedical applications. *Bioconjugate Chemistry, 26*, 975–986.

Evans, E., & Ritchie, K. (1997). Dynamic strength of molecular adhesion bonds. *Biophysical Journal, 72*, 1541–1555.

Fernandez, J. M., & Li, H. (2004). Force-clamp spectroscopy monitors the folding trajectory of a single protein. *Science (New York, N.Y.), 303*, 1674–1678.

Finkelstein, A. V. (2018). 50+ Years of protein folding. *Biochemistry-Moscow, 83*, S3–S18.

Gosse, C., & Croquette, V. (2002). Magnetic tweezers: Micromanipulation and force measurement at the molecular level. *Biophysical Journal, 82*, 3314–3329.

Gruber, S., Löf, A., Sedlak, S. M., Benoit, M., Gaub, H. E., & Lipfert, J. (2020). Designed anchoring geometries determine lifetimes of biotin-streptavidin bonds under constant load and enable ultra-stable coupling. *Nanoscale, 12*, 21131–21137.

Guo, Z. L., & Chen, H. (2021). Single-molecule manipulation by magnetic tweezers. In S. Huang (Ed.). *Single-molecule tools for bioanalysis*. Jenny Stanford Publishing.

Guo, Z. L., Hong, H. Y., Sun, H., Zhang, X. F., Wu, C. X., Li, B., et al. (2021). SpyTag/SpyCatcher tether as a fingerprint and force marker in single-molecule force spectroscopy experiments. *Nanoscale, 13*, 11262–11269.

Guo, Z. L., Hong, H. Y., Yuan, G. H., Qian, H., Li, B., Cao, Y., et al. (2020). Hidden intermediate state and second pathway determining folding and unfolding dynamics of GB1 protein at low forces. *Physical Review Letters, 125*, 198101.

Gursky, O., & Atkinson, D. (1996). High- and low-temperature unfolding of human high-density apolipoprotein A-2. *Protein Science, 5*, 1874–1882.

Hong, H. Y., Guo, Z. L., Sun, H., Yu, P., Su, H. H., Ma, X. N., et al. (2021). Two energy barriers and a transient intermediate state determine the unfolding and folding dynamics of cold shock protein. *Communications Chemistry, 4*, 156.

Huang, J. R., Craggs, T. D., Christodoulou, J., & Jackson, S. E. (2007). Stable intermediate states and high energy barriers in the unfolding of GFP. *Journal of Molecular Biology, 370*, 356–371.

Hughes, M. L., & Dougan, L. (2016). The physics of pulling polyproteins: A review of single molecule force spectroscopy using the AFM to study protein unfolding. *Reports on Progress in Physics, 79*, 076601.

Jagannathan, B., Elms, P. J., Bustamante, C., & Marqusee, S. (2012). Direct observation of a force-induced switch in the anisotropic mechanical unfolding pathway of a protein. *Proceedings of the National Academy of Sciences of the United States of America, 109*, 17820–17825.

Janissen, R., Berghuis, B. A., Dulin, D., Wink, M., van Laar, T., & Dekker, N. H. (2014). Invincible DNA tethers: Covalent DNA anchoring for enhanced temporal and force stability in magnetic tweezers experiments. *Nucleic Acids Research, 42*, e137.

Jha, S. K., & Marqusee, S. (2014). Kinetic evidence for a two-stage mechanism of protein denaturation by guanidinium chloride. *Proceedings of the National Academy of Sciences of the United States of America, 111*, 4856–4861.

Jumper, J., Evans, R., Pritzel, A., Green, T., Figurnov, M., Ronneberger, O., et al. (2021). Highly accurate protein structure prediction with AlphaFold. *Nature, 596*, 583–589.

Keeble, A. H., Banerjee, A., Ferla, M. P., Reddington, S. C., Anuar, I. N. A. K., & Howarth, M. (2017). Evolving accelerated amidation by SpyTag/SpyCatcher to analyze membrane dynamics. *Angewandte Chemie-International Edition, 56*, 16521–16525.

Keeble, A. H., Yadav, V. K., Ferla, M. P., Bauer, C. C., Chuntharpursat-Bon, E., Huang, J., et al. (2022). DogCatcher allows loop-friendly protein-protein ligation. *Cell Chemical Biology, 29*, 339–350.

Kellermayer, M. S. Z., Smith, S. B., Granzier, H. L., & Bustamante, C. (1997). Folding-unfolding transitions in single titin molecules characterized with laser tweezers. *Science (New York, N.Y.), 276*, 1112–1116.

Knowles, T. P. J., Vendruscolo, M., & Dobson, C. M. (2014). The amyloid state and its association with protein misfolding diseases. *Nature Reviews. Molecular Cell Biology, 15*, 384–396.

Le, S., Hu, X., Yao, M., Chen, H., Yu, M., Xu, X., et al. (2017). Mechanotransmission and mechanosensing of human alpha-actinin 1. *Cell Reports, 21*, 2714–2723.

Le, S., Yu, M., & Yan, J. (2019). Direct single-molecule quantification reveals unexpectedly high mechanical stability of vinculin-talin/α-catenin linkages. *Science Advances, 5*, eaav2720.

Le, S., Yu, M., & Yan, J. (2021). Mechanical regulation of tension-transmission supramolecular linkages. *Current Opinion in Solid State and Materials Science, 25*, 100895.

Lei, H., Zhang, J. S., Li, Y., Wang, X., Qin, M., Wang, W., et al. (2022). Histidine-specific bioconjugation for single-molecule force spectroscopy. *ACS Nano, 16*, 15440–15449.

Li, B., Wang, X., Li, Y., Paananen, A., Szilvay, G. R., Qin, M., et al. (2018). Single-molecule force spectroscopy reveals self-assembly enhanced surface binding of hydrophobins. *Chemistry—A European Journal, 24*, 9224–9228.

Löf, A., Walker, P. U., Sedlak, S. M., Gruber, S., Obser, T., Brehm, M. A., et al. (2019). Multiplexed protein force spectroscopy reveals equilibrium protein folding dynamics and the low-force response of von Willebrand factor. *Proceedings of the National Academy of Sciences of the United States of America, 116*, 18798–18807.

Los, G. V., Encell, L. P., McDougall, M. G., Hartzell, D. D., Karassina, N., Zimprich, C., et al. (2008). HatoTag: A novel protein labeling technology for cell imaging and protein analysis. *ACS Chemical Biology, 3*, 373–382.

Ma, X. N., Sun, H., Hong, H. Y., Guo, Z. L., Su, H. H., & Chen, H. (2022). Free-energy landscape of two-state protein acylphosphatase with large contact order revealed by force-dependent folding and unfolding dynamics. *Physical Review E, 106*, 024404.

Majumder, P., & Baumeister, W. (2020). Proteasomes: Unfoldase-assisted protein degradation machines. *Biological Chemistry, 401*, 183–199.

Matouschek, A. (2003). Protein unfolding — An important process in vivo? *Current Opinion in Structural Biology, 13*, 98–109.

Min, D., Arbing, M. A., Jefferson, R. E., & Bowie, J. U. (2016). A simple DNA handle attachment method for single molecule mechanical manipulation experiments. *Protein Science, 25*, 1535–1544.

Neuman, K. C., & Nagy, A. (2008). Single-molecule force spectroscopy: Optical tweezers, magnetic tweezers and atomic force microscopy. *Nature Methods, 5*, 491–505.

Oliveberg, M., & Wolynes, P. G. (2005). The experimental survey of protein-folding energy landscapes. *Quarterly Reviews of Biophysics, 38*, 245–288.

Ott, W., Durner, E., & Gaub, H. E. (2018). Enzyme-mediated, site-specific protein coupling strategies for surface-based binding assays. *Angewandte Chemie-International Edition, 57*, 12666–12669.

Paik, D. H., Roskens, V. A., & Perkins, T. T. (2013). Torsionally constrained DNA for single-molecule assays: An efficient, ligation-free method. *Nucleic Acids Research, 41*.

Parker, M. J., Spencer, J., & Clarke, A. R. (1995). An integrated kinetic-analysis of intermediates and transition-states in protein-folding reactions. *Journal of Molecular Biology, 253*, 771–786.

Peterman, E. J. G., Gittes, F., & Schmidt, C. F. (2003). Laser-induced heating in optical traps. *Biophysical Journal, 84*, 1308–1316.

Popa, I., Berkovich, R., Alegre-Cebollada, J., Badilla, C. L., Rivas-Pardo, J. A., Taniguchi, Y., et al. (2013). Nanomechanics of HaloTag tethers. *Journal of the American Chemical Society, 135*, 12762–12771.

Popa, I., Rivas-Pardo, J. A., Eckels, E. C., Echelman, D. J., Badilla, C. L., Valle-Orero, J., et al. (2016). A HaloTag anchored ruler for week-long studies of protein dynamics. *Journal of the American Chemical Society, 138*, 10546–10553.

Rief, M., Gautel, M., Oesterhelt, F., Fernandez, J. M., & Gaub, H. E. (1997). Reversible unfolding of individual titin immunoglobulin domains by AFM. *Science (New York, N.Y.), 276*, 1109–1112.

Rivas-Pardo, J. A., Eckels, E. C., Popa, I., Kosuri, P., Linke, W. A., & Fernández, J. M. (2016). Work done by titin protein folding assists muscle contraction. *Cell Reports, 14*, 1339–1347.

Sedlak, S. M., Schendel, L. C., Gaub, H. E., & Bernardi, R. C. (2020). Streptavidin/biotin: Tethering geometry defines unbinding mechanics. *Science Advances, 6*, eaay5999.

Seol, Y., Carpenter, A. E., & Perkins, T. T. (2006). Gold nanoparticles: Enhanced optical trapping and sensitivity coupled with significant heating. *Optics Letters, 31*, 2429–2431.

Shon, M. J., Rah, S. H., & Yoon, T. Y. (2019). Submicrometer elasticity of double-stranded DNA revealed by precision force-extension measurements with magnetic tweezers. *Science Advances, 5*, eaav1697.

Smith, S. B., Cui, Y. J., & Bustamante, C. (1996). Overstretching B-DNA: The elastic response of individual double-stranded and single-stranded DNA molecules. *Science (New York, N.Y.), 271*, 795–799.

Stigler, J., Ziegler, F., Gieseke, A., Gebhardt, J. C. M., & Rief, M. (2011). The complex folding network of single calmodulin molecules. *Science (New York, N.Y.), 334*, 512–516.

Strick, T. R., Allemand, J. F., Bensimon, D., Bensimon, A., & Croquette, V. (1996). The elasticity of a single supercoiled DNA molecule. *Science (New York, N.Y.), 271*, 1835–1837.

Sun, H., Guo, Z. L., Hong, H. Y., Yu, P., Xue, Z., & Chen, H. (2021). Protein folding mechanism revealed by single-molecule force spectroscopy experiments. *Biophysics Reports, 7*, 399–412.

Tapia-Rojo, R., Eckels, E. C., & Fernández, J. M. (2019). Ephemeral states in protein folding under force captured with a magnetic tweezers design. *Proceedings of the National Academy of Sciences of the United States of America, 116*, 7873–7878.

Wong, W. P., & Halvorsen, K. (2006). The effect of integration time on fluctuation measurements: Calibrating an optical trap in the presence of motion blur. *Optics Express, 14*, 12517–12531.

Yan, J., Skoko, D., & Marko, J. F. (2004). Near-field-magnetic-tweezer manipulation of single DNA molecules. *Physical Review, E: Statistical, Nonlinear, and Soft Matter Physics, 70*, 011905.

Yao, M., Goult, B. T., Chen, H., Cong, P., Sheetz, M. P., & Yan, J. (2014). Mechanical activation of vinculin binding to talin locks talin in an unfolded conformation. *Scientific Reports, 4*, 4610.

Yao, M., Qiu, W., Liu, R. C., Efremov, A. K., Cong, P. W., Seddiki, R., et al. (2014). Force-dependent conformational switch of α-catenin controls vinculin binding. *Nature Communications, 5*, 4525.

Yao, M. X., Goult, B. T., Klapholz, B., Hu, X., Toseland, C. P., Guo, Y. J., et al. (2016). The mechanical response of talin. *Nature Communications, 7*, 11966.

Yuan, G. H., Le, S. M., Yao, M. X., Qian, H., Zhou, X., Yan, J., et al. (2017). Elasticity of the transition state leading to an unexpected mechanical stabilization of titin immunoglobulin domains. *Angewandte Chemie-International Edition, 56*, 5490–5493.

Zakeri, B., Fierer, J. O., Celik, E., Chittock, E. C., Schwarz-Linek, U., Moy, V. T., et al. (2012). Peptide tag forming a rapid covalent bond to a protein, through engineering a bacterial adhesin. *Proceedings of the National Academy of Sciences of the United States of America, 109*, E690–E697.

CHAPTER TEN

Single-molecule tethering methods for membrane proteins

Daehyo Lee[a] **and Duyoung Min**[a,b,*]
[a]Department of Chemistry, Ulsan National Institute of Science and Technology, Ulsan, Republic of Korea
[b]Center for Wave Energy Materials, Ulsan National Institute of Science and Technology, Ulsan, Republic of Korea
*Corresponding author. e-mail address: dymin@unist.ac.kr

Contents

Methods in Enzymology, Volume 694
ISSN 0076-6879, https://doi.org/10.1016/bs.mie.2023.12.013

Abstract

Molecular tethering of a single membrane protein between the glass surface and a magnetic bead is essential for studying the structural dynamics of membrane proteins using magnetic tweezers. However, the force-induced bond breakage of the widely-used digoxigenin-antidigoxigenin tether complex has imposed limitations on its stable observation. In this chapter, we describe the procedures of constructing highly stable single-molecule tethering methods for membrane proteins. These methods are established using dibenzocyclooctyne click chemistry, traptavidin-biotin binding, SpyCatcher-SpyTag conjugation, and SnoopCatcher-SnoopTag conjugation. The molecular tethering approaches allow for more stable observation of structural transitions in membrane proteins under force.

1. Introduction

Magnetic tweezers have been utilized as a powerful single-molecule tool over broad research areas, including DNA mechanics, DNA-protein interactions, protein-protein interactions, and soluble protein folding (Choi, Kim et al., 2022; De Vlaminck & Dekker, 2012). Recently, this versatile technique was successfully adapted for studies on membrane protein folding, observing the interactions between transmembrane (TM) helices of membrane proteins within lipid bilayers (Hong, Choi, & Yoon, 2022; Wijesinghe & Min, 2023). The reconstitution of a lipid bilayer environment is crucial for membrane protein studies, as it significantly impacts their stability, fold, and function. Thus, long molecular spacers linked to a target membrane protein, such as DNA handles, have been employed to accommodate lipid bilayer mimetics between solid supports, i.e., the glass surface and a magnetic bead (Kim, Lee, Wijesinghe, & Min, 2023). To exert mechanical force to the target molecule, both ends of the protein-DNA hybrid construct are then affixed to the solid supports. Although various covalent tethering approaches have been developed (Kim et al., 2023), the single-molecule pulling methods using DNA handles have primarily relied on the weak, noncovalent linkage between digoxigenin (dig) and antidigoxigenin (antidig) (Choi et al., 2019; Choi, Kang et al., 2022; Min, Jefferson, Bowie, & Yoon, 2015; Min et al., 2018) (Fig. 1; tethering modes 1 and 2). However, the dig-antidig tether exhibits a short lifetime of ~1 s at 25 pN and ~0.1 s at 50 pN (Van Patten et al., 2018), which limits stable force applications to a single molecule.

Here, we present the procedures for single-molecule tethering methods with high stability, developed for investigating the structural dynamics of

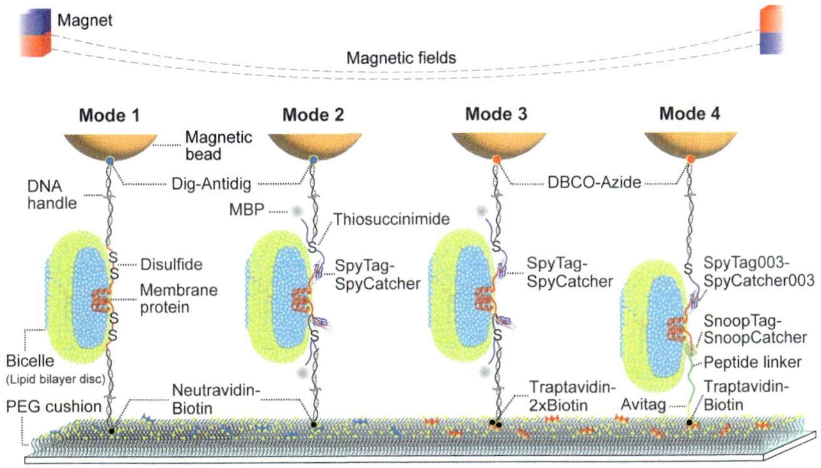

Fig. 1 Schematics depicting four single-molecule tethering methods for membrane proteins. Tethering modes 1–3 were established in previous studies. Tethering mode 4 is presented here for the first time. This chapter provides the detailed procedures pertaining to the tethering modes 3 and 4, since they are the most advanced methods regarding the tethering stability.

membrane proteins (Fig. 1; tethering modes 3 and 4). In tethering mode 3 (Kim et al., 2023), we utilize three orthogonal tethering strategies: covalent dibenzocyclooctyne (DBCO)-azide crosslinking (Kim & Koo, 2019), non-covalent yet exceptionally strong traptavidin-biotin binding (Chivers et al., 2010), and SpyCatcher-SpyTag complex which forms a spontaneous iso-peptide bond upon binding (Min, Arbing, Jefferson, & Bowie, 2016; Zakeri et al., 2012). In tethering mode 4, the biotin-modified DNA handle is replaced with a biotinylated SnoopCatcher (SnoopCatcher-bt), conjugated to SnoopTag at the end of the target membrane protein. This approach using the SnoopCatcher handle improves the attachment efficiency and enhances the yield of the hybrid molecular construct. Our single-molecule tethering methods can also be employed for studying other types of biomolecules, such as soluble proteins and DNA hairpins (Kim et al., 2023).

2. Overall scheme

Our single-molecule tethering systems are constructed for pulling an *E. coli* intramembrane protease GlpG as an example case (Fig. 2 for the overall system assembly). In tethering mode 3, a 13-residue peptide tag, SpyTag, is

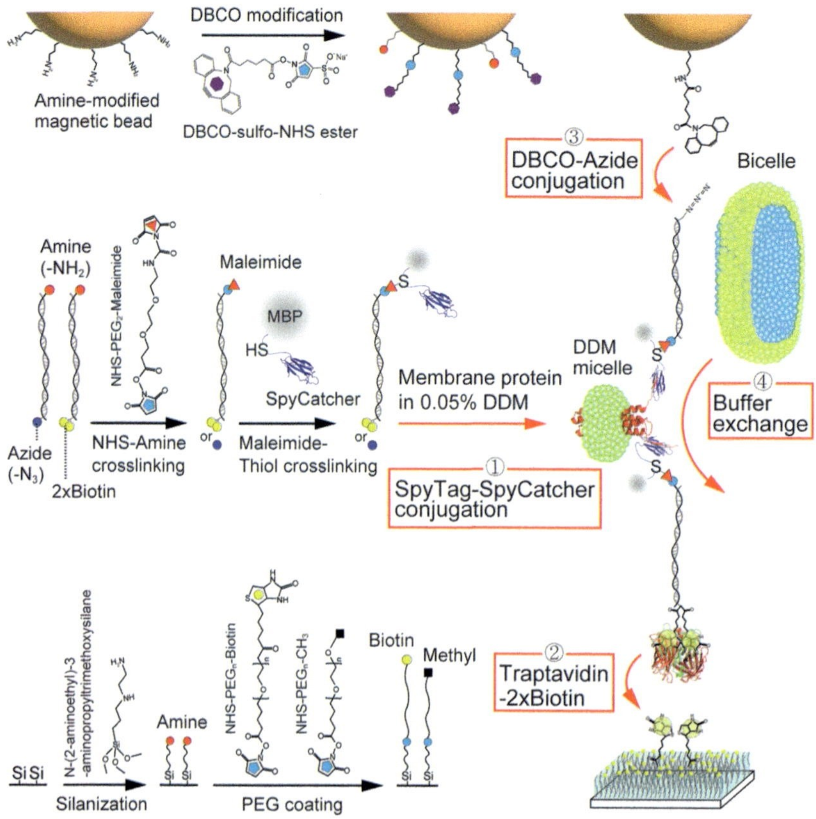

Fig. 2 Procedures for the single-molecule tethering and system assembly. This diagram pertains to the tethering mode 3. In the tethering mode 4, the biotin-modified DNA handle is substituted with biotinylated SnoopCatcher (refer to the main text for more details). *Adapted from our earlier work with permission (Kim et al., 2023).*

engineered at both the N- and C-termini of GlpG for DNA handle attachment (Fig. 3). To attach the DNA handles, a ~1000-bp long DNA is first linked to maltose-binding protein (MBP)-fusion SpyCatcher (mSpyCatcher) through thiol-maleimide crosslinking (Fig. 4A). The mSpyCatcher-linked DNA handles are then attached to both ends of GlpG via SpyCatcher-SpyTag conjugation (Fig. 4B). The other ends of the DNA handles are modified with azide or 2× biotin. The azide-modified end is attached to a magnetic bead through DBCO-azide conjugation. The 2× biotin-modified end is affixed to the glass surface through traptavidin binding. The glass surface is passivated using polyethylene glycol (PEG) to prevent nonspecific adhesion of magnetic beads and the tethered molecular constructs. The successful tethering is verified by monitoring the unfolding events of GlpG (Fig. 4C).

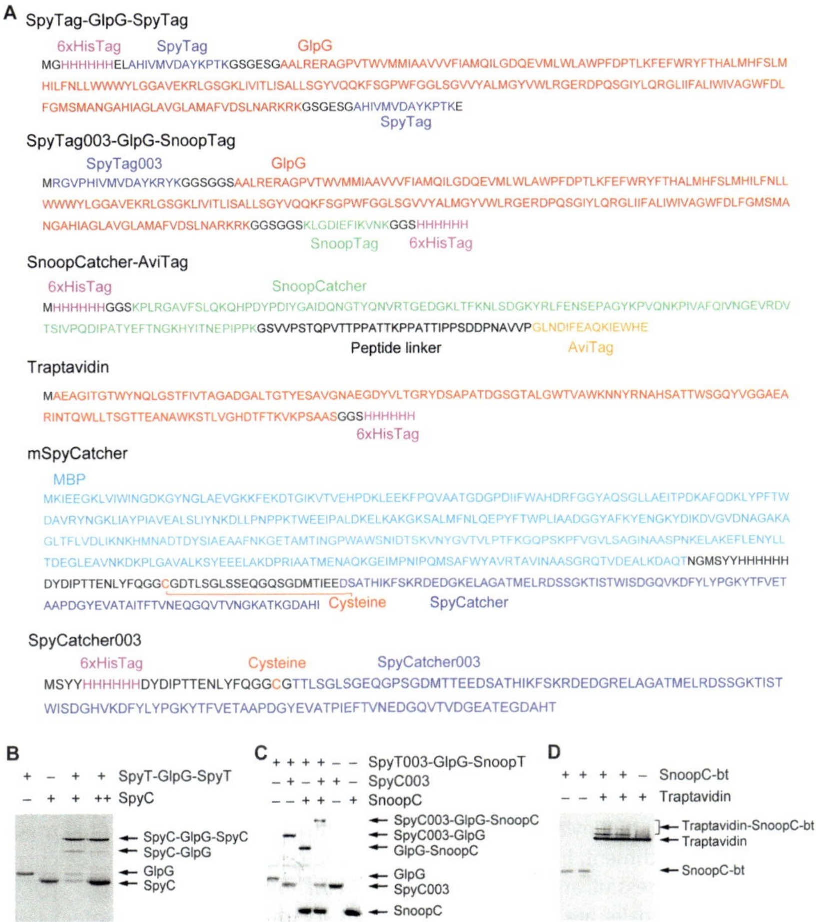

Fig. 3 Protein constructs and purification results. (A) Amino acid sequence of purified protein constructs of GlpG, SnoopCatcher-AviTag, traptavidin, mSpyCatcher, and SpyCatcher003. mSpyCatcher indicates maltose-binding protein (MBP)-fusion SpyCatcher. (B–D) Gel results obtained from 12% SDS-PAGE demonstrating successful protein purifications. The bindings between relevant protein partners are also confirmed. In panel D, the two gel lanes for SnoopC-bt and the mixture of SnoopC-bt and traptavidin correspond to those for different sample fractions of SnoopC-bt after purification using size exclusion chromatography. SpyC, SpyC003, SpyT, SpyT003, and SnoopC-bt are abbreviated notations for SpyCatcher, SpyCatcher003, SpyTag, SpyTag003, and biotinylated SnoopCatcher, respectively. The ++ symbol indicates a higher concentration.

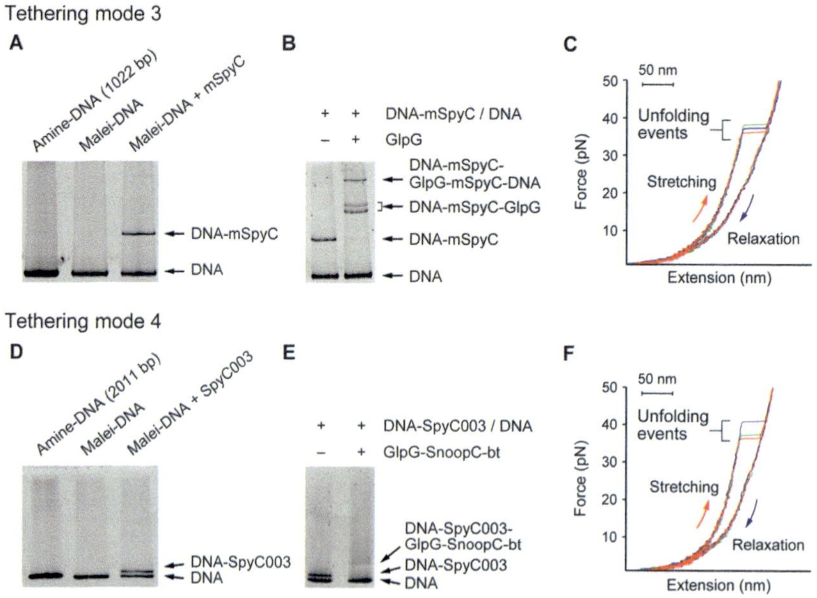

Fig. 4 Attachment of DNA handle to GlpG membrane protein and its unfolding results. (A–F) Gel results obtained from 6% SDS-PAGE demonstrating the successful protein-DNA attachment and representative force-extension curves of GlpG for tethering mode 3 (A–C) and tethering mode 4 (D–F). Panels A and D show the attachment of DNA to SpyCatcher constructs, while panels B and E show the attachment of SpyCatcher-linked DNA to GlpG. In tethering mode 3, the two gel bands in panel B correspond to two molecular constructs where DNA is attached to the N- or C-terminus of GlpG. In tethering mode 4, biotinylated SnoopCatcher is initially attached to GlpG, followed by DNA handle attachment. In panel E, the shift of the gel band for DNA-SpyCatcher003 signifies the successful attachment of the DNA handle to GlpG. mSpyC, SpyC003, SnoopC-bt, and malei are abbreviated notations for maltose-binding protein (MBP)-fusion SpyCatcher, SpyCatcher003, biotinylated SnoopCatcher, and maleimide, respectively. The stepwise signals in panels C and F indicate the unfolding events of GlpG at a magnet speed of 0.3 mm/s. The extension, i.e., the end-to-end distance of the hybrid constructs, was median-filtered with a 6-Hz window.

In tethering mode 4 presented here for the first time, SpyTag003 and SnoopTag are engineered at the N- and C-termini of GlpG, respectively (Fig. 3). SpyTag003 is a mutant of SpyTag with faster binding rate (Keeble et al., 2019), while SnoopTag is a 12-residue peptide tag that binds and conjugates to SnoopCatcher (Veggiani et al., 2016). In this tethering approach, the SnoopTag end of GlpG is first attached to the Snoop-Catcher-bt. The biotinylation of SnoopCatcher is carried out using the

BirA enzyme at AviTag engineered in the SnoopCatcher (SnoopCatcher-AviTag) (Fairhead & Howarth, 2015). A ~2000-bp long DNA handle is then attached to GlpG via SpyCatcher003-SpyTag003 conjugation (Fig. 4D,E). The successful tethering is verified by monitoring the unfolding events of GlpG (Fig. 4F), as in the case of tethering mode 3. The peptide linker and AviTag of SnoopCatcher-bt before the biotin-modified site and the PEG polymer on the glass surface are estimated as ~15 nm and ~32 nm in contour length (Ma et al., 2014), respectively. According to the unfolding results, these lengths appear to be sufficiently long for the reconstitution of membrane mimetics, such as bicelles.

3. Protein expression and purification

All necessary proteins in constructing the tethering systems are expressed in *E. coli* competent cells using isopropyl β-D-1-thiogalactopyranoside (IPTG) induction. The overexpressed proteins are subsequently purified using a 6× HisTag, which binds to nickel-iminodiacetic acid (IDA) resin. Amylose resin is employed for the MBP-fusion protein. Further purification is performed using size exclusion chromatography (SEC) to enhance the level of purity. Following the purification, the affinity of the peptide tags is confirmed using sodium dodecyl sulfate-polyacrylamide gel electrophoresis (SDS-PAGE). Refer to Fig. 3A for amino acid sequences of the protein constructs and Fig. 3B–D for gel data confirming the protein purity and tag affinity.

3.1 Equipment

- Multi-chamber shaking incubator JSSI-202 (JSR)
- Micro centrifuge Smart R17 Plus (Hanil Scientific)
- High-speed refrigerated centrifuge VS-24SMTi (Vision Scientific)
- High-speed centrifuge Supra R22 (Hanil Scientific)
- High-pressure homogenizer Emulsiflex-C3 (Avestin)
- ÄKTA pure™ 25 L (Cytiva)
- Vortex mixer VM1 (LABTron)

3.2 Materials

- *E. coli* BL21 (DE3) (Thermo Fisher Scientific)
- Cloned plasmids for protein constructs (using pTrcHisA vector for SpyTag-GlpG-SpyTag, pET24a vector for SpyTag003-GlpG-SnoopTag, SpyCatcher003, SnoopCatcher-AviTag, and traptavidin, and pDEST14 vector for mSpyCatcher)

- Sterile Luria-Bertani (LB, Merck) agar plate with 100 mg/L ampicillin (agar plate A)
- Sterile LB (Merck) agar plate with 25 mg/L kanamycin (agar plate B)
- Sterile LB medium (Merck, 25 g/L)
- Sterile LB medium (Merck, 25 g/L with 100 mg/L ampicillin; LB-amp medium)
- Sterile LB medium (Merck, 25 g/L with 25 mg/L kanamycin; LB-kan medium)
- Sterile Terrific broth (TB) medium (Novagen, 47 g/L with 4 mL/L glycerol and 100 mg/L ampicillin; TB-amp medium)
- Buffer A (50 mM Tris-HCl, pH 7.4, 300 mM NaCl)
- Buffer B (100 mM Tris-HCl, pH 8.0, 400 mM NaCl)
- PBS Buffer A (10 mM Na_2HPO_4, 2 mM KH_2PO_4, pH 7.4, 137 mM NaCl, 2.7 mM KCl)
- Lysis Buffer A (25 mM Tris-HCl, pH 7.4, 150 mM NaCl, 10% glycerol, 1 mM TCEP, 1 mM PMSF)
- Lysis Buffer B (50 mM Tris-HCl, pH 8.0, 200 mM NaCl, 1 mM DTT, 1 mM PMSF, 1 μg/ml DNase I, 5 mM $MgCl_2$)
- Lysis Buffer C (PBS Buffer A, 150 mM NaCl, 5 mM EDTA, 1% Triton X-100)
- Wash Buffer A (25 mM Tris-HCl, pH 7.4, 150 mM NaCl, 10% glycerol, 1 mM TCEP, 40 mM imidazole; 80 mM imidazole and additional 0.1% DDM only for GlpG)
- Wash Buffer B (50 mM Tris-HCl, pH 8.0, 200 mM NaCl, 1 mM TCEP)
- Wash Buffer C (PBS Buffer A, 150 mM NaCl, 30 mM imidazole)
- Elution Buffer A (25 mM Tris-HCl, pH 7.4, 150 mM NaCl, 10% glycerol, 1 mM TCEP, 300 mM imidazole; additional 0.1% DDM only for GlpG)
- Elution Buffer B (50 mM Tris-HCl, pH 8.0, 200 mM NaCl, 1 mM TCEP, 300 mM maltose)
- Elution Buffer C (PBS Buffer A, 150 mM NaCl, 300 mM imidazole)
- Storage Buffer A (25 mM Tris-HCl, pH 7.4, 150 mM NaCl, 10% glycerol, 1 mM TCEP, 0.1% DDM)
- Extraction Buffer (PBS Buffer A, 150 mM NaCl, 0.5% Triton X-100)
- Dilution Buffer (PBS Buffer A, 10 mM imidazole)
- 0.2 μm polyethersulfone (PES) syringe filter (HYUNDAI MICRO)
- Nickel-IDA resin (Takara Bio)
- Amylose resin (NEB)
- Superdex 200 Increase 10/300 GL (Cytiva)

3.3 Procedure for protein expression

1. Add 1–2 μL of 100–200 ng/μL plasmid into 25 μL of *E. coli* BL21 (DE3) competent cells and gently mix the sample tube by finger tapping. Incubate for 10 min on ice.
2. For bacterial transformation using heat shock, incubate for 40 s at 42 °C and then place the tube on ice for 2 min. Add 1 mL sterile LB medium and incubate for 1 hr at 37 °C while shaking at 150 rpm.
3. Centrifuge for 30 s at 4 °C (13,000 rpm). Discard the supernatant with ~100 μL left and then resuspend the cell pellet by pipetting.
4. Spread the cell suspension on an agar plate using a cell spreader (agar plate A for SpyTag-GlpG-SpyTag and mSpyCatcher, or agar plate B for SpyTag003-GlpG-SnoopTag, SpyCatcher003, SnoopCatcher-AviTag, and traptavidin). Incubate the plate for 16 hr at 37 °C. Seal the agar plate with parafilm and store at 4 °C.
5. For cell culture, take one colony from the agar plate with a pipette tip and put the tip into 20 mL LB-amp medium for SpyT-GlpG-SpyT, 20 mL LB-kan medium for SpyT003-GlpG-SnoopT, SpyCatcher003, Snoop Catcher-AviTag, and traptavidin, or 20 mL TB-amp medium for mSpy Catcher. Culture the cells for 16 hr at 37 °C while shaking at 180 rpm.
6. Put the cell culture into 1 L of each LB or TB medium. Incubate the larger curtures at 37 °C until reaching $OD_{600\ nm} = 0.7$ for SpyT-GlpG-SpyT, SpyT003-GlpG-SnoopT, SpyCatcher003, and SnoopCatcher-AviTag, 0.9 for traptavidin, or 1.0 for mSpyCatcher.
7. For protein overexpression, add 400 μL of 1 M IPTG and incubate for 3 hr at 37 °C for SpyT-GlpG-SpyT, SpyT003-GlpG-SnoopT, SpyCatcher003, and SnoopCatcher-AviTag, 500 μL and 4 hr at 37 °C for traptavidin, or 800 μL and 4 hr at 30 °C for mSpyCatcher.
8. Centrifuge for 10 min at 4 °C (5700 rpm) and discard the supernatant. For later use, store the cell pellet at –80 °C.

3.4 Procedure for protein purification: GlpG, SpyCatcher, SnoopCatcher

1. Suspend the cell pellet in 50 mL Lysis Buffer A. Lyse the cells with Emulsiflex-C3 for 5 min at 4 °C (15,000–17,000 psi). In case of GlpG, add 5.6 mL of 10% DDM into the cell lysate and incubate for 1 hr at 4 °C with slow rotation. Centrifuge the lysate for 30 min at 4 °C (18,000 rpm) and collect the supernatant.
2. Add 1130 μL imidazole and 2 mL Ni-IDA resin reconstituted with Buffer A into the supernatant. Incubate for 1 hr at 4 °C with slow rotation.

3. Load the mixture on a gravity column, wash the column with 10 mL Wash Buffer A (×3), and then elute the protein with 1 mL Elution Buffer A (×6). In case of GlpG, perform additional purificaiton using SEC with Superdex 200 Increase 10/300 GL. Elute with Storage Buffer A, aliquot, and store the purified protein at −80 °C.

3.5 Procedure for protein purification: mSpyCatcher

1. Suspend the cell pellet in 50 mL Lysis Buffer B. Lyse the cells with Emulsiflex-C3 for 5 min at 4 °C (15,000–17,000 psi). Centrifuge the cell lysate for 30 min at 4 °C (18,000 rpm) and collect the supernatant. Filter the supernatant with 0.2 μm PES syringe filter.
2. Add 2 mL amylose resin reconstituted with Buffer B into the supernatant. Incubate for 2 hr at 4 °C with slow rotation.
3. Load the mixture on a gravity column, collect the flow-through, wash the column with 10 mL Wash Buffer B (×3), and then elute the protein with 1 mL Elution Buffer B (×6).
4. Repeat the steps 2 and 3 with the flow-through once again. Aliquot and store the purified protein at −80 °C.

3.6 Procedure for protein purification: Traptavidin

1. Suspend the cell pellet in 50 mL Lysis Buffer C. Lyse the cells with Emulsiflex-C3 for 5 min at 4 °C (15,000–17,000 psi). Centrifuge the cell lysate for 30 min at 4 °C (18,000 rpm) and discard the supernatant.
2. Suspend the cell pellet in 20 mL Extraction Buffer. Centrifuge for 10 min at 4 °C (16,000 rpm) and discard the supernatant (×3). For purification in a denaturation condition, suspend the cell pellet in 6 mL guanidinium hydrochloride.
3. For protein refolding, add 5 mL of the suspension slowly (drop by drop) into 200 mL Dilution Buffer with stirring at 400 rpm. Incubate the diluted sample for 16 hr at 4 °C with slow rotation.
4. Centrifuge the sample for 15 min at 4 °C (13,000 rpm) and collect the supernatant. Add 2 mL Ni-IDA resin reconstituted with PBS Buffer A into the supernatant. Incubate for 1 hr at 4 °C with slow rotation.
5. Load the mixture on a gravity column, wash the column with 10 mL Wash Buffer C (×3), and then elute the protein with 1 mL Elution Buffer C (×6). Aliquot and store the purified protein at −80 °C.

3.7 Notes

1. All the LB or TB media, glasses, and flasks must be autoclaved before the protein expression process. Keep the samples on ice during the whole time of protein purification.
2. To make a glycerol stock, add 200 μL of 75% glycerol into 500 μL cell culture after the step 5 of protein expression, and store it at −80 °C. In this step, a frozen piece can be taken from the glycerol stock, instead of a colony cultured on an agar plate.
3. If the concentration of GlpG or mSpyCatcher is low after protein purification, concentrate the protein sample at least up to 5–10 μM or 100 μM, respectively, using 10 kDa or 3 kDa Amicon® Ultra centrifugal filter unit. These concentration levels are required for the DNA handle attachment described in the following section.
4. In making Lysis Buffer C used at the step 1 of traptavidin purification, dissolve Triton X-100 using a spatula instead of vortexing or pipetting to prevent the generation of foams that can damage the high-pressure homogenizer Emulsiflex-C3.

4. DNA handle attachment to membrane proteins

The SpyCatcher-SpyTag conjugation is utilized to attach DNA handles to GlpG. To this end, the DNA handles are first linked to mSpyCatcher in tethering mode 3 or SpyCatcher003 in tethering mode 4, using the thiol-maleimide crosslinking. In the tethering mode 3, the DNA handles linked to mSpyCatcher are then attached to both the N- and C-termini of GlpG through the SpyCatcher-SpyTag conjugation. In the tethering mode 4, the DNA handle linked to SpyCatcher003 is attached to the N-terminus of GlpG through the SpyCatcher003-SpyTag003 conjugation, following the conjugation of SnoopCatcher-bt to the C-terminus of GlpG.

4.1 Equipment

- Mini-PROTEAN® Tetra vertical electrophoresis cell (Bio-Rad)
- High speed centrifuge Supra R22 (Hanil Scientific)
- ÄKTA pure™ 25 L (Cytiva)
- NanoDrop 2000 spectrophotometer (Thermo Fisher Scientific)
- Vortex mixer VM1 (LABTron)

4.2 Materials

- 250 mM SM(PEG)$_2$ (Thermo Fisher Scientific) in DMSO
- 1 M and 50 mM bicine (pH 8.3)
- 1 M $NaHCO_3$
- PBS Buffer B (0.1 M sodium phosphate, pH 7.2–7.4, 150 mM NaCl)
- Buffer C (1.24% DDM with 12 mM TCEP)
- Buffer D (50 mM Tris-HCl, pH 7.5, 150 mM NaCl, 1.5% bicelle, 1 mM TCEP)
- Storage Buffer B (25 mM Tris-HCl, pH 7.4, 150 mM NaCl, 10% glycerol, 1 mM TCEP)
- Start Buffer (20 mM Tris-HCl, pH 7.5)
- End Buffer (20 mM Tris-HCl, pH 7.5, 1 M NaCl)
- 20× Reaction Buffer (200 mM ATP, 200 mM MgOAc, 20 mM d-Biotin)
- 10× Taq Buffer (750 mM Tris-HCl, pH 8.8, 200 mM $(NH_4)_2SO_4$, 20 mM $MgCl_2$, 0.1% tween20)
- Lambda DNA (NEB)
- 100 μM PCR oligonucleotides (purchased from IDT; Table 1)
- Taq DNA polymerase with 10 mM dNTP mixture (Biofact)
- 5–10 μM SpyTag-GlpG-SpyTag, 5–10 μM SpyTag003-GlpG-SnoopTag, ~100 μM mSpyCatcher, ~100 μM SpyCatcher003, ~111 μM Snoop Catcher-AviTag, and 1 mg/mL BirA
- Hispeed plasmid maxi kit (Qiagen)
- PD-10 desalting column (Cytiva)
- HITRAP Q HP 5 mL (Cytiva)
- Amicon® Ultra-4 centrifugal filter unit (10 kDa) (Merck)
- Minitrap G-25 desalting column (Cytiva)
- Superdex 200 Increase 10/300 GL (Cytiva)

Table 1 Primer list for making DNA handles.

Primers in tethering mode 3 (1022 bp)

Antisense primer	/5AmMC12/ACA GAA AGC CGC AGA GCA
Sense primer 1	/52-Bio/TCG CCA CCA TCA TTT CCA
Sense primer 2	/5AzideN/TCG CCA CCA TCA TTT CCA
Primers in tethering mode 4 (2011 bp)	
Antisense primer	/5AmMC12/CGC CCG TGA AGA GAA AGA
Sense primer	/5AzideN/TCG CCA CCA TCA TTT CCA

4.3 Procedure for making SpyCatcher-conjugated DNA handles

1. Prepare a mixture for polymerase chain reaction (PCR) shown in Table 2 and then perform PCR using the thermocycling condition shown in Table 3. Purify the PCR product using Hispeed plasmid maxi kit. Elute with 1 mL $NaHCO_3$ equilibrated at room temperature (RT).
2. For maleimide modification at the amine-end of DNA, add 4 μL of 250 mM $SM(PEG)_2$ into the PCR product, vortex, and incubate the mixture for 20 min at RT. Purify the mixture using PD-10 desalting column. Elute with 1.5 mL PBS Buffer B equilibrated at RT.

Table 2 Composition of PCR mixture.

Volume	Component	Final concentration
800 μL	10× Taq Buffer	1×
40 μL	Taq DNA polymerase	1×
160 μL	dNTP mixture	0.2 mM
40 μL	Lambda DNA	2.5 μg/mL
40 μL	Antisense primer	0.5 μM
20 μL	Sense primer 1	0.25 μM
20 μL	Sense primer 2	0.25 μM
6880 μL	Milli-Q water	

Table 3 Thermocycling condition for PCR.

Step	Temperature	Time
1	95 °C	2 min
2	95 °C	20 s
3	52 °C	30 s
4	72 °C	2 min for 1022 bp 2.5 min for 2011 bp
	Repeat the steps 2–4 for 40 cycles	
5	72 °C	5 min
6	10 °C	10 min

3. For the conjugation of SpyCatcher to DNA, add 100 μL mSpyCatcher for tethering mode 3 or SpyCatcher003 for tethering mode 4 into the eluted sample. Incubate for 2 hr at RT. Here, the cysteine thiol groups of the SpyCatcher constructs are crosslinked to the maleimide moiety of DNA (Fig. 3A for the amino acid sequence of the SpyCatchers). Add 3.1 mL Start Buffer. Perform anion exchange chromatography using HITRAP Q HP column to remove unconjugated proteins (see Table 4 for details).
4. Measure the DNA concentration of eluted fractions (blank: a mixture of 20% Start Buffer and 80% End Buffer). Collect two fractions of the highest and second highest DNA concentrations. Concentrate the sample up to 500–600 nM as DNA concentration using Amicon® Ultra-4 centrifugal filter unit (10 kDa).
5. Estimate the conjugation yield using 6% SDS-PAGE (typically, 20–40%). The SpyCatcher-conjugated DNA handle (denoted as DNA-mSpyCatcher for tethering mode 3 or DNA-SpyCatcher003 for tethering mode 4) is estimated as 100–150 nM. Aliquot in 10 μL and store at −80 °C. Refer to Fig. 4A,D for gel data.

4.4 Procedure for DNA handle attachment to GlpG: Tethering mode 3

1. Mix 10 μL DNA-mSpyCatcher, 1.1 μL Buffer C, and 2 μL GlpG for tethering mode 3 (SpyTag-GlpG-SpyTag). Incubate for 2 hr at RT and then keep the sample at 4 °C.
2. Using 6% SDS-PAGE, estimate the conjugation yield for GlpG with distinct DNA handles at both the N- and C-terminus (i.e., azide-modified DNA handle at one end and 2× biotin-modified DNA handle at the other end).

Table 4 Condition for anion exchange chromatography.

Step	Volume
Equilibrium	5 mL Start Buffer (2 mL/min)
Sample injection	6 mL mixture for SpyCatcher-conjugated DNA handles (1 mL/min)
Washing	10 mL Start Buffer (2 mL/min)
Elution	60 mL End Buffer of 0–100% gradient (2 mL/min; 2 mL per fraction)
Cleaning	25 mL End Buffer (2 mL/min)

3. Dilute the mixture with Buffer D to make the target construct to be ~200 pM. Aliquot in 10 μL and store at −80 °C. Refer to Fig. 4B for gel data.

4.5 Procedure for DNA handle attachment to GlpG: Tethering mode 4

1. For the biotinylation of SnoopCatcher, prepare a mixture containing SnoopCatcher-AviTag, biotin, and BirA enzyme (see Table 5), and then incubate the mixture for 3 hr at 30 °C. Remove free biotins using SEC with Superdex 200 Increase 10/300 GL and then elute with Storage Buffer B. Aliquot SnoopCatcher-bt and store at −80 °C.
2. Mix GlpG for tethering mode 4 (SpyTag003-GlpG-SnoopTag) and SnoopCatcher-bt at approximately 1:1 ratio (total 10 μL at final ~2 μM each; see Notes). Incubate for 16 hr at 4 °C.
3. Mix 10 μL DNA-SpyCatcher003, 1 μL Buffer C, and 1 μL GlpG-SnoopCatcher-bt complex. Incubate for 3 hr at RT and then keep the sample at 4 °C.
4. Using 6% SDS-PAGE, confirm whether the gel band of DNA-SpyCatcher003 is completely shifted by the attachment to GlpG-SnoopCatcher-bt. Dilute the mixture with Buffer D to make the target construct to be ~500 pM. Aliquot in 10 μL and store at −80 °C. Refer to Fig. 4E for gel data.

4.6 Notes

1. Run the gel electrophoresis for 60 min at 150 V for the 1022-bp DNA handle or 120 min at 150 V on ice for the 2011-bp DNA handle. After the gel running, stain the DNA handles, not proteins, to confirm the successful attachments.
2. For a higher biotinylation efficiency, exchange the buffer for SnoopCatcher-AviTag to 50 mM bicine (pH 8.3) using Minitrap G-25 desalting column.

Table 5 Mixture components for protein biotinylation.

Volume	Component
540 μL	~111 μM SnoopCatcher-AviTag
30 μL	1 M bicine (pH 8.3)
30 μL	20× Reaction Buffer
6 μL	1 mg/mL BirA

3. In the SEC chromatogram for removing free biotins, the largest and second largest peaks correspond to the biotin and biotinylated protein, respectively.
4. The 1:1 ratio condition for the mixture of GlpG and SnoopCatcher-bt means a concentration ratio at which the yield of the protein complex becomes a possible maximum from the gel analysis, not 1:1 molar ratio based on the measured protein concentrations.

5. Passivation of coverslip surfaces

The coverslip surfaces to which the molecular constructs are tethered are passivated using a mixture of biotin-PEG and methyl-PEG (mPEG) at 1:100 as a mass ratio. The biotin-PEG allows for specific binding of target constructs to the surface, while the mPEG prevents nonspecific binding of magnetic beads and molecular constructs to the surface. The surface passivation procedure includes KOH etching, piranha cleaning, silanization for amine functionalization, and finally, PEG coating.

5.1 Equipment

- Bransonic® CPXH ultrasonic cleaning bath (Branson)
- High speed centrifuge Supra R22 (Hanil Scientific)
- Vortex mixer VM1 (LABTron)

5.2 Materials

- Methanol
- 95% sulfuric acid
- 30% hydrogen peroxide
- Acetic acid
- N-(2-Aminoethyl)-3-aminopropyltrimethoxysilane (amino-silane) (Merck)
- 1 M KOH
- 1 M $NaHCO_3$
- mPEG-SVA, MW 5000 (Laysan bio) (mPEG)
- Biotin-PEG-SVA, MW 5000 (Laysan bio) (biotin-PEG)
- Micro cover glass (24 × 40 mm, No 1.5) (VWR)
- Micro cover glass (24 × 50 mm, No 1.5) (VWR)
- Customized Teflon tank
- Customized Teflon jig
- Teflon tweezers

5.3 Procedure

1. Put 100 coverslips of 24 × 40 mm and 24 × 50 mm, each into two Teflon jigs. Pour 800 mL methanol each into two Teflon tanks and put the coverslip-containing jigs containing the coverslips. Slowly move the jigs inside back and forth (×10). After incubating for 10 min, wash the coverslips and Teflon tanks with water.
2. Pour 800 mL of 1 M KOH into the tanks for KOH etching. Sonicate the coverslips for 30 min. Wash the coverslips and tanks with water. Remove the jigs from the tanks.
3. For piranha cleaning of 24 × 50 mm coverslips, pour 600 mL sulfuric acid and 300 mL of 30% hydrogen peroxide (equilibrated at RT) into a Teflon tank. Mix the solution using Teflon tweezers and put the jig containing 24 × 50 mm coverslips into the tank immediately. After incubating for 30 min, wash the jig and tank with water. Remove the jig from the tank.
4. For silanization, pour 760 mL methanol, 38 mL acetic acid, and 7.6 mL amino-silane each into the two Teflon tanks. Stir the mixture and put the two coverslip-containing jigs into the tanks, immediately. Incubate for 30 min. Remove the jigs from the tanks and dump the mixture.
5. Pour 800 mL methanol and then put the jigs back into the tanks. Slowly move the jigs inside back and forth (×10). After incubating for 2 min, wash the jigs and tanks with water. Put the cleaned coverslips into 50 mL conical tubes (2 coverslips per one tube). Centrifuge for 5 min at 20 °C (2000 rpm).
6. Prepare a PEG solution for PEGylation on 24 × 50 mm coverslips. Add 200 mg mPEG and 4 mg biotin-PEG into 800 μL and 16 μL $NaHCO_3$, respectively (here, PEGylation for 20 coverslips). Vortex until transparent. Add 10 μL biotin-PEG solution into the mPEG solution, and vortex again.
7. Put ~500 μL PEG solution into two 1.5 mL tubes, centrifuge for 3 min at 13,000 rpm, and combine ~420 μL solution of the upper side each into a fresh 1.5 mL tube.
8. Put 80 μL PEG solution on the center of each coverslip. Pile another coverslip onto the coverslip (total 10 coverslip pairs). Incubate for 4–16 hr at RT.
9. Separate the two piled coverslips and then put them into a Teflon jig. Wash the jig with water for 2 min. Put the coverslips into 50 mL conical tubes (2 coverslips per one tube). Centrifuge for 5 min at 20 °C (2000 rpm). Store the PEGylated coverslips at −20 °C.

5.4 Notes

1. During the piranha cleaning, always wear acid-resistant gloves and face shield. Do not pour sulfuric acid into hydrogen peroxide (the reverse direction) as it could cause an explosion. The temperature will go up over 100 °C and do not decrease the temperature on purpose. Cool down the piranha solution for 2 hr after the piranha cleaning. Dilute the piranha solution with 1 L water and throw out.
2. During the piling of coverslips for PEGylation, air bubbles can be made between the coverslips. In this case, remove the bubbles by moving a coverslip on the other one back and forth using a pipette tip. During the PEGylation, keep the humidity by incubating the piled coverslips in a microtube storage box containing water to readily separate the coverslips.

6. Making single-molecule sample chambers

The hybrid protein-DNA constructs are initially attached to the glass surface through the traptavidin-(2×)biotin binding and then linked to magnetic beads through the DBCO-azide conjugation. Non-magnetic polystyrene beads are also affixed to the glass surface through streptavidin-biotin binding to correct for microscope stage drift. Bovine serum albumin (BSA) is briefly incubated in the sample chamber to further passivate the glass surface. Unoccupied biotin-binding pockets of traptavidin are blocked with a biotin-modified DNA oligo to prevent possible nonspecific adhesion. Following all the molecular tethering procedures, detergent micelles are exchanged with bicelles for a lipid bilayer reconstitution for membrane proteins.

6.1 Equipment

- Micro centrifuge Smart R17 Plus (Hanil Scientific)
- Vortex mixer VM1 (LABTron)

6.2 Materials

- 1,2-dimyristoyl-sn-glycero-3-phosphocholine (DMPC) (Avanti)
- 1,2-dimyristoyl-sn-glycero-3-phospho-(1′-rac-glycerol) (DMPG) (Avanti)
- 3-([3-Cholamidopropyl] dimethylammonio)-2-hydroxy-1-propanesulfonate (CHAPSO) (Merck)
- 1 mM dibenzocyclooctyne-sulfo-N-hydroxysuccinimidyl ester (DBCO-sulfo-NHS ester; Merck) in DMSO
- 100 μM biotin-modified DNA oligo (30 nt)
- PBS Buffer B (0.1 M sodium phosphate, pH 7.2–7.4, 150 mM NaCl)

- Storage Buffer C (PBS Buffer B, 0.01% Tween 20)
- PBST0.1 (PBS Buffer B, 0.1% Tween 20)
- Buffer E (50 mM Tris-HCl, pH 7.4, 150 mM NaCl)
- Buffer F (50 mM Tris-HCl, pH 7.4, 150 mM NaCl, 0.1% DDM)
- Buffer G (50 mM Tris-HCl, pH 7.4, 150 mM NaCl, 0.05% DDM)
- Buffer H (50 mM Tris-HCl, pH 7.4, 150 mM NaCl, 2% bicelle)
- 100 mg/mL BSA (Merck) in PBS Buffer B
- 2 μM traptavidin
- Dynabeads™ M-270 Amine (Thermo Fisher Scientific)
- Streptavidin-coated polystyrene bead (Spherotech SVP-10-5)
- DynaMag™-2 Magnet (InvitrogenTM)
- Surface-passivated coverslips prepared in the previous section

6.3 Procedure for bicelle stock preparation

1. Add 213.6 mg DMPC (or 216.9 mg DMPG) and 79.5 mg CHAPSO into 2.7 mL milli-Q water (lipid:detergent = 2.5:1 as a molar ratio), and vortex until transparent.
2. Cool the sample on ice for 10 min and vortex. Freeze the sample at –80 °C for 10 min. Thaw it at 42 °C for 10 min and vortex. Repeat this cooling-freezing-thawing cycle (more than three times).
3. Cool the sample on ice for 10 min and vortex. Add milli-Q water to make total 3 mL bicelle solution (~10% total amphiphile). Put 1 mL solution each into three 1.5 mL tubes. Centrifuge for 3 min at 4 °C (13,000 rpm).
4. Combine ~950 μL solution of the upper side each into a 15 mL conical tube. Aliquot in 30 μL and store at –80 °C.

6.4 Procedure for surface modification of magnetic beads

1. Add 35 μL amine-modified M-270 magnetic bead into a 1.5 mL tube. Wash the bead slurry with 1 mL Storage Buffer C with DynaMag™-2 Magnet (×3).
2. Add 100 μL Storage Buffer C and 10 μL DBCO-sulfo-NHS. Mix the bead slurry by pipetting and vortex. Incubate for 3 hr at RT with slow rotation.
3. Wash the bead slurry with 1 mL Storage Buffer C with the magnet (×5). Add 35 μL Storage Buffer C, vortex, and store at 4 °C.

6.5 Procedure for sample chamber assembly

1. To reconstitute the streptavidin-coated polystyrene bead, add 1 μL polystyrene bead stock into 500 μL PBST0.1, vortex briefly, centrifuge

for 3 min at 13,000 rpm, and then discard the supernatant (×2). Add 50 μL PBST0.1, vortex, and store at 4 °C.

2. Add 1 μL traptavidin into 9 μL Buffer F (0.2 μM traptavidin). For the traptavidin binding to the target protein-DNA construct, mix 10 μL of ~200 pM protein-DNA construct (prepared in Sections 4.4, 4.5) and 1 μL diluted traptavidin. Incubate for 15 min at RT.
3. Construct a sample chamber using a pair of 24 × 40 mm and 24 × 50 mm coverslips and double-sided tape. The channel volume (CV) is ~10 μL.
4. Inject 1 CV of the reconstituted polystyrene bead and incubate for 2 min at RT. Inject 5 CV of 100 mg/mL BSA and incubate for 5 min at RT.
5. Wash with 5 CV Buffer E and then 5 CV Buffer F. Inject 1 CV mixture of the protein-DNA construct and traptavidin. Incubate for 5 min in tethering mode 3 or 10 min in tethering mode 4 at RT.
6. Add 1 μL of 100 μM biotin-modified DNA oligo into 9 μL Buffer F (10 μM DNA oligo). Inject 1 CV of the DNA oligo and incubate for 5 min at RT.
7. To reconstitute the DBCO-modified magnetic bead, add 1 μL magnetic bead stock into 500 μL Buffer E, vortex briefly, and discard the supernatant using the magnet. Add 50 μL Buffer G, vortex briefly, and discard the supernatant using the magnet. Add 20 μL Buffer G and vortex.
8. Wash the sample chamber by injecting 5 CV Buffer G. Inject 1 CV of the reconstituted magnetic bead and incubate for 1 hr at 25 °C. Slowly wash the sample chamber with 5–10 CV of Buffer H. For long measurements lasting tens of hours, apply epoxy at channel openings to prevent water evaporation in the sample chamber.

6.6 Notes

1. In dissolving lipids and detergents for the bicelle preparation, it is helpful to break big lipid/detergent chunks using pipetting and cover the tube with ice at −80 °C in the freezing step.
2. Keep all samples on ice during the whole time. In the final 1-hr incubation, keep the humidity by incubating the sample chamber in a microtube storage box containing water to prevent the evaporation of sample buffer.
3. The DBCO-azide crosslinking for magnetic bead attachment is highly sensitive to the lipids and detergents contained in the buffer. As in the described procedure, reconstitute the magnetic bead exactly with 0.05% DDM.

7. Conclusion

This chapter provides the detailed procedures in constructing single-molecule tethering systems for membrane protein studies. The stability of the tethering mode 3 surpasses that of the conventional tethering relying on dig-antidig binding by ~100-fold, as shown in the next chapter. It is reasonable to expect that the tethering mode 4 will also exhibit the high stability, given that all the tethering points are covalent except for traptavidin, same as in the tethering mode 3. These robust molecular tethering methods will allow us to conduct more stable investigation of structural dynamics of a single membrane protein under force. In the next chapter, we will present an application example of the tethering methods, focusing on membrane protein folding studies.

Acknowledgments

This work was supported and funded by the National Research Foundation of Korea (NRF-2020R1C1C1003937 to D.M.).

References

Chivers, C. E., Crozat, E., Chu, C., Moy, V. T., Sherratt, D. J., & Howarth, M. (2010). A streptavidin variant with slower biotin dissociation and increased mechanostability. *Nature Methods, 7*(5), 391–393.

Choi, H. K., Kang, H., Lee, C., Kim, H. G., Phillips, B., Park, S., ... Yoon, T. Y. (2022a). Evolutionary balance between foldability and functionality of a glucose transporter. *Nature Chemical Biology, 18*(7), 713–723.

Choi, H. K., Kim, H. G., Shon, M. J., & Yoon, T. Y. (2022b). High-resolution single-molecule magnetic tweezers. *Annual Review of Biochemistry, 91*, 33–59.

Choi, H. K., Min, D., Kang, H., Shon, M. J., Rah, S. H., Kim, H. C., ... Yoon, T. Y. (2019). Watching helical membrane proteins fold reveals a common N-to-C-terminal folding pathway. *Science (New York, N. Y.), 366*(6469), 1150–1156.

De Vlaminck, I., & Dekker, C. (2012). Recent advances in magnetic tweezers. *Annual Review of Biophysics, 41*, 453–472.

Fairhead, M., & Howarth, M. (2015). Site-specific biotinylation of purified proteins using BirA. *Methods in Molecular Biology, 1266*, 171–184.

Hong, H., Choi, H. K., & Yoon, T. Y. (2022). Untangling the complexity of membrane protein folding. *Current Opinion in Structural Biology, 72*, 237–247.

Keeble, A. H., Turkki, P., Stokes, S., Anuar, I. N. A. K., Rahikainen, R., Hytonen, V. P., & Howarth, M. (2019). Approaching infinite affinity through engineering of peptide-protein interaction. *Proceedings of the National Academy of Sciences of the United States of America, 116*(52), 26523–26533.

Kim, E., & Koo, H. (2019). Biomedical applications of copper-free click chemistry: In vitro, in vivo, and ex vivo. *Chemical Science, 10*(34), 7835–7851.

Kim, S., Lee, D., Wijesinghe, W. C. B., & Min, D. (2023). Robust membrane protein tweezers reveal the folding speed limit of helical membrane proteins. *Elife, 12*, e85882.

Ma, Z., LeBard, D. N., Loverde, S. M., Sharp, K. A., Klein, M. L., Discher, D. E., & Finkel, T. H. (2014). TCR triggering by pMHC ligands tethered on surfaces via poly (ethylene glycol) depends on polymer length. *PLoS One, 9*(11), e112292.

Min, D., Arbing, M. A., Jefferson, R. E., & Bowie, J. U. (2016). A simple DNA handle attachment method for single molecule mechanical manipulation experiments. *Protein Science: A Publication of the Protein Society, 25*(8), 1535–1544.

Min, D., Jefferson, R. E., Bowie, J. U., & Yoon, T. Y. (2015). Mapping the energy landscape for second-stage folding of a single membrane protein. *Nature Chemical Biology, 11*(12), 981–987.

Min, D., Jefferson, R. E., Qi, Y., Wang, J. Y., Arbing, M. A., Im, W., & Bowie, J. U. (2018). Unfolding of a ClC chloride transporter retains memory of its evolutionary history. *Nature Chemical Biology, 14*(5), 489–496.

Van Patten, W. J., Walder, R., Adhikari, A., Okoniewski, S. R., Ravichandran, R., Tinberg, C. E., ... Perkins, T. T. (2018). Improved free-energy landscape quantification illustrated with a computationally designed protein-ligand interaction. *Chemphyschem: A European Journal of Chemical Physics and Physical Chemistry, 19*(1), 19–23.

Veggiani, G., Nakamura, T., Brenner, M. D., Gayet, R. V., Yan, J., Robinson, C. V., & Howarth, M. (2016). Programmable polyproteams built using twin peptide superglues. *Proceedings of the National Academy of Sciences of the United States of America, 113*(5), 1202–1207.

Wijesinghe, W. C. B., & Min, D. (2023). Single-molecule force spectroscopy of membrane protein folding. *Journal of Molecular Biology*, 167975.

Zakeri, B., Fierer, J. O., Celik, E., Chittock, E. C., Schwarz-Linek, U., Moy, V. T., & Howarth, M. (2012). Peptide tag forming a rapid covalent bond to a protein, through engineering a bacterial adhesin. *Proceedings of the National Academy of Sciences of the United States of America, 109*(12), E690–E697.

CHAPTER ELEVEN

Robust magnetic tweezers for membrane protein folding studies

Seoyoon Kim[a] and Duyoung Min[a,b,*]
[a]Department of Chemistry, Ulsan National Institute of Science and Technology, Ulsan, Republic of Korea
[b]Center for Wave Energy Materials, Ulsan National Institute of Science and Technology, Ulsan, Republic of Korea
*Corresponding author. e-mail address: dymin@unist.ac.kr

Contents

Abstract

Single-molecule magnetic tweezers have recently been adapted for monitoring the interactions between transmembrane helices of membrane proteins within lipid bilayers. In this chapter, we describe the procedures of conducting studies on membrane protein folding using a robust magnetic tweezer method. This tweezer method is capable of observing thousands of (un)folding transitions over extended periods of several to tens of hours. Using this approach, we can dissect the folding pathways of membrane proteins, determine their folding time scales, and map the

Methods in Enzymology, Volume 694
ISSN 0076-6879, https://doi.org/10.1016/bs.mie.2023.12.014

folding energy landscapes, with a higher statistical reliability. Our robust magnetic tweezers also allow for estimating the folding speed limit of helical membrane proteins, which serves as a link between the kinetics and barrier energies.

1. Introduction

Magnetic tweezers have emerged as a powerful approach to membrane protein folding studies at nm-scale precision (Hong, Choi, & Yoon, 2022; Wijesinghe & Min, 2023). The key advantage lies in the ability to perform near-equilibrium measurements in a low force range, enabling the observation of transmembrane (TM) helix interactions within lipid bilayers (Choi, Kim, Shon, & Yoon, 2022; Wijesinghe & Min, 2023). However, conventional approaches face challenges, particularly with the large mechanical hysteresis exhibited by membrane proteins, even at very low force-loading rates like 1 pN/s (Choi et al., 2019; Choi, Kang, et al., 2022; Lu et al., 2018; Min, Jefferson, Bowie, & Yoon, 2015; Min et al., 2018). The large hysteresis hampers the repetitive observation of numerous and sustained transitions under constant forces. Recent attempts have captured the folding transitions for some membrane proteins at a low force level of 5–7 pN, albeit limited to short time scales, typically hundreds of seconds (Choi et al., 2019; Choi, Kang, et al., 2022). This low-force range strongly favors the native state and remains quite narrow, thus limiting the extensive sampling of intermediate states and hindering reliable zero-force extrapolation analysis.

In this context, we have recently developed a robust magnetic tweezer method to address these limitations in membrane protein folding studies (Kim, Lee, Wijesinghe, & Min, 2023). Our tweezer system exhibits improved stability with a survival time of ~12 h at 50 pN, enabling th extended observation of repetitive transitions across a wide force spectrum. This method offers several advantages over previous tweezer approaches to membrane protein folding, including: (1) enhanced clarity in observing (un)folding steps owing to reduced bead fluctuations at higher forces, (2) more reliable characterization of (un) folding properties due to the increased number of transitions from a single molecule, (3) higher stability in using a model-free deconvolution method for energy landscape reconstruction, (4) improved reliability of zero-force extrapolation analysis for the kinetics and thermodynamics due to a broad force coverage, and (5) systematic monitoring of changes in folding properties over an extended period (e.g., impact from oxidative damage).

In this chapter, we present how this method can be utilized to investigate various aspects of membrane protein folding, including folding pathways, kinetics, energy landscapes, and the folding speed limit.

2. Robust membrane protein tweezers

2.1 Instrumentation

Magnetic tweezer apparatuses are constructed on inverted microscopes (Olympus, IX73) along with motorized XY stages (ASI, MS-2000FT or Prior Scientific, H117P1), as previously described (Kim et al., 2023; Min et al., 2015; Min, Arbing, Jefferson, & Bowie, 2016; Min et al., 2018). A light-emitting diode (Thorlabs, M455L4, λ = 447 nm) is utilized to illuminate the magnetic beads tethered to target molecules and the reference non-magnetic beads affixed to the sample chamber surface. Diffraction patterns of the beads are monitored at a rate of 60–90 Hz using a charge-coupled device camera (JAI, CM-040GE or CM-030GE). Heights of a magnetic bead from the chamber surface is calibarated from its diffraction pattern changes by moving the focal plane of an objective lens (Olympus, UPLFLN100XO2 or UPLXAPO100XO) in known increments with a piezo nanopositioner (Mad City Labs, Nano-F100S). The extension (end-to-end distance) of a tethered molecule, represented by the bead height, is tracked via chi-square analysis with the calibration data. The reference beads immobilized on the chamber surface are also tracked to correct for XY stage drift. Magnetic field gradients are established using a pair of permanent neodymium magnets configured with antiparallel magnetic moments (each 10×10×12 mm; separated by 1 mm). The vertical and rotational movements of the magnets are controlled using a translational motor (PI, M-126. PD1 or M-126. PD2) and a rotational motor (PI, DT-50 or DT-34). The mechanical force applied to the tethered molecule is calibrated as a function of magnet position using the formula, $F = k_B T \cdot L/\delta x^2$, which is derived from the inverted pendulum model (Sarkar & Rybenkov, 2016). In this equation, F is the applied force, k_B is the Boltzmann constant, T is the absolute temperature, L is the extension, and δx^2 is the magnitude of lateral fluctuations. The magnetic tweezers are mounted on pneumatic optical tables (Daeil Systems, DVIO-I-1512M-300t-800H) to mitigate environmental vibrations. The imaging rooms are maintained at a temperature range of 20–22 °C.

2.2 Single membrane protein tethering

A designer single-chained TM homodimer (scTMHC2) is adopted as a model protein for our membrane protein folding studies (Kim et al., 2023; Lu et al., 2018). Individual membrane proteins are tethered to both the glass surface and magnetic beads within a sample chamber through ~1000-bp DNA handles (Fig. 1A; see tethering mode 3 in the previous chapter). The bicelles consisting of DMPC lipids and CHAPSO detergents are adopted to reconstitute a lipid bilayer environment for membrane proteins. All the detailed procedures, necessary materials, and buffer conditions for the molecular tethering can be found in the previous chapter. The time-resolved extension trace of a tethered molecule is mointored using a calibrated extension profile. Real-time force-extension curves are generated from the extension trace and calibrated force profile (Fig. 1B). The bead tracking and data visualization are achieved using custom-built

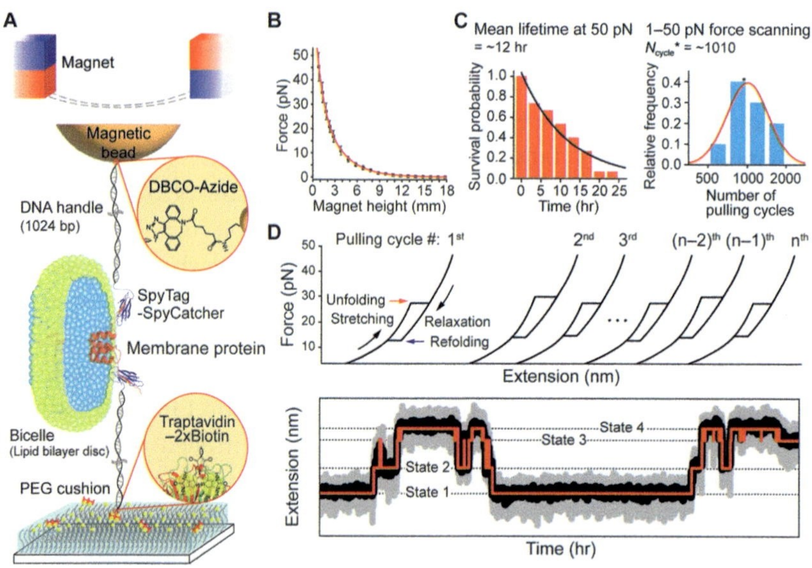

Fig. 1 Robust membrane protein tweezers for membrane proteins. (A) Schematics of our single-molecule tethering system. (B) Calibrated force as a function of magnet height. (C) Robustness evaluation of the tweezer system. The left panel shows the survival probability of single-molecule tethering at a constant force of 50 pN over time ($N = 15$ molecules). The right panel shows the distribution of repetitive pulling numbers until bead detachment (magnet speed: 0.5 mm/s, force scan: 1–50 pN, $N = 10$ molecules). (D) Schematic diagrams depicting the experiments using the constant-speed mode (upper) and the constant-force mode (lower). *Adapted for panels A–C from our earlier work with permission, Kim et al., (2013).*

LabVIEW programs. The membrane protein (un)folding experiments are conducted exclusively for the molecules that are singly tethered to the magnetic beads, following these criteria:

1. During a force scan ranging from 1 pN to 50 pN at a constant magnet speed of 0.1–0.5 mm/s, the force-extension curves should match those expected from the extensible worm-like chain (eWLC) polymer model for the sub-μm DNA handles (Shon, Rah, & Yoon, 2019).
2. The unfolding step sizes should also match those expected from a combined polymer model of the worm-like chain (WLC) model (Marko & Siggia, 1995) and the Kessler-Rabin (KR) model (Kessler & Rabin, 2004) for a membrane protein. For helical membrane proteins, the helix-coil transition is observed to occur at ~18 pN (Kim et al., 2023). Thus, at forces above 18 pN, the unfolded polypeptide state with all helices fully unstructured (the unfolded coil state; U_c) is solely modeled by the WLC model. At forces below 18 pN, the unfolded helical state with all helices fully structured (U_h) is modeled by the KR model for the helices and the WLC model for the polypeptide linkers in between (Kim et al., 2023). At forces close to 18 pN, the unfolding step sizes appear in the helix-coil transition zone bounded by the two unfolded states.

2.3 Evaluation of system robustness

The robustness of the magnetic tweezers is assessed using two key metrics: the survival probability at a constant force and the number of pulling cycles at a constant magnet speed. The survival probability is determined by measuring the total duration at a constant force until bead detachment. The number of pulling cycles is determined by counting the repetitive stretching-relaxation cycles at a constant magnet speed until bead detachment, with a waiting time of 1 s between the cycles (enough for refolding of scTMHC2; 1–50 pN force scanning). For the first robustness test, a constant force of 50 pN is chosen as it exceeds the typical unfolding-force range observed in optical/magnetic tweezers (Kim et al., 2023). For the second robustness test, a constant magnet speed of 0.5 mm/s is chosen as it is relatively fast while still ensuring clear identification of unfolding steps. From the metrics obtained under these experimental conditions, our single-molecule magnetic tweezers exhibit a mean lifetime of ~12 h at 50 pN and a mean pulling-cycle number of ~1000 at 0.5 mm/s (Fig. 1C). These results indicate that the tweezer system is >100 times more stable than conventional systems previously used in membrane protein folding studies (Kim et al., 2023).

3. Two modes of (un)folding experiments

Two types of (un)folding experiments are mainly conducted using the robust membrane protein tweezers: the constant-speed mode (Fig. 1D, upper) and the constant-force mode (Fig. 1D, lower). Using the constant-speed mode, where the magnet speed remains constant, we monitor the force-dependent (un)folding steps and analyze the (un) folding force distribution to characterize zero-force (un)folding kinetics. This mode can be transformed into the constant force-loading mode by adjusting the magnet speed at different magnet heights. Using the constant-force mode, we monitor the (un)folding steps at constant forces, more clearly probing potential intermediate states. Under this mode, the (un)folding kinetics and energy landscapes can be characterized at selected applied forces. The zero-force kinetics and landscapes are obtained by extrapolation analysis from the data. This mode also allows us to estimate the so-called folding speed limit of helical membrane proteins. Detailed procedures and relevant analyses in each mode are outlined in the following sections.

4. (Un)folding studies using constant-speed mode

4.1 Structural state analysis

1. The stretching and relaxation cycles are repetitively performed for a singly tethered molecule in a force range of 1–50 pN at a constant magnet speed. Sufficient waiting times between the cycles are given to allow for refolding (Fig. 2A). The waiting time is set as 1 s, which is enough for refolding of scTMHC2. The appropriate duration of waiting time depends on target membrane proteins, for example, hundreds of seconds for a rhomboid protease GlpG (Min et al., 2015). Typically, a range of 0.1–0.5 mm/s is adopted for the magnet speed. Force-extension curves of the protein-DNA hybrid molecule are generated in real-time based on the extension and force values.
2. To enhance clarity in the observation, the force-extension curves are median-filtered for the extension data with a 6–9 Hz window and smoothed for the force data with a 3–4.5 Hz window. The clear observation of refolding steps largely depends on the magnet speed and specific protein. In constrast, unfolding steps are typically well resolved at the magnet speed of 0.1–0.5 mm/s.

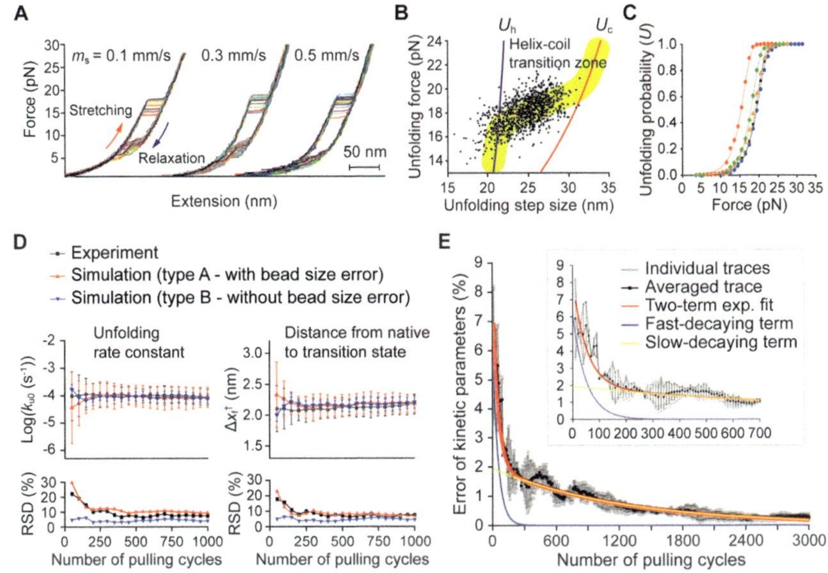

Fig. 2 Membrane protein (un)folding and kinetics characterization using constant-speed mode. (A) Force-extension curves of scTMHC2 attached to DNA handles at constant magnet speeds (m_S, 0.1–0.5 mm/s). (B) Scatter plot of unfolding forces and step sizes of scTMHC2 with theoretical lines of polymer models ($N = 5$ molecules, $n = 1000$ data points). (C) Unfolding probability as function of force ($N = 5$ molecules). (D) Trends of kinetic parameters with increasing pulling cycles. RSD, denoting relative standard deviation, is defined as the ratio of the standard deviation to the mean. (E) Error in the kinetic parameters during progression of pulling cycles. *Adapted from our earlier work with permission, Kim et al., (2013).*

3. The scatter-plotted data of unfolding forces and step sizes are compared with the expected 'protein' force-extension curves for the U_c and U_h states of helical membrane proteins (Fig. 2B; see above for each state for more details). The U_c state is modeled using the Marko–Siggia formula of the WLC model (Kim et al. 2023; Marko & Siggia, 1995), $FP_c/k_BT = l_c/C_c + (1 - l_c/C_c)^{-2}/4 - 1/4$. Here, F is the applied force, l_c is the protein extension of the U_c state, P_c is the polypeptide persistence length of 0.4 nm, and C_c is the polypeptide contour length, estimated as the product of the number of unfolded residues from tertiary structure and the average residue-residue distance of 0.36 nm.
4. For the U_h state, the helices are modeled using the Kessler-Rabin (KR) formula (Kessler & Rabin, 2004; Kim et al., 2023), $l_h = -1/2f - \alpha/(f\cdot\tanh(2\alpha)) + C_h/\tanh(fC_h) - (2\alpha^2/3f)\cdot(1/\tanh(fC_h) - fC_h/\sinh^2(fC_h) - 1)$,

$f = F/k_BT$, $\alpha = \sqrt{fC_h^2/4P_h}$. Here, l_h is the protein extension of the U_h state, P_h is the helix persistence length of 9.17 nm, and C_h is the helix contour length, estimated as the product of the number of helix residues and the average helical rise per residue of 0.16 nm. The intervening peptide linkers in the U_h state are modeled using the Marko–Siggia formula.

5. The protein force-extension curves are adjusted for the end-to-end distance of the pulling points of tertiary structure (Δd), since the extension changes during unfolding are measured smaller by the Δd.
6. The TM helical portions of the U_h state can be inserted into the lipid membrane at low forces (Choi et al., 2019; Kim et al., 2023). However, in the U_c state, the corresponding polypeptide portions are unlikely to remain within the lipid membrane due to the energetic cost associated with the polar backbone in contact with nonpolar lipid tails (Kim et al., 2023). Therefore, the unfolding at high forces above the helix-coil (U_h-U_c) transition zone (~18 pN) will be a combined process involving the unfolding and extraction from the membrane.
7. The refolding transitions typically occur at low forces below 18 pN. In this case, the reference unfolded state just before the folding is the U_h state, requiring the polymer model analysis for the U_h state, as described above.

4.2 Characterization of (un)folding kinetics

1. The cumulative plots of unfolding force distributions are normalized to the unfolding probabilities as a function of force. To characterize the unfolding kinetics (k_{u0} and $\Delta x_f^\dagger$), the unfolding probability profiles are fitted using the formula, $U = 1 - \exp(\int dF(-k_{u0}\cdot\exp(F\Delta x_f^\dagger/k_BT)/\dot{F}))$, which is derived from the first-order rate equation, $dN/dt = -k_uN$, and the Bell equation, $k_u = k_{u0}\cdot\exp(-F\Delta x_f^\dagger/k_BT)$ (Fig. 2C). Here, F is the force, U is the unfolding probability, N (=1−U) is the folding probability, k_u (k_{u0}) is the rate constant at an arbitrary (zero) force, $\Delta x_f^\dagger$ is the distance between the native and transition states, and $\dot{F}$ (=dF/dt) is the force-loading rate. The folding kinetics can be characterized in a similar manner.
2. The complex formula can be simplified as follow (Kim et al., 2023). In our tweezer setup, dF/dt (=$\dot{r}dF/dz$) can be approximated as a linear function of the force (aF+b) within the force range of $U = 0$ to 1. Here, dF/dz is the first derivative of force with respect to magnet height and $\dot{r}$ (=dz/dt) is the constant magnet speed. The constants of a and b are determined by fitting dF/dt as a function of F. To further simplify the formula, the exponential integral function ($\int \exp(x)/x dF$) is approximated as the first term of series

expansion ($\exp(x)/x$). This approximation is valid within an error of ~5% because the higher-order terms contribute only ~5% of the first term within the unfolding force range. The fitting formula is finally simplified to $U = 1 - \exp(-k_{u0}k_B T \cdot \exp(F\Delta x_f^{\dagger}/k_B T)/(\Delta x_f^{\dagger}(aF+b)))$.

4.3 Error estimation for (un)folding kinetics

1. Magnetic beads have an error in their diameters (~3% for M-270 bead used here as per the manual), leading to inaccuracies in the unfolding forces/probabilities and thus the characterized kinetics. Another error source in the kinetics is a low number of data points (i.e., a low number of pulling cycles). To estimate the kinetic error, the parameters of kinetics are obtained every 50 pulling cycles from multiple molecules. Although the relative standard deviation (RSD; defined as the ratio of the standard deviation to the mean) ranges from a few to tens of %, the mean values of the kinetic parameters converge after the initial hundred cycles, indicating that the bead size effect is mitigated by the averaging (Fig. 2D). This trend is further confirmed by Monte Carlo (MC) simulations (Kim et al., 2023).
2. The mean values of the kinetic parameters are still subject to the other error from the number of pulling cycles. The MC simulations for probability profiles with various curve shapes yield possible error estimates relevant to the error source (Fig. 2E). The two-term exponential model shows a better fit, compared to the reciprocal square root model (Kim et al., 2023). The regression model employed in our analysis is a phenomenological model that provides a more accurate description of the error decay curve.
3. From the regression analysis, the kinetic error curve is separated into two components: the fast and slow decays, with $\tau = 52$ and 1223 cycles, respectively (Fig. 2E). With correction of instrumental errors (Cossio, Hummer, & Szabo, 2015; Jacobson & Perkins, 2020; Neupane & Woodside, 2016), the kinetic error can be estimated as approximately 3.7%, 1.8%, and 0.9% at 60, 200, and 1000 pulling cycles, respectively (see the next section for more details of the correction). Based on this error estimation, it is evident that after a few initial hundred cycles, increasing the number of pulling cycles up to 1000 does not significantly enhance the accuracy; it only marginally reduces the error by ~1%. This suggests that blindly performing a large number of stretching-relaxation cycles on single molecules does not provide substantial benefits in a practical sense.

5. (Un)folding studies using constant-force mode

5.1 Structural state analysis

1. To induce complete unfolding to an unstructured random coil (i.e., the U_c state), a relatively high force is initially applied at a rapid magnet speed of 15–25 mm/s. For scTMHC2, 30 pN is sufficient for the complete unfolding. Once unfolded, the force is quickly reduced to a lower value to enable its refolding. The force is kept constant to observe repetitive unfolding and refolding events over an hours-long duration (Fig. 3A). The (un)folding transitions are manifested as stepwise changes in extension.
2. To correct for a long-term drift, the 90-Hz extension trace, measured over 9 h at 12 pN in our example case, is first smoothed with a 1.8-Hz window. The long-term drift is eliminated by subtracting a polynomial-fit baseline for regions of the native state (N) in the smoothed trace.
3. The original trace with the drift corrected is processed by median-filtering with a 4.5-Hz window to more clearly identify the (un)folding transitions (Fig. 3A). The extension trace is then transformed into an extension distribution. The Gaussian peaks in the extension distribution are compared to the extension values expected from particular structural states, which are estimated by the WLC and KR polymer models (Fig. 3A, right, and Fig. 3B; see above for the polymer model analysis for more details).
4. The four structural states matching the extension positions are identifed for the example case of scTMHC2 (Kim et al., 2023): the two unfolded states (U_c and U_h), the intermediate helical state with one monomer unfolded and the other folded ($I_{1,2}$), and the N state. The I_1 and I_2 states are indistinguishable as their step sizes are identical.
5. The penetration angles and depths of the TM helices within lipid membrane vary depending on the applied force (Kim et al., 2023) (Fig. 3C). The average penetration angle ($<\theta_p>$) is estimated using the formula, $L_{h,exp} = 4D_{helix} \cdot \cos <\theta_p> + \Sigma L_{linker} - \Delta d$, where $L_{h,exp}$ is the measured step size at the constant force for the U_h state, D_{helix} is the contour length of the helices, L_{linker} is the peptide linker extension at the constant force obtained from the WLC model, and Δd is the end-to-end distance of the pulling points of tertiary structure.
6. The average penetration depth ($<d_p>$) is estimated by $<d_p> = (D_{helix} - <D_{helix,w}>) \cdot \sin <\theta_p>$, where $<D_{helix,w}>$ is the average contour length of the soluble parts of helices bounded by RK ring on the pulling side

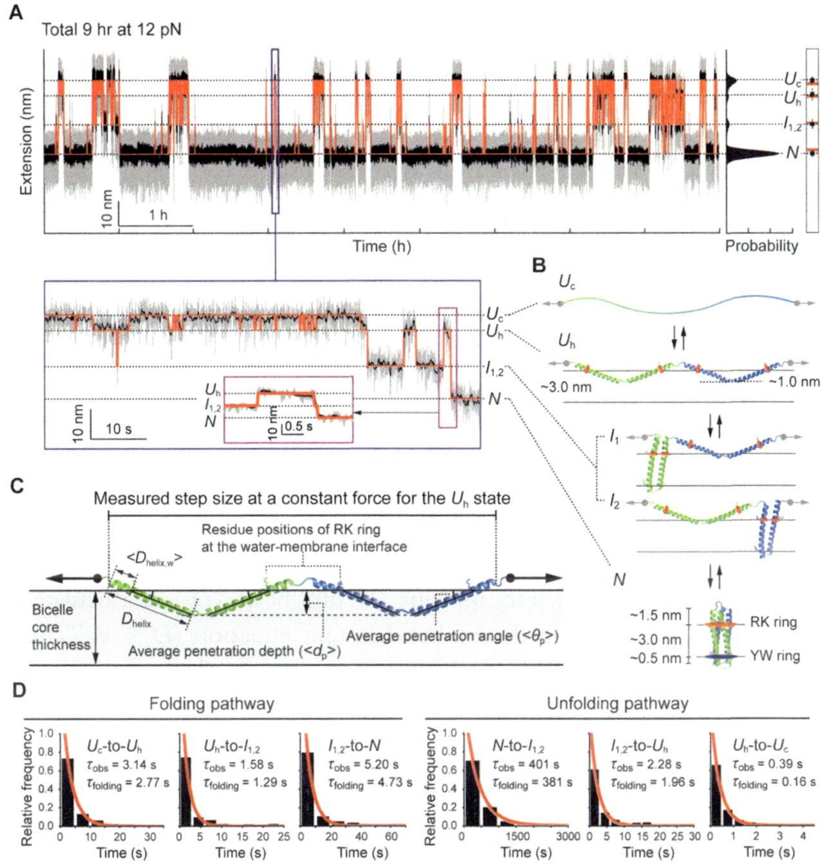

Fig. 3 Membrane protein (un)folding and kinetics characterization using constant-force mode. (A) Time-resolved extension trace of scTMHC2 at a constant force of 12 pN. The gray and black traces represent the raw data and median-filtered data with a 4.5-Hz window, respectively. The median-filtered trace is analyzed using the hidden Markov modeling (red). The red lines on the right represent the extension values expected for the structural states from polymer model analysis. (B) Predicted structural states of scTMHC2 for the four extension positions at the constant force. The RK and YW ring are designed to be positioned at the water-membrane interface (Lu et al., 2018). (C) Estimation of average penetration angles and depths of TM helices at the constant force. (D) Characterization of (un)folding transition times with instrumental errors corrected. The corresponding (un)folding rate constants are determined as the reciprocal of the (un)folding times. *Adapted from our earlier work with permission, Kim et al., (2013).*

(Lu et al., 2018). The lipid tail portion of DMPC/CHAPSO bicelle (the bicelle core) is estimated as ~3.0 nm, the average of previously measured values (Giudice et al., 2022; Leite, Wu, Pingali, Urban, & Lieberman, 2022; Murugova et al., 2022).

5.2 Characterization of (un)folding times and kinetics

1. A hidden Markov modeling (HMM) analysis (Jiao, Rebane, Ma, & Zhang, 2017; Zhang, Jiao, & Rebane, 2016) is applied to the median-filtered extension trace to obtain the observed (un)folding time (τ_{obs}) for each transition at the constant force (Fig. 3D). The τ_{obs} represents the dwell time constant in one state before transitioning to another state.
2. The τ_{obs} values are corrected for limited temporal resolution using the formula, $\tau_{corr} = \tau_{obs} - t_r$, where τ_{corr} is the corrected (un)folding time and t_r is the instrument time resolution (Jacobson & Perkins, 2020).
3. To characterize the molecular (un)folding time ($\tau_{(un)folding}$), the τ_{corr} is further multiplied by a correction factor of 0.95, accounting for the tethered bead-handle effect (Cossio et al., 2015). This kinetic error mainly arises from a relatively large size of the tethered bead in magnetic tweezers (Neupane & Woodside, 2016). The procedure for determining the correction factor is outlined as below.
4. The diffusion coefficient for our 2.8 μm bead (D_q) is calculated as 1.5×10^5 nm^2/s using the Stokes–Einstein equation ($D = k_B T / 6\pi\eta R$) (Hummer & Szabo, 2010), where η represents the dynamic viscosity of the solvent and R is the radius of the spherical particle. The water viscosity can be used for this calculation, since the addition of bicelles does not significantly change the solution viscosity in the range of 0.1–12% (w/v) total amphiphiles (Lu et al., 2012).
5. The diffusion coefficient for (un)folding of a soluble helical protein on the barrier top (D_x) is estimated as 10^6 nm^2/s (Yu et al., 2012). With a conservative estimate for the viscosity ratio of DMPC lipid membranes to water ($\sim 10^2$) (Kim et al., 2023), the D_x is reduced to $\sim 10^4$ nm^2/s from the Einstein relation ($D = k_B T / \eta$) (Chung & Eaton, 2013).
6. The ratio of D_x/D_q is then estimated as $\sim 10^{-1}$. The bead-handle effect in this low D_x/D_q regime is approximated by a factor of 0.95, according to a Langer theory-based prediction (Cossio et al., 2015).
7. The (un)folding rate constants are obtained as the reciprocal of the (un) folding times. The zero-force kinetics (k_0 and $\Delta x^\dagger$) can also be obtained using force-dependent kinetic models, such as the Bell-Evans model (Bell, 1978; Evans, 2001; Evans & Ritchie, 1997) or Dudko-Hummer-Szabo (DHS) model (Dudko, Hummer, & Szabo, 2006). The DHS model yields an additional energy parameter, i.e., the energy barrier height ($\Delta G^\dagger$) at zero force.

5.3 Reconstruction of folding energy landscapes

1. The Boltzmann relation, $\Delta G(l) = -k_B T \cdot \ln(p(l)/p(l = 0))$, allows us to characterize the folding energy landscape at a constant force (Gebhardt, Bornschlogla, & Rief, 2010; Woodside et al., 2006). Here, l is the protein extension, $p(l)$ is the probability density of the protein extension, and $\Delta G(l)$ is the free energy difference from $l = 0$. The $p(l)$ is obtained by applying deconvolution to the probability density of detected extension , $P(l)$, which corrects for the bead-handle effect.
2. A nonlinear constrained iterative deconvolution method (Gebhardt et al., 2010; Woodside et al., 2006) is employed in our case, respresented by this formula: $p^{(n+1)}(l) = p^{(n)}(l) + r[p^{(n)}(l)] \cdot (P(l) - S(l) {}^* p^{(n)}(l))$, $r[p^{(n)}(l)] = r_0(1 - 2|p^{(n)}(l) - 1/2|)$, where r_0 is the amplitude, $S(l)$ is the point spread function (PSF) obtained from the N state at the constant force, and n is the index of iterations. In our example, the parameters of r_0 and n_{max} are set as $r_0 = 2$ and $n_{max} = 5000$.
3. The validity of the deconvolution method is confirmed using the MC simulations where generated extension traces are convoluted and subsequently deconvoluted by a generated PSF (Fig. 4A). The parameters of the simulated traces, such as time constants, noise strength, and total duration, are chosen to mimic the experimental data. The dwell times at each state and thermal extension noises are randomly generated from the exponential and normal distributions, respectively.
4. The deconvoluted probability density of extension trace corresponds to $p(l)$ in the Boltzmann relation (Fig. 4B). The free energy differences between the energy wells (ΔG) and the barrier heights ($\Delta G^{\dagger}$), along with their corresponding positions, are determined by fitting the vertices with the Gaussian function. The zero-force energy values and positions can also be estimated by extrapolation from data at various forces.

5.4 Estimation of the folding speed limit

1. The pre-exponential factor (τ_{ω}) for each folding transition can be characterized using our magnetic tweezers, serving as an estimate for the so-called folding speed limit (Chung & Eaton, 2018; Kubelka, Hofrichter, & Eaton, 2004). The τ_{ω} values are obtained from the determined folding energy barriers ($\Delta G^{\dagger}$) and folding times ($\tau_{folding}$) using the Kramers rate equation ($\tau_{folding} = \tau_{\omega} \cdot \exp(\Delta G^{\dagger}/k_B T)$) (Hanggi, Talkner, & Borkovec, 1990) (Fig. 4C).
2. The τ_{ω} for each structural transition in scTMHC2 is as follows (Kim et al., 2023): 184 ± 0.06 ms for the coil-to-helix transition, 21.4 ± 0.09 ms for

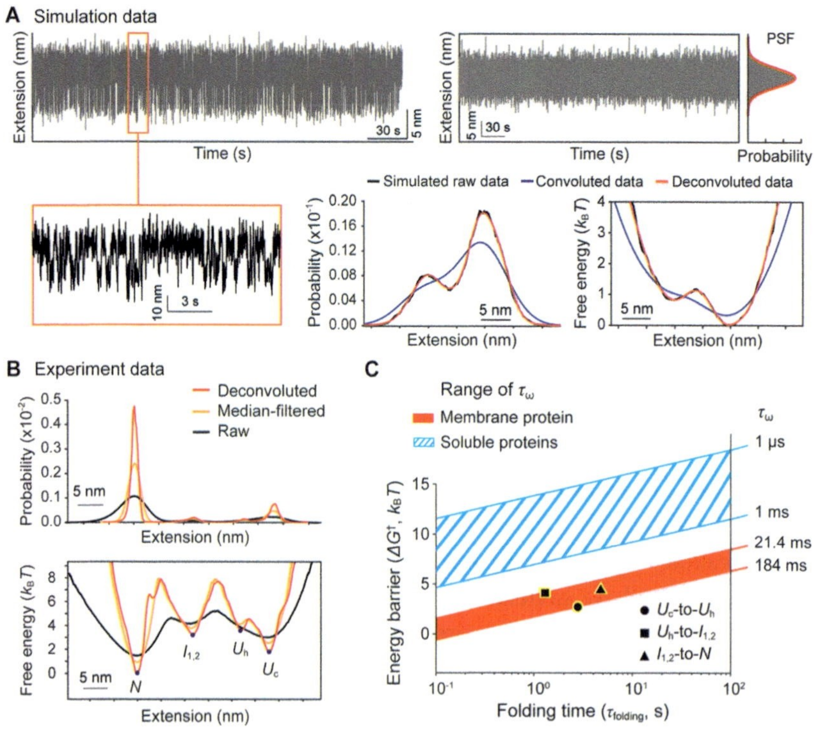

Fig. 4 Characterization of energy landscape and folding speed limit using constant-force mode. (A) Validation of deconvolution method using Monte Carlo (MC) simulations. This process involves convoluting and subsequently deconvoluting the probability density of a two-state extension trace using an extension trace of point spread function (PSF). The free energy landscapes are obtained using the Boltzmann relation. The original energy landscape is restored by the deconvolution. (B) Application of the deconvolution method to the measured extension trace shown in Fig. 3A. (C) Characterization of pre-exponential factor (τ_ω) for each folding transition, serving as an estimate for the folding speed limit. The τ_ω values are obtained from the energy barrier heights ($\Delta G^\dagger$) and folding transition times ($\tau_{folding}$) using the Kramers equation. In the $\Delta G^\dagger$–$\tau_{folding}$ plane, the blue- and red-shaded regions represent the τ_ω range for soluble proteins previously reported and for the membrane protein, respectively. *Adapted from our earlier work with permission, Kim et al., (2013).*

the helical hairpin formation, and 48.7 ± 0.11 ms for the final folding step, comprising the other helical hairpin formation and dimerization.

3. A measurement error for the energy barrier heights and folding times can arise due to the variation in magnetic bead size, likely the largest error source (Neupane & Woodside, 2016). However, this type of error does not impact the τ_ω estimation, as both the barrier heights and folding times are characterized under the same bead.

4. All three widely used kinetic models assume that τ_ω is independent of force (Hane, Attwood, & Leonenko, 2014). For this case, the obtained τ_ω value for each transition remain valid as a rough estimate even at zero force. Readers are referred to a previous report for more details on another potential source of error, concerning the membrane-water transfer event of helices and soluble linkers (Kim et al., 2023).
5. The solvent viscosity of lipid bilayers is likely the determining factor for τ_ω (Kim et al., 2023). Thus, the estimated τ_ω values could be non-negligibly altered by different lipid compositions and membrane reconstitutions.
6. The obtained τ_ω range is considerably different from that for soluble proteins (Fig. 4C), leading to more accurate reconstruction of the folding energy landscapes of helical membrane proteins using the Kramers rate framework.

6. Conclusion

This chapter provides the detailed procedures in characterizing the folding features of membrane proteins using our robust magnetic tweezers. Membrane proteins with more complex structures and larger sizes exhibit lower foldability (Wijesinghe & Min, 2023), resulting in a lower number of sampling for on-pathway folding transitions under the same experimental conditions. In such cases, the extended observations are more essential to increase the sampling numbers and ensure the statistical reliability of characterizations. The mangetic tweezers with a highly stable molecular tethering fulfills the technical needs, since it can sample a large number of structural transitions in a single membrane protein over tens of hours. We anticipate that our tweezer approach will broaden the scope of more accurate, mechanistic studies on membrane protein folding.

Acknowledgments

This work was supported and funded by the National Research Foundation of Korea (NRF-2020R1C1C1003937 to D.M.).

References

Bell, G. I. (1978). Models for the specific adhesion of cells to cells. *Science (New York, N. Y.), 200*(4342), 618–627. https://doi.org/10.1126/science.347575.

Choi, H. K., Kang, H., Lee, C., Kim, H. G., Phillips, B., Park, S., ... Yoon, T. Y. (2022). Evolutionary balance between foldability and functionality of a glucose transporter. *Nature Chemical Biology, 18*(7), 713. https://doi.org/10.1038/s41589-022-01002-w.

Choi, H. K., Kim, H. G., Shon, M. J., & Yoon, T. Y. (2022). High-resolution single-molecule magnetic tweezers. *Annual Review of Biochemistry, 91*, 33–59. https://doi.org/10.1146/annurev-biochem-032620-104637.

Choi, H. K., Min, D., Kang, H., Shon, M. J., Rah, S. H., Kim, H. C., ... Yoon, T. Y. (2019). Watching helical membrane proteins fold reveals a common N-to-C-terminal folding pathway. *Science (New York, N. Y.), 366*(6469), 1150–1156. https://doi.org/10.1126/science.aaw8208.

Chung, H. S., & Eaton, W. A. (2013). Single-molecule fluorescence probes dynamics of barrier crossing. *Nature, 502*(7473), 685. https://doi.org/10.1038/nature12649.

Chung, H. S., & Eaton, W. A. (2018). Protein folding transition path times from single molecule FRET. *Current Opinion in Structural Biology, 48*, 30–39. https://doi.org/10.1016/j.sbi.2017.10.007.

Cossio, P., Hummer, G., & Szabo, A. (2015). On artifacts in single-molecule force spectroscopy. *Proceedings of the National Academy of Sciences of the United States of America, 112*(46), 14248–14253. https://doi.org/10.1073/pnas.1519633112.

Dudko, O. K., Hummer, G., & Szabo, A. (2006). Intrinsic rates and activation free energies from single-molecule pulling experiments. *Physical Review Letters, 96*(10), https://doi.org/ARTN10810110.1103/PhysRevLett.96.108101.

Evans, E. (2001). Probing the relation between force—Lifetime—And chemistry in single molecular bonds. *Annual Review of Biophysics and Biomolecular Structure, 30*, 105–128. https://doi.org/DOI 10.1146/annurev.biophys.30.1.105.

Evans, E., & Ritchie, K. (1997). Dynamic strength of molecular adhesion bonds. *Biophysical Journal, 72*(4), 1541–1555. https://doi.org/10.1016/S0006-3495(97)78802-7.

Gebhardt, J. C. M., Bornschlogla, T., & Rief, M. (2010). Full distance-resolved folding energy landscape of one single protein molecule. *Proceedings of the National Academy of Sciences of the United States of America, 107*(5), 2013–2018. https://doi.org/10.1073/pnas.0909854107.

Giudice, R. D., Paracini, N., Laursen, T., Blanchet, C., Roosen-Runge, F., & Cardenas, M. (2022). Expanding the toolbox for bicelle-forming surfactant-lipid mixtures. *Molecules (Basel, Switzerland), 27*(21), https://doi.org/ARTN 762810.3390/molecules27217628.

Hane, F. T., Attwood, S. J., & Leonenko, Z. (2014). Comparison of three competing dynamic force spectroscopy models to study binding forces of amyloid-beta (1-42). *Soft Matter, 10*(12), 1924–1930. https://doi.org/10.1039/c3sm52257a.

Hanggi, P., Talkner, P., & Borkovec, M. (1990). Reaction-rate theory—50 years after kramers. *Reviews of Modern Physics, 62*(2), 251–341. https://doi.org/10.1103/revmodphys.62.251.

Hong, H., Choi, H. K., & Yoon, T. Y. (2022). Untangling the complexity of membrane protein folding. *Current Opinion in Structural Biology, 72*, 237–247. https://doi.org/10.1016/j.sbi.2021.11.013.

Hummer, G., & Szabo, A. (2010). Free energy profiles from single-molecule pulling experiments. *Proceedings of the National Academy of Sciences of the United States of America, 107*(50), 21441–21446. https://doi.org/10.1073/pnas.1015661107.

Jacobson, D. R., & Perkins, T. T. (2020). Correcting molecular transition rates measured by single-molecule force spectroscopy for limited temporal resolution. *Physical Review E, 102*(2), https://doi.org/ARTN 02240210.1103/PhysRevE.102.022402.

Jiao, J., Rebane, A. A., Ma, L., & Zhang, Y. (2017). Single-molecule protein folding experiments using high-precision optical tweezers. *Methods in Molecular Biology, 1486*, 357–390. https://doi.org/10.1007/978-1-4939-6421-5_14.

Kessler, D. A., & Rabin, Y. (2004). Distribution functions for filaments under tension. *Journal of Chemical Physics, 121*(2), 1155–1164. https://doi.org/10.1063/1.1760743.

Kim, S., Lee, D., Wijesinghe, W. C. B., & Min, D. (2023). Robust membrane protein tweezers reveal the folding speed limit of helical membrane proteins. *Elife, 12*, e85882. https://doi.org/10.7554/eLife.85882.

Kubelka, J., Hofrichter, J., & Eaton, W. A. (2004). The protein folding speed limit. *Current Opinion in Structural Biology, 14*(1), 76–88. https://doi.org/10.1016/j.sbi.2004.01.013.

Leite, W. C., Wu, Y. Q., Pingali, S. V., Urban, V. S., & Lieberman, R. L. (2022). Change in morphology of dimyristoylphosphatidylcholine/bile salt derivative bicelle assemblies with dodecylmaltoside in the disk and ribbon phases. *Journal of Physical Chemistry Letters, 13*(42), 9834–9840. https://doi.org/10.1021/acs.jpclett.2c02445.

Lu, P., Min, D., DiMaio, F., Wei, K. Y., Vahey, M. D., Boyken, S. E., ... Baker, D. (2018). Accurate computational design of multipass transmembrane proteins. *Science (New York, N. Y.), 359*(6379), 1042–1046. https://doi.org/10.1126/science.aaq1739.

Lu, Z. W., Van Horn, W. D., Chen, J., Mathew, S., Zent, R., & Sanders, C. R. (2012). Bicelles at low concentrations. *Molecular Pharmaceutics, 9*(4), 752–761. https://doi.org/10.1021/mp2004687.

Marko, J. F., & Siggia, E. D. (1995). Stretching DNA. *Macromolecules, 28*(26), 8759–8770 https://doi.org/DOI 10.1021/ma00130a008.

Min, D., Arbing, M. A., Jefferson, R. E., & Bowie, J. U. (2016). A simple DNA handle attachment method for single molecule mechanical manipulation experiments. *Protein Science, 25*(8), 1535–1544. https://doi.org/10.1002/pro.2952.

Min, D., Jefferson, R. E., Bowie, J. U., & Yoon, T. Y. (2015). Mapping the energy landscape for second-stage folding of a single membrane protein. *Nature Chemical Biology, 11*(12), 981–987. https://doi.org/10.1038/nchembio.1939.

Min, D., Jefferson, R. E., Qi, Y., Wang, J. Y., Arbing, M. A., Im, W., & Bowie, J. U. (2018). Unfolding of a ClC chloride transporter retains memory of its evolutionary history. *Nature Chemical Biology, 14*(5), 489–496. https://doi.org/10.1038/s41589-018-0025-4.

Murugova, T. N., Ivankov, O. I., Ryzhykau, Y. L., Soloviov, D. V., Kovalev, K. V., Skachkova, D. V., ... Gordeliy, V. I. (2022). Mechanisms of membrane protein crystallization in 'bicelles'. *Scientific Reports, 12*(1), https://doi.org/10.1038/s41598-022-13945-0.

Neupane, K., & Woodside, M. T. (2016). Quantifying instrumental artifacts in folding kinetics measured by single-molecule force spectroscopy. *Biophysical Journal, 111*(2), 283–286. https://doi.org/10.1016/j.bpj.2016.06.011.

Sarkar, R., & Rybenkov, V. V. (2016). A guide to magnetic tweezers and their applications. *Frontiers in Physics, 4* https://doi.org/ARTN 4810.3389/fphy.2016.00048.

Shon, M. J., Rah, S. H., & Yoon, T. Y. (2019). Submicrometer elasticity of double-stranded DNA revealed by precision force-extension measurements with magnetic tweezers. *Science Advances, 5*(6) https://doi.org/ARTN eaav169710.1126/sciadv.aav1697.

Wijesinghe, W. C. B., & Min, D. Y. (2023). Single-molecule force spectroscopy of membrane protein folding. *Journal of Molecular Biology, 435*(11) https://doi.org/ARTN 16797510.1016/j.jmb.2023.167975.

Woodside, M. T., Anthony, P. C., Behnke-Parks, W. M., Larizadeh, K., Herschlag, D., & Block, S. M. (2006). Direct measurement of the full, sequence-dependent folding landscape of a nucleic acid. *Science (New York, N. Y.), 314*(5801), 1001–1004. https://doi.org/10.1126/science.1133601.

Yu, H., Gupta, A. N., Liu, X., Neupane, K., Brigley, A. M., Sosova, I., & Woodside, M. T. (2012). Energy landscape analysis of native folding of the prion protein yields the diffusion constant, transition path time, and rates. *Proceedings of the National Academy of Sciences of the United States of America, 109*(36), 14452–14457. https://doi.org/10.1073/pnas.1206190109.

Zhang, Y. L., Jiao, J. Y., & Rebane, A. A. (2016). Hidden Markov modeling with detailed balance and its application to single protein folding. *Biophysical Journal, 111*(10), 2110–2124. https://doi.org/10.1016/j.bpj.2016.09.045.

CHAPTER TWELVE

Magnetic nano-tweezer for interrogating mechanosensitive signaling proteins in space and time

Minsuk Kwak[a,b,*]
[a]Center for Nanomedicine, Institute for Basic Science, Seoul, Republic of Korea
[b]Department of Nano Biomedical Engineering, Advanced Science Institute, Yonsei University, Seoul, Republic of Korea
*Corresponding author. e-mail address: minsuk.kwak@yonsei.ac.kr

Contents

Abstract

Spatiotemporal interrogation of signal transduction at the single-cell level is necessary to understand how extracellular cues are converted into biochemical signals and differentially regulate cellular responses. Using single-cell perturbation tools such as optogenetics, specific biochemical cues can be delivered to selective molecules or cells at any desired location and time. By measuring cellular responses to provided

Methods in Enzymology, Volume 694
ISSN 0076-6879, https://doi.org/10.1016/bs.mie.2024.01.009

perturbations, investigators have decoded and deconstructed the working mechanisms of a variety of neuroelectric and biochemical signaling processes. However, analogous methods for deciphering the working mechanisms of mechanosensitive signaling by regulating mechanical inputs to cell receptors have remained elusive. To address this unmet need, we have recently developed a nanotechnology-based single-cell and single-molecule perturbation tool, termed mechanogenetics, that enables precise spatial and mechanical control over genetically encoded cell-surface receptors in live cells. This tool combines a magnetofluorescent nanoparticle (MFN) actuator, which provides precise spatial and mechanical signals to receptors via target-specific one-to-one interaction, with a micromagnetic tweezers that remotely controls the force exerted on a single nanoparticle. This chapter provides comprehensive experimental protocols of mechanogenetics consisting of four stages: (i) chemical synthesis of MFNs, (ii) bio-conjugation and purification of monovalent MFNs, (iii) establishment of cells with genetically encoded mechanosensitive proteins, and (iv) modular targeting and control of cell-surface receptors in live cells. The entire procedure takes up to 1 week. This mechanogenetic tool can be generalized to study many outstanding questions related to the dynamics of cell signaling and transcriptional control, including the mechanism of mechanically activated receptor.

1. Introduction

Many biological processes are regulated by force. Mechanosensitive cell-surface receptors enable cells to perceive external physical cues, such as mechanical forces from cell-matrix and cell-cell interactions and elicit corresponding cellular responses (Vogel & Sheetz, 2006). They integrate mechanical, chemical, spatial, and temporal cues, activating downstream signaling pathways crucial in development, homeostasis, and disease (Iskratsch, Wolfenson, & Sheetz, 2014). The development of physical tools for applying mechanical stimulation in space and time to mechanoreceptors allows us to interrogate and understand the operating mechanisms of the mechanotransduction pathways.

Single-molecule force spectroscopy such as optical and magnetic tweezers are powerful and versatile tools to study forces and motions associated with biomolecules and biological processes with their capability to generate and deliver a strong force (~nanonewton, nN) to specific biomolecules (Block, Goldstein, & Schnapp, 1990; Neuman & Nagy, 2008; Svoboda, Schmidt, Schnapp, & Block, 1993). Magnetic particles ranging in size from 0.5 to 5 μm generate and deliver the force to the attached biomolecules in the direction of, and proportional to, an external magnetic field gradient (Choi, Kim, Shon, & Yoon, 2022; De Vlaminck & Dekker, 2012). Magnetic tweezer techniques have been particularly useful for revealing the effects of physical perturbations (i.e. mechanical tension) in conformation transitions of biomolecules that play critical roles in transcriptional control, mechanotransduction of cell signaling, protein biogenesis, and homeostasis (Choi et al., 2022; Smith, Finzi, & Bustamante, 1979, 1992). However, these techniques are not ideally suitable for spatial and mechanical control of individual membrane proteins in live cells because microprobes tend to induce protein clustering upon binding due to their large size (~μm) and multivalent character (Dufrêne et al., 2011).

To address this need, we have developed nanotechnology-driven single cell perturbation tools, or mechanogenetics (i.e. genetically encoded mechanically control of cell signaling), using multifunctional magnetic nanoparticles capable of mechanically activating, localizing, and imaging mechanoreceptors at the single-cell or -molecule level (Kim et al., 2017; Seo et al., 2016). With their characteristic sizes comparable to those of cell signaling molecules, magnetic nanoparticles have provided an unprecedented means of selective and quantitative delivery of physical stimulations to targeted receptors (Cho et al., 2012; Etoc et al., 2013; Hoffmann et al., 2013; Jeong et al., 2023). These nanoparticles can be further assembled with other functional modules for targeting and imaging into a single nanosystem. A modular nanoprobe based on magnetoplasmonic nanoparticles (MPNs), comprising a Zn-doped ferrite magnetic core, a plasmonic shell, and an oligonucleotide-based targeting module, was developed. MPNs could precisely label receptors with target-specific one-to-one engagement and minimize nonspecific perturbation as compared with traditional magnetic tweezers. Another critical advantage of this MPN-based mechanogenetic platform is the capacity to provide both spatial and mechanical control of the targeted receptors using a micromagnetic tweezer (μMT). The probes can be designed with targeting modules which can label multiple different proteins or different sites in the same proteins, allowing for chemical control of receptor activity. With these capabilities, nanoprobes enable the study of differential roles of external cues in regulating downstream cell signaling (Fig. 1).

Recently, we developed an alternative mechanogenetic nanoprobe with improved usability, applicability, and versatility based on small, clickable, and monovalent magnetofluorescent nanoparticles (MFNs) (Kwak et al., 2019). Through organic-phase surface PEGylation, click chemistry, and charge-based valency discrimination principles, monovalent MFNs are compact (<30 nm), allowing for a significantly enhanced labeling of target receptors in crowded surface microenvironments, and

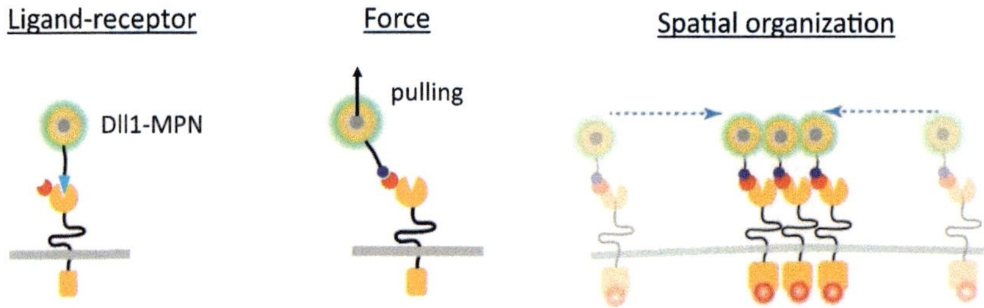

Fig. 1 Chemical, mechanical, and spatial regulation of cell-surface receptors using multifunctional magnetic nanoparticles.

fluorescent for interrogation of cellular responses using 3D confocal imaging. Thus, MFNs represent robust and versatile probes for mechanogenetic interrogation of a wide range of mechanosensitive cell signaling systems. In this chapter, we provide detailed description of the chemical synthesis and functionalization of MFNs, specific labeling of target receptors, and mechanical and spatial perturbation of the targeted receptors using the μMT. This chapter describes the protocol of mechanogenetic interrogation of Notch receptor and epithelial cadherin (E-cadherin) surrounded by a highly crowded microenvironment.

2. Chemical synthesis and functionalization of magneto-fluorescent nanoparticles

The key components that comprise the mechanogenetic nanoparticle probes are (1) a force-generating domain that delivers a controlled physical stimulation to the target protein, (2) an imaging domain to visualize spatiotemporal distribution of the probes, and (3) a targeting domain that specifically binds to the receptor. The MFNs integrate all three components, where $Zn_{0.4}Fe_{2.6}O_4$ magnetic core (13 nm) coated with a thin (2 nm) SiO_2 layer (M@SiO_2), a PEGylated layer for surface conjugation of ligands, and an oligonucleotide tether serve as the force-generating, imaging, and targeting domains, respectively. Since the plasmonic gold shell as the imaging module comprises a significant portion of previous MPNs, we developed a method to directly conjugate surface ligands such as fluorescent dyes to M@SiO_2 nanoparticles, while providing the imaging capability of the gold shell. While multivalent probes (e.g., probes containing multiple targeting ligands) result in uncontrolled receptor cross-linking and are unable to deliver a defined force to a single receptor, monovalent probes ensure one-to-one probe-receptor engagement, eliminating problems associated with multivalency (Farlow et al., 2013; Howarth et al., 2008).

2.1 Equipment

- 50-ml Three-neck round-bottom flask (KJ Scientific Instruments, cat. no. KJ-F15-3)
- Reflux condenser (KJ Scientific Instruments, cat. no. KJ-C9-5)
- Vacuum pump (Edwards, model no. RV8)
- Schlenk line connected to a vacuum pump and supply of argon
- Top-loading balance (0.0001-g readability; Mettler, cat. no. AE240)
- 20 Vials (50 ml; VWR, cat. no. 66012-066)
- Glass jars (100 ml, 1 liter; Pyrex, cat. nos. 1395-100, 1395-1L)
- Thermo controller (Poong Lim, cat. no. MC-NP200)
- Hot plate with magnetic stir bar (Corning, cat. no. PC420D)
- Bench-top centrifuge (Eppendorf, model no. 5424)
- Centrifuge (Beckman, model no. J2-21M)
- Shaker (Wisd Laboratory Instruments, model no. SHO-1D)
- MidiMACS Separator (Miltenyi Biotec, cat. no. 130-042-302)
- MACS separation columns (25 LS columns, Miltenyi Biotec, cat. no. 130-042-401)
- Amicon Ultra-15 centrifugal filter (10 K MWCO, Merck Millipore, cat. no. UFC801008)
- Illustra NAP-10 column (GE Healthcare, cat. no. 17-0854-02)
- Transmission electron microscope (TEM; JEOL, model no. JEM-2100)
- TEM grid (Carbon on 400 mesh; Ted Pella, cat. no. 01754-F)
- Plasma cleaner (Harrick Plasma, cat no. PDC-32G)
- UV–visible absorption spectrophotometer (Shimazu, model no. UV-1650(PC)s)
- Zetasizer (Malvern Instruments, model no. ZEN3600)
- Zeba spin desalting columns (7 K MWCO, 0.5 ml; Thermo Fisher Scientific, cat. no. 89882)
- HPLC system (Agilent Technologies, model no. 1220 Infinity II LC)
- DNAPac PA100 oligonucleotide column (4 × 250 mm; Thermo Fisher Scientific, cat. no. 043010)

2.2 Materials

- Iron (III) acetylacetonate (Strem Chemicals, cat. no. 26-2300)

- Zinc chloride (Sigma-Aldrich, cat. no. 208086)
- Oleic acid (Sigma-Aldrich, cat. no. 364525)
- Oleylamine (Sigma-Aldrich, cat. no. O7805)
- Trioctylamine (Sigma-Aldrich, cat. no. T81000)
- Igepal CO-520 (Sigma-Aldrich, cat. no. 238643)
- Ammonium hydroxide solution, 28.0–30.0% (vol/vol) (NH_4OH; Sigma-Aldrich, cat. no. 221228)
- Tetraethyl orthosilicate (TEOS; Sigma-Aldrich, cat. no. 333859)
- (3-Aminopropyl)trimethoxysilane (APTMS; Sigma-Aldrich, cat. no. 281778)
- [3-(2-Aminoethylamino)propyl]trimethoxysilane (AEAPTMS; Sigma-Aldrich, cat. no. 440302)
- Tetramethylammonium hydroxide in methanol solution (25% (wt/vol), TMAOH in methanol solution; Sigma-Aldrich, cat. no. 334901)
- Tris base (Thermo Fisher Scientific, cat. no. BP152-1)
- Sodium chloride (NaCl; Thermo Fisher Scientific, cat. no. BP358-1)
- Toluene (≥99.9% (vol/vol); Sigma-Aldrich, cat. no. 650579)
- Ethanol (≥99.9% (vol/vol); Sigma-Aldrich, cat. no. 362808)
- Cyclohexane (≥99.7% (vol/vol); Sigma-Aldrich, cat. no. 34855)
- Methanol (≥99.8% (vol/vol); Sigma-Aldrich, cat. no. 179337)
- DMSO (Sigma-Aldrich, cat. no. D5879)
- Methoxy- and azido-PEG_{24}-TFP ester (Quanta BioDesign (USA))
- HS-$(CH_2)_{11}$-EG_{12}-OCH_2-COOH, HS-$(CH_2)_{11}$-EG_{12}-OH custom-synthesized from Prochimia Surfaces (Poland).
- BG-GLA-NHS (New England BioLabs, cat. no. S9151S)
- HaloTag succinimidyl ester ligand building block (cat. no. P6751, Promega)
- HEPES (Thermo Fisher Scientific, cat. no. BP310-1)
- DBCO-modified DNA: 5′ - DBCO - T_{60} CAGT_{10} - 3′ (DBCO-DNA1)
- Amino-functionalized oligonucleotides (Amine-DNA2): 5′ - Amino Modifier C6 - T_{60} ACTG_{10} - 3′
- BG-modified oligonucleotides (BG-DNA2): BG - T_{60} ACTG_{10} – 3′
- Halo tag-modified oligonucleotides (Halo tag-DNA2): Halo Tag - T_{60} ACTG_{10} – 3′

2.3 Protocol

2.3.1 Synthesis of 13-nm Zn-doped ferrite ($Zn_{0.4}Fe_{2.6}O_4$) core nanoparticles

(1) Add 423.6 mg of iron (III) acetylacetonate and 120 mg of zinc chloride to a 50-ml three-neck round-bottom flask.

(2) Add 1.2 ml of oleic acid, 4.8 ml of oleylamine and 4.8 ml of trioctylamine to the flask.
(3) Connect the flask to a Schlenk line and stir the solution for 30 min at 300 r.p.m. under vacuum conditions.
(4) Heat the solution using a temperature controller under argon gas flow at 90 ml/min. Stir the solution at 300 r.p.m. during the reaction according to the five-step process shown below.
(5) Add 1.92 ml of toluene and 30 ml of ethanol to the solution. Transfer the solution to a 50-ml centrifuge tube.
(6) Centrifuge the solution at 1600 g for 5 min at RT.
(7) Discard the supernatant, re-disperse the black precipitate in 8 ml of toluene and add 30 μL of oleylamine.
(8) Centrifuge the solution at 650 g for 3 min at RT and collect the supernatant.
(9) Add 4 ml of ethanol to the supernatant and centrifuge again at 1600 g for 5 min at RT.
(10) Discard the supernatant and re-disperse the precipitate (final product = 13-nm $Zn_{0.4}Fe_{2.6}O_4$ nanoparticles) in 4 ml of toluene. Perform TEM analysis under an acceleration voltage of 200 kV.
(11) Measure absorption of the nanoparticle solution at 400 nm using a UV–visible absorption spectrophotometer and determine the nanoparticle concentration using the extinction coefficient for 13-nm $Zn_{0.4}Fe_{2.6}O_4$ nanoparticles: 5.6×10^7/M/cm at 400 nm.

2.3.2 Preparation of silica-coated $Zn_{0.4}Fe_{2.6}O_4$ nanoparticles ($M@SiO_2$)

(1) Add 12.6 ml of cyclohexane to a 50-ml vial.
(2) Add 780 mg of Igepal CO-520 and shake until it is completely dissolved.
(3) Add 1 mg (100 μL of the 2.75 μM) $Zn_{0.4}Fe_{2.6}O_4$ nanoparticle solution from step 1 to the solution from step 13.
(4) Add 105 μL of NH_4OH solution and shake for 1 min in a fume hood. Upon addition of NH_4OH, the solution will become turbid and soon will turn transparent with shaking.
(5) Add 2 μL of TEOS under vigorous vortex and the emulsion was incubated at room temperature for 2 days to form silica-coated magnetic nanoparticles ($M@SiO_2$).
(6) Add 6 μL of 3-aminopropyltrimethoxysilane (APTMS) and shaken for 2 h to functionalize the silica surface of $M@SiO_2$ with amine groups

(7) Add 4 ml of 50 mM tetramethylammonium hydroxide (TMAOH) in MeOH solution into the mixture, and vigorously shaken for 20 s

(8) In 30 s, phase separation between the methanol solution containing nanoparticles (dark brown color) and cyclohexane becomes evident. Carefully remove the dark brown layer (~4.5 ml) using a pipette and transfer to a new vial.

(9) Mount a MACS LS column on a MidiMACS separator and load 3.0 ml of solution from Step 19. Allow the nanoparticle solution to enter the column completely. Then apply 8 ml of 50 mM TMAOH in methanol solution to wash out the by-products.

(10) Allow 1 ml of DMSO to enter the column completely. Apply an additional 1.5 ml of DMSO and elute the M-SiO_2 by dismounting the column from the MidiMACS separator.

(11) Determine the particle solution concentration using a UV–visible absorption spectrophotometer. Dilute the particle solution to 400 nM using DMSO.

2.3.3 Organic-phase PEGylation of silica-coated $Zn_{0.4}Fe_{2.6}O_4$ nanoparticles (M@SiO_2)

(1) Prepare PEG solutions containing 20 μmol of methoxy-PEG_{24}-TFP ester and 0.8 μmol of azido-PEG_{24}-TFP ester (OCH_3:N_3 = 25:1) in 250 μL anhydrous DMSO.

(2) Mix 100 pmol of amine-functionalized $Zn_{0.4}Fe_{2.6}O_4$ @SiO_2 nanoparticles with PEG solutions from above.

(3) Add 30 μmol of trimethylamine and shake the mixture overnight to complete the reaction.

(4) Add 50 μmol of succinic anhydride and 50 μmol of triethylamine dissolved in 50 μL anhydrous DMSO and incubate for 8 h or longer to remove remaining amine groups on the silica surface.

(5) Mount a MACS® MS column mounted to a MiniMACS™ separator and load with 300 μL of 1 mM NaCl solution to equilibrate the column.

(6) Add succinylated M@SiO_2-PEG into the column and allow to enter the column completely. Wash the byproducts with 2 ml of 1 mM NaCl solution.

(7) Load 1 ml of DI water and collect the solution by dismounting the column from separator when the nanoparticles start to leak.

2.3.4 Monovalent conjugation of M@SiO2-PEG with a single-stranded DBCO-DNA1

(1) Prepare 5′-DBCO modified oligonucleotides (DBCO-DNA1; $T_{60}CAGT_{10}$) in clean DW.

(2) Mix 10 pmole of the PEGylated nanoparticles and 300 pmol of DBCO-DNA1. The ratio of DBCO-DNA1 and the particles can vary.

(3) Add clean, filtered 1xPBS and deionized water to a total particle concentration of 80 nM in a total reaction volume of 500 μL.

(4) Incubate the mixture at room temperature overnight.

(5) After overnight incubation, add a copious amount of clean DW to dilute the reaction.

(6) Concentrate the particles to a final volume of 110 μL by centrifugation using Amicon (100kD) on a tabletop centrifuge (3000 g, 10 min).

(7) Inject 100 μL of the reaction solution into an HPLC system to separate the monovalent MPNs (DNAPac PA100 oligonucleotide column; $4 \times 250\ \mathrm{mm}^2$). The following AE-HPLC method was used: salt concentration gradient was 40 - 1500 mM NaCl in 25 mM Tris-HCl buffer (pH 8.0) for 30 min at $1\ \mathrm{ml\ min}^{-1}$ of flow rate with detection wavelength of 580 nm.

(8) Collect a fraction containing monovalent MFNs (1 ml) and dilute to a total volume of 20 ml with clean DW.

(9) Dilute sequentially the salt concentration in the collected nanoparticles using an Amicon-15 column (100kD) with repeated washing.

(10) Concentrate the particles using an Ultra centricon (100kD) to a final concentration ≥100 nM.

(11) Measure absorbance of the nanoparticle solution at λ_{max} using a UV–visible spectrophotometer and determine the concentration using the extinction coefficients.

2.3.5 DBCO-fluorescent dye conjugation of monovalent magnetic nanoparticles

1. Add 10^4 molar excess amount of DBCO-modified AF647 dyes into monovalently conjugated M@SiO2-PEG
2. Incubate in the dark at room temperature overnight.
3. Desalt the excess dyes and exchange the buffer using a 7k Zeba 0.5 ml column

4. Dilute the mixture into 14 ml of DW and wash using an an Amicon-15 column (100kD) to remove remaining dyes.
5. Concentrate the particles using an Ultra centricon (100kD).

3. Efficient and specific labeling of target cell-surface receptors with magneto-fluorescent nanoparticles

Using a two-step targeting and hybridization procedure, we achieve efficient and specific labeling of target receptors with MFNs. First, cells expressing receptors are treated with an oligonucleotide conjugated with the receptor's ligand. Second, monovalent MFNs bearing an oligonucleotide with complementary sequence were targeted to hybridize with the DNA-conjugated receptors. This strategy provides the unique advantages for high specificity, modularity, and a rapid binding rate. The DNA tether length is critical to improving accessibility of MFNs and henace the labeling efficiency. While the optical length depends on the cell type and accessibility of the receptor, we recommend an oligonucleotide with linker lengths of at least 60 bases or longer for efficient labeling. Another important consideration for optimization of cell labeling is blocking conditions, which are necessary to minimize non-specific interactions of MFNs with cell-surface. In this section, we recommend the optimal labeling conditions for U2OS cells expressing SNAP-tagged Notch or Halo-tagged E-cadherin to use 10%.

3.1 Equipment

- An inverted microscope (Nikon ECLIPSE Ti-E) equipped with a TIRF system (Nikon)
- A Nikon Ti-E microscope equipped with the A1 confocal laser scanning microscope system (Nikon)
- White light source (SuperK Compact supercontinuum laser; NKT Photonics). Other collimated white light sources, such as mercury, xenon and LED lamps, may be used instead.
- Three OBIS solid state laser lines (488, 552, 642 nm)
- EM charge-coupled device (CCD) camera (Princeton Instruments, model no. ProEM 512)
- A motorized stage (ASI, MS-2000).
- Color CCD (Nikon, model no. DS-Ri2)
- pH meter (Fisher Scientific, cat. no. AB15)

- Disposable smartSpatula (anti-static; Sigma-Aldrich, cat. no. Z561762)
- 35-mm Dish with 7-mm glass-bottom dish (MatTeK, cat. no. P35G-1.5-7-C)
- Neon transfection system (Thermo Fisher Scientific, cat. no. MPK5000)
- ImageJ (https://imagej.nih.gov/ij/)

3.2 Materials

- Cell line of interest. We describe here the use of U2OS cells (ATCC, cat. no. HTB-96) or U2OS-Trex cells (gift from S. Blacklow (Harvard University))
- McCoy's 5A (modified) medium (Thermo Fisher Scientific, cat. no. 16600082)
- HyClone McCoy's 5A medium, without phenol red (GE Healthcare Life Sciences, cat. no. SH30270.01)
- FBS (heat-inactivated FBS; Gibco, cat. no. 10500)
- Trypsin–EDTA solution (0.25% (wt/vol) trypsin/1 mM EDTA-4Na (1×), liquid; Thermo Fisher Scientific, cat. no. 25200-072)
- BSA (Thermo Fisher Scientific, cat. no. BP9703100)
- 5% (wt/vol) Alkali-soluble casein (Merck Millipore, cat. no. 70955)
- Collagen type I, rat tail (Corning, cat. no. 354236)
- Doxycycline hydrochloride (Sigma-Aldrich, cat. no. D3447)
- BG-GLA-NHS (New England BioLabs, cat. no. S9151S)
- HaloTag succinimidyl ester ligand building block (cat. no. P6751, Promega)
- HEPES (Thermo Fisher Scientific, cat. no. BP310-1)
- Amino-functionalized oligonucleotides (Amine-DNA2): 5′ - Amino Modifier C6 - T_{60} $ACTG_{10}$ - 3′
- BG-modified oligonucleotides (BG-DNA2): BG - T_{60} $ACTG_{10}$ – 3′
- Halo tag-modified oligonucleotides (Halo tag-DNA2): Halo Tag - T_{60} $ACTG_{10}$ – 3′

3.3 Protocol

3.3.1 Generation of U2OS stable cell line co-expressing Halo-Ecadherin-GFP (Halo-Ecad-GFP) and SNAP-human Notch1-mCherry (S-hN1-mC)

(1) U2OS stable cell line expressing S-hN1-mC was created as previously described [ref].

(2) Detach 5×10^5 U2OS cells from a cell culture plate (70–100% confluency on a 60-mm dish or T25 flask) with 1 ml of 0.05% (wt/vol) trypsin–EDTA solution. Neutralize the lifted cells with 2 ml of McCoy's 5A medium with 10% FBS and transfer the cells to a 35-mm dish.

(3) Transfect 2 μg Halo-Ecad-GFP plasmid of high purity (A260/280 ratio > 1.9) into a stable S-hN1-mC U2OS cell line using a Neon electroporation transfection kit following the electroporation condition recommended by the manufacturer.

(4) To investigate the effects of spatial localization of Ecadherin on F-actin polymerization, transfect the above U2OS cells with the plasmid encoding Lifeact7-mC

(5) After transfection, wash the cells three times with 15 ml of PBS after 6 h of incubation to remove cellular debris and inclusion bodies.

(6) Culture the transfected cell lines with selection media containing 400 μg/ml hygromycin for S-hN1-mC and 800 μg/ml neomycin for Halo-Ecad-GFP for 2 weeks.

(7) Label the cells for membrane expression of S-hN1-mC or Halo-Ecad-GFP using SNAP- or Halo Tag- surface dyes.

(8) Perform FACS to detect and enrich the cell subpopulations co-stained for Halo-Ecad-GFP and S-hN1-mC with the top 2% membrane expressions of both Halo- and SNAP-fluorescent signals.

(9) Perform FACS sorting into a 96 well plate by a single cell. Each well would contain a monoclonal stable cell line with homogeneous protein expression.

(10) After 2-3 weeks of selection, identify the cell populations showing the strongest fluorescence signals for S-hN1-mC and Halo-Ecad-GFP and cryo-preserve them until use.

3.3.2 Preparation of Halo ligand- or BG-modified DNA

1. Dissolve 2 mg of BG-GLA-NHS or HaloTag succinimidyl ester ligand building block in 200 μL of anhydrous DMSO.
2. Add 20 μL of 1 mM Amino-functionalized oligonucleotides (Amine-DNA2) to 100 μL of 1.0 M HEPES buffer (pH 8.5).
3. Mix the two solutions and make up the volume to 1 ml with deionized water.
4. Incubate the reaction solution at RT for 1 h.
5. Purify Halo ligand- or BG-DNA2 with Illustra NAP-10 columns.
6. Detect and purify the conjugated oligonucleotides by reverse-phase high performance liquid chromatography.

3.3.3 Labeling of SNAP-tagged notch receptors and halo-tagged E-cadherin receptors with MFNs

(1) Treat the cells expressing S-hN1-mC and Halo-Ecad-GFP receptors with 2 μM of receptor-targeting DNA (BG-$T_{60}ACTG_{60}$ for Notch and Halo Tag-$T_{60}ACTG_{60}$ for E-cadherin). Incubate for 45 mins in 37 °C and 5% CO_2.

(2) Wash the cells five times with 5 ml of McCoy's 5A culture medium containing 10% (wt/vol) BSA. Alexa 647-$(CAGT)_5$ (10 nM) can be used to confirm the binding of BG- or Halo Tag- $T_{60}ACTG_{60}$. A nontransfected control cell line can be used as a negative control.

(3) Prepare the MFN solution by quickly mixing the stock solution of MFNs monovalently conjugated with the oligonucleotides ($T_{60}CAGT_{10}$) with McCoy's 5A culture medium containing 0.5% (wt/vol) alkaline casein to make the final nanoparticle concentrations of 10-20 nM. Typical concentration of monovalent MFN stock solution is in the range of 200–250 nM.

(4) Treat the cells with 10–20 nM MFN and incubate for 10 min

(5) Wash the cells three times with 10 ml of McCoy's 5A culture medium containing 10% FBS.

(6) Add phenol-red-free McCoy's 5 A culture medium with 10% FBS for imaging.

3.3.4 Quantitative assessment of cell labeling efficiency using flow cytometry analysis

(1) After labeling with MFNs, detach the cells using 40 μL of 0.05% Trypsin.

(2) Add 60 μL media and used vigorous pipetting at least 15 times to completely detach all cells from the substrate, maximizing the number of cells for analysis.

(3) Transfer the detached cells into a 1.5 ml Eppendorf tube, add 200 μL additional media, centrifuge at 300 x g for 5 min. Resuspend the cells in 300 μL PBS containing 2% FBS (FACS running buffer) filtered in a 0.22 μm membrane syringe-filter.

(4) Acquire single-cell fluorescence data on a BD FACSAria Fusion using FACSDiva software. Analyze samples in uniformly processed batches containing negative controls for protein (S-hN1-mC or Halo-Ecad-GFP) expression using plain U2OS and negative controls for nanoparticle labeling using cells treated with non-targeting DNA.

(5) Analyze total of 10,000 cells after initial gating in each measurement. Determine the gating for nanoparticle labeling based on AF647 signals to make the fraction of labeling-positive cells in negative controls to be zero %.
(6) Detect GFP fluorescence with a 488 nm excitation laser and a 530/30-nm FITC emission filter for E-cadherin expression.
(7) Detect AF647 fluorescence with a 640 nm excitation laser and a 670/30-nm APC, AF647 emission filter for evaluating MFN labeling on cells.

4. Spatial and mechanical control of targeted receptors and cell signaling using a micro-magnetic tweezer

To quantitatively deliver a defined force, we utilize the distance-dependent force generation capability of MFNs coupled with a custom-built micro-magnetic tweezer (μMT). Force delivery per MFN as a function of distance (*d*) between the nanoparticle and the magnetic tip end was calibrated previously. The calibration curve for MFNs with a 13-nm $Zn_{0.4}Fe_{2.6}O_4$ is $F_{pN} = 0.48{d_{\mu m}}^{-4.86} + 3.07{d_{\mu m}}^{-1.43}$, where a single MFN can generate forces from 100 fN to 50 pN by changing *d* (Kim et al., 2017; Seo et al., 2016). Because spatial control of receptors can be induced with sub-pN force and mechanical activation of receptors requires a few to tens of pN per receptor, we can achieve either spatial or mechanical perturbation modes by changing *d*.

4.1 Equipment

- An inverted microscope (Nikon ECLIPSE Ti-E) equipped with a TIRF system (Nikon)
- A Nikon Ti-E microscope equipped with the A1 confocal laser scanning microscope system (Nikon)
- White light source (SuperK Compact supercontinuum laser; NKT Photonics) Other collimated white light sources, such as mercury, xenon and LED lamps, may be used instead.
- Three OBIS solid state laser lines (488, 552, 642 nm)
- EM CCD camera (Princeton Instruments, model no. ProEM 512)
- A motorized stage (ASI, MS-2000).
- Color CCD (Nikon, model no. DS-Ri2)
- Steel needle (length = 15 mm; diameter = 0.8 mm; tapered angle = 15°; Semprex, custom order)

- NdFeB permanent magnet (length = 25 mm; diameter = 10 mm; K&J Magnetics, cat. no. N52)
- Magnet adaptor, custom-made (cone shape; length = 5 mm; upper diameter = 1 mm; bottom diameter = 5 mm; hole size = 0.8 mm; custom order)
- xyz translation stage (Physik Instrument, cat. nos. VT-80 250-DC, VT-80 25-DC, and M-227.50 for x, y and z translation stages, respectively)
- UV–visible spectroscopy system (Agilent, model no. 8453)
- μManager (https://www.micro-manager.org/)

4.2 Materials

- Cell line of interest
- McCoy's 5A (modified) medium (Thermo Fisher Scientific, cat. no. 16600082)
- Polycaprolactone (Sigma-Aldrich, cat. no. 440744)

4.3 Protocols

4.3.1 μMT setup

(1) Integrate an *xyz* controller system by assembling three single-axis motorized translation stages, or utilize a microinjection controller system that is routinely used for gene transfection.
(2) Install an *xyz* translation stage on the inverted microscope.
(3) Assemble a steel needle onto a magnet adaptor.
(4) Place the steel needle–magnet adaptor assembly on a hot plate and heat for 100 °C for 3 min
(5) Apply 50 mg of polycaprolactone to the top of the needle tip, using a tweezer to coat and prevent oxidation of the steel needle. Apply heat to melt polycaprolactone and cover the entire surface of needle–magnet adaptor assembly.
(6) Cool the polycaprolactone-coated needle–magnet adaptor assembly at RT for 10 min
(7) Magnetically attach the steel needle to the NdFeB magnet and mount it on the *z* translation stage.
(8) Control the horizontal position of the needle by linking the xy stage to a computer control program such as μmanager.

4.3.2 Preparation for mechanogenetic control of cell-surface receptors

(1) Equilibrate the temperature of a microscope cell culture chamber to 37 °C before starting the experiment.

(2) Align the μMT with the microscope objective lens. Position the end of the μMT tip at the center of the objective oculus with bright-field illumination. Perform this step sequentially with 20×, 40× and 100× objective lenses and set the height of the μMT tip to ~5 cm above the 100× objective lens.
(3) Apply a small drop of immersion oil onto the 100× objective lens, and place the bottom dish containing the cells on the microscope stage.
(4) Locate a subcellular region of interest by visualizing the MPNs using dark-field microscopy.
(5) Lower the μMT into the medium and place the tip of the μMT above the target cell or area.
(6) Determine the distance (*d*) between the tip of the μMT and the MPNs on a cell membrane and adjust the *d* to 10 μm.

4.3.3 Spatial control of receptors

(1) Move the tip of the μMT so that its tip is right above the cell or the subcellular location where receptors are to be mechanogeneically regulated.
(2) Place the tip of the μMT 2 μm above a target subcellular location and leave for 5–20 min
(3) Monitor the spatial localization of the MFNs at the μMT by imaging MFN fluorescence (λ_{em} = 647 nm).
(4) Monitor the fluorescence signals of S-hN1-mC (λ_{em} = 610 nm) or Halo-Ecad-GFP (λ_{em} = 509 nm) within the same region of the accumulated MFN signals.
(5) After 20 min of stimulation, move the μMT up (d > 100 μm) and immediately proceed to confocal z-stack 3D imaging.
(6) To monitor F-actin polymerization during E-cadherin spatial control, monitor fluorescence signal changes of F-actin (Lifeact7-mC) and Ecadherin (GFP) every 3 min for 10–15 min during mechanogenetic stimulation.

4.3.4 Mechanical control of receptors

(1) Move the tip of the μMT so that its tip is right above the cell or the subcellular location where receptors are to be mechanogeneically regulated.
(2) Place the tip of the μMT 0.7 μm above a target subcellular location and leave for 5–20 min
(3) Monitor the spatial localization of the MFNs at the μMT by imaging MFN fluorescence (λ_{em} = 647 nm).

(4) Monitor the spatial distribution and activation of the receptors by imaging fluorescence signals of S-hN1-mC (λ_{em} = 610 nm) or Halo-Ecad-GFP (λ_{em} = 509 nm) within the same region.

5. Summary

We have provided here a detailed description of the methods for synthesis of MFNs, bio-conjugation for targeting and fluorescent imaging, specific labeling of cell-surface receptors, and mechanogenetic control of the targeted receptors. Small, monovalent, and modular MFNs demonstrate the utility as an improved mechanogenetic nanoprobe and provide many unique and powerful features to interrogate the mechanism of mechanosensitive signaling proteins.

Funding source

This work was supported by Institute for Basic Science (IBS-R026-D1) and National Research Foundation of Korea (NRF-2021R1F1A1063378).

Declarations of interest

The authors declare no competing financial interest.

References

Block, S. M., Goldstein, L. S. B., & Schnapp, B. J. (1990). Bead movement by single kinesin molecules studied with optical tweezers. *Nature, 348*, 348–352.

Cho, M. H., et al. (2012). A magnetic switch for the control of cell death signalling in in vitro and in vivo systems. *Nature Materials, 11*, 1038–1043.

Choi, H.-K., Kim, H. G., Shon, M. J., & Yoon, T.-Y. (2022). High-resolution single-molecule magnetic tweezers. *Annual Review of Biochemistry*. https://doi.org/10.1146/annurev-biochem-032620.

De Vlaminck, I., & Dekker, C. (2012). *Recent advances in magnetic tweezers. Annual Review of Biophysics*453–472.

Dufrêne, Y. F., et al. (2011). Five challenges to bringing single-molecule force spectroscopy into living cells. *Nature Methods, 8*, 123–127.

Etoc, F., et al. (2013). Subcellular control of Rac-GTPase signalling by magnetogenetic manipulation inside living cells. *Nature Nanotechnology, 8*, 193–198.

Farlow, J., et al. (2013). Formation of targeted monovalent quantum dots by steric exclusion. *Nature Methods, 10*, 1203–1205.

Hoffmann, C., et al. (2013). Spatiotemporal control of microtubule nucleation and assembly using magnetic nanoparticles. *Nature Nanotechnology, 8*, 199–205.

Howarth, M., et al. (2008). Monovalent, reduced-size quantum dots for imaging receptors on living cells. *Nature Methods, 5*, 397–399.

Iskratsch, T., Wolfenson, H., & Sheetz, M. P. (2014). Appreciating force and shape—The rise of mechanotransduction in cell biology. *Nature Reviews. Molecular Cell Biology, 15*, 825–833.

Jeong, S., et al. (2023). Hydrogel magnetomechanical actuator nanoparticles for wireless remote control of mechanosignaling in vivo. *Nano Letters, 23*, 5227–5235.
Kim, J., et al. (2017). Single-cell mechanogenetics using monovalent magnetoplasmonic nanoparticles. *2017 Nature Protocols, 12*(9 12), 1871–1889.
Kwak, M., et al. (2019). Small, clickable, and monovalent magnetofluorescent nanoparticles enable mechanogenetic regulation of receptors in a crowded live-cell microenvironment. *Nano Letters, 19*, 3761–3769.
Neuman, K. C., & Nagy, A. (2008). Single-molecule force spectroscopy: Optical tweezers, magnetic tweezers and atomic force microscopy. *Nature Methods, 5*, 491–505.
Seo, D., et al. (2016). A mechanogenetic toolkit for interrogating cell signaling in space and time. *Cell, 165*, 1507–1518.
Smith, S. B., Finzi, L., & Bustamante, C. (1992). Direct mechanical measurements of the elasticity of single dna molecules by using magnetic beads. *Science (New York, N. Y.) (1979), 258*, 1122–1126.
Svoboda, K., Schmidt, C. F., Schnapp, B. J., & Block, S. M. (1993). Direct observation of kinesin stepping by optical trapping interferometry. *Nature, 365*, 721–727.
Vogel, V., & Sheetz, M. (2006). Local force and geometry sensing regulate cell functions. *Nature Reviews. Molecular Cell Biology, 7*, 265–275.

CHAPTER THIRTEEN

Magnetic tweezers in cell mechanics

Claudia Tanja Mierke*
Faculty of Physics and Earth System Sciences, Peter Debye Institute for Soft Matter Physics, Biological Physics Division, Leipzig University, Leipzig, Germany
*Corresponding author. e-mail address: claudia.mierke@uni-leipzig.de

Contents

Abstract

The chapter provides an overview of the applications of magnetic tweezers in living cells. It discusses the advantages and disadvantages of magnetic tweezers technology with a focus on individual magnetic tweezers configurations, such as electromagnetic tweezers. Solutions to the disadvantages identified are also outlined. The specific role of magnetic tweezers in the field of mechanobiology, such as mechanosensitivity, mechano-allostery and mechanotransduction are also emphasized. The specific usage of magnetic tweezers in mechanically probing cells via specific cell surface receptors, such as mechanosensitive channels is discussed and why mechanical probing has revealed the opening and closing of the channels. Finally, the future direction of magnetic tweezers is presented.

1. Introduction

Cells can be described as mechanosensitive objects. They are able to perceive mechanical cues from their outside environment and react to them.

Methods in Enzymology, Volume 694
ISSN 0076-6879, https://doi.org/10.1016/bs.mie.2023.12.007

Thereby, cells can transfer these mechanical cues into biochemical cues and generate mechanical signaling and memory. During the last two decades a growing number of proteins have been identified that utilized mechanical forces to conduct their functions (Ingber, 2006; Iskratsch et al., 2014). Consequently, the field of mechanobiology has emerged and expanded in a rapid manner. There have been major breakthroughs in the field of structural biology, such as the protein structures of several mechanical proteins at near-atomic resolution. However, a number of experiments of cellular proteins have mostly been conducted in an acellular environment. In addition, the single time-points obtained by X-ray crystallography and cryo-electron microscopy are static, since the proteins are fixed and hence are devoid of many dynamics part, such as alterations of the protein conformation (Wang & Wang, 2017).

Mechanosensitive proteins are located in living cells and are subject to many forces *in vivo* that are bidirectional (Mierke, 2021; Ravi et al., 2022), such as extracellular forces (outside in signaling) and intracellular forces (inside out signaling). Consequently, these mechanosensitive proteins can modify their transient structures and interactions.

For a comprehend knowledge of how these mechano-proteins react, their force-based characteristics have to be analyzed with high spatio-temporal resolution, preferably by probing a single molecule.

Therefore, at least two biophysical techniques are well suited, such as magnetic tweezers and optical tweezers, which have both been shown to probe individual biomolecules (Neuman & Nagy, 2008). Traditionally these techniques have been employed to measure slow and huge alterations within proteins. Currently, magnetic tweezers provide a resolution on a nanometer length scale with sub-millisecond precision in time that can even probe rapid dynamical remodeling of proteins on the length scale of a few amino acid residues. Besides molecular measurements, magnetic tweezers have been used to measure living cells (Aermes et al. 2020, 2021; Kollmannsberger et al., 2011; Mierke et al. 2008, 2010, 2020). Magnetic tweezers have been utilized for cellular rheology examinations (Bausch et al., 1999) or for investigating the binding strengths of specific membrane proteins (Mierke et al., 2010).

The force range of electro-magnetic tweezers is generally on the order of 10^{-3}–10^{4} pN (Neuman et al., 2007), which is several orders of magnitude higher than that of optical tweezers (Ashkin et al., 1986; Guck et al., 2001) and dielectrophoresis-based tweezers (Kodama et al., 2013). It is important to note that usage of piconewton forces in tweezer setups can uncover the hidden, load-dependent sites of biomolecules and induce a specific function of them, such as tethering of interacting molecules. While the short-lived conformations and their fine modifications are not amenable to traditional methods, single-molecule tweezers "force" the targeted molecules toward a new thermodynamic equilibrium to monitor the force-induced conditions so that their stabilities can be deduced.

In this chapter, it is presented how singular magnetic tweezers can be used to unravel the molecular mechanisms underlying mechanosensitivity found in living cells. The emphasis will be on typical examples such as cytoskeletal proteins, cell surface receptors and mechanosensitive ion channels, and then whole cell measurements will be discussed, which will be contextualized in mechanobiology. Finally, high-resolution tweezers appear to be essential instruments that can encourage fascinating discoveries in the new discipline of nanoprecision mechanobiology.

2. History and principles of the magnetic tweezer

Magnetic tweezers utilize the magnetic field that can be either generated by permanent or electromagnets to exert a force and/or rotate magnetic particles (referred to as beads) attached to a biological material, such as cells. Electromagnetics comprise temporary magnets that provide a strong magnetic field. The advantage of electromagnetics is that the field intensity can be controlled by regulating the flow of the electric current in the coil. Permanent magnets permit controlling the position and angle of the magnetic tweezers toward the placed bead by a simple motorized stage (Strick et al., 1998) to stretch and twist single molecules, such as DNA (Gosse & Croquette, 2002). A major advantage of electromagnets is that they offer a fast regulation of the applied tension, which is needed for application of different probing magnetic tweezer modes (Gosse & Croquette, 2002). Thereby a mechanical stress is triggered. Magnetic tweezers follow a number of magnetic beads attached to the surface in parallel, whereby the beads are brought under a pair of magnets and pulled to the top (De Vlaminck et al., 2012). With magnetic tweezers, the force is adjusted through the location of the magnet, and the subsequent movements of the beads positions are recorded, which is referred to as force clamp type of measurement. The latest progress in magnetic tweezers, in both hardware and software, has considerably optimized the resolution of measurements, so that the most modern devices have an impressive level of performance similar to that of optical tweezers (Dulin et al., 2015; Huhle et al., 2015; Lansdorp et al., 2013; Shon et al., 2019). Magnetic tweezers can be applied in two key fields of biophysics, such as cellular mechanics and single-molecule biophysics. There are huge differences in the forces exerted for these two types of applications. The force used in cellular mechanics is typically fairly large (<<1 nN) (Kah et al., 2020; Kollmannsberger et al., 2011; Mierke et al. 2010, 2017, 2020) in comparison to single-molecule analyses (<<100 pN) (Dulin et al., 2013). Hence the respective magnetic tweezer instruments required for these two applications vary largely. The focus of the chapter is on cellular magnetic tweezer devices. The measurement noise of magnetic tweezers varies according to several parameters, among others the geometry of the bead tether and the amount of force exerted. However, as a rough estimate, modern devices can accurately identify structural modifications of around 1 nm with a temporal resolution of about 1 ms in the piconewton force range.

2.1 Single electromagnetic tweezer

Single-molecule magnetic tweezers have been introduced in the 1990s by two different laboratories, the Bustamante laboratory and the Croquette and Bensimon laboratory, which pioneered the modern type of magnetic tweezers that can generate a constant force (even less than 1 pN) and a constant torque, permitting the investigation of the mechanical characteristics of DNA at the single-molecule scale (Smith et al., 1992; Strick et al., 1996). These days, magnetic tweezers are available in multiple laboratories all over the world. Apart from single-molecule studies on various protein-nucleic acids systems, cellular and viral RNA polymerases, nucleoprotein filaments and mechanical stability of protein folding and protein-ligand interactions, cellular mechanical characteristics have become the focus of mechanobiological research using nano-precision magnetic tweezers. The ease and the stability of the biophysical instrument renders the single-molecule force spectroscopy as a useful technique that is spread over multiple laboratories.

A disadvantage of permanent magnets to be used as magnetic tweezer is the requirement that they have to be mechanically positioned to apply temporal control of the magnetic field at a specific point in space. Thus, fast alterations cannot be achieved and vibration-free movement is also not easy to implement. In contrast, electromagnets utilized as magnetic tweezers can cause dissipation of heat, magnetic hysteresis, and lower field magnitudes, which may restrict their usage for every kind of measurement.

However, for cellular applications, the usage of electromagnetics has been preferred.

The straight forward design for magnetic tweezers comprises an electromagnet in the shape of a solenoid with a core material that possesses high magnetic permeability, for instance, iron alloys. The core tapers on one side to a sharp spike with a radius of typically under 10 μm. A magnetic field develops at the apex with a high gradient, drawing superparamagnetic beads into the vicinity (Fisher et al., 2006). A major disadvantage, though, is the impact of the magnetic hysteresis of the core compound, which is generated due to remanent magnetization (Basoli et al., 2018; Bessalova et al., 2016; Neuman & Nagy, 2008; Sarkar & Rybenkov, 2016) (Table 1). Magnetic hysteresis means that the magnetization of the core matter relies not solely on this magnetic current, but also on its progression. In reality, this implies that the relation of the magnetic current to the produced force is not easy to forecast and is normally typically only calibrated for a single, specific type of current protocol, for example for rising currents following core demagnetization, which is referred to as degaussing. Therefore, the residual magnetization of the magnetic core needs to be removed following every measurement through the generation of a sinusoidal alternating current with diminishing amplitude (Fisher et al., 2006; Tanase et al., 2007). However, demagnetization invariably leads to a high, abrupt, distance-dependent force effect on close-by beads, even when it is carried out very rapidly. This may pre-stimulate these beads before they are measured.

The residual magnetic field of the core material can be balanced with a closed-loop control circuit to regulate the magnetic induction in place of the magnetic current. A system of this type has actually previously been used in a four-pole tweezer design (Amblard et al., 1996). The primary purpose of the forced induction feedback is to decrease the magnetic interference between the magnets, but is not intended to offset the hysteresis of the core compound. The maximum force of this device amounted to about 1 pN for beads with a diameter of 2.8 μm remains too small for numerous uses.

A high-force single-pole magnetic tweezer setup has been introduced with a triggered induction feedback that involves a Hall probe (Kah et al., 2020). In specific, as an option, a lower current can be supplied in the reverse direction to balance the residual magnetic field of the core, although the level of this reverse current needs to be individually calibrated for every force application (Kollmannsberger & Fabry, 2007). Another prevalent mitigation technique for the issues related to magnetic hysteresis

Table 1 Advantages and disadvantages of magnetic tweezers for cell mechanical analyses. Possible solutions for weaknesses are also provided.

Advantages	Disadvantages	Solutions
Specific receptor targeting	Hysteresis issues with the magnetic core of the magnetic tweezer (Kah et al., 2020)	Force-feedback control (Kah et al., 2020)
Multiple cells can be measured in parallel	Measurement noise via bead-tether geometry (Yang et al., 2022)	To scan proteins, DNA handles are covalently linked to the two amino acid moieties of individual protein molecules. Thus, small variations in end-to-end expansion arising from conformational modifications are accurately mirrored in the motion of the bead (Yang et al., 2022)
Measurement of non-equilibrium responses to force	Measurement noise via applied force (Yang et al., 2022)	The exerted force can be altered to explore thermodynamics due to mechanical disturbance (Yang et al., 2022).
Dynamical probing of cells, organelles or cellular constituents	Temperature variations between experiments of cells and proteins (Aermes et al., 2021; Seifert et al., 2020)	Measurement at various temperatures using an environmental chamber (Aermes et al., 2021; Seifert et al., 2020; Zhou et al., 2023)
Different kinds of force-exertion procedures for elastic characteristics, intermediate conformations and dynamics of a transition measures	A crucial issue for these experiments is the precise calibration of the forces acting on the typically paramagnetic beads in the external inhomogeneous magnetic field (Sarkar & Rybenkov, 2016).	Management of the precise calculation of forces (Bausch et al., 1999; Kollmannsberger & Fabry, 2007; Mierke et al., 2008)

Advancements of analysing particle movements within living cells and organelles	Creation and control of localized magnetic field gradients, which are responsible for force generation (Timonen & Grzybowski, 2017)	Precise engineering of magnetic tweezers that are usually produced in the shape of extremely sharp magnetic tips (Timonen & Grzybowski, 2017)
Magnetic fields permeate biological tissue unhindered		
In contrast to optical stretching, for magnetic tweezers, the refractive index of the cells plays no role, for example, melanoma cells can be measured without overheating or cell rupture that occur in optical stretching approaches.		

is the usage of a mu-metal core, a nickel-iron-molybdenum alloy that is engineered for low hysteresis (Fig. 1). Nonetheless, Mu metal has a markedly lower maximum magnetic induction than conventional iron alloys (Landgraf et al., 2008) and can revert to hysteretic properties after machining, such as sharpening the spike with a grinding machine. The center of the magnetic tweezers comprises a 100 mm long cylinder having a diameter of 4.5 mm, which consists of either St37 steel (Kah et al., 2020) or Mu-metal (Aermes et al. 2020, 2021; Mierke et al., 2020) (Fig. 1A). As the cylinder is milled using a milling machine (Fig. 1B), one end of the cylinder is narrowed to a sharp point with an opening angle of 60° and a point radius less than 5 μm using a precision drilling machine. A Hall sensor (for example, SS495A; Honeywell, Charlotte, NC) for sensing magnetic induction is attached to the back end of the tweezer needle. The needle is fitted in a solenoid with ~200 turns (24 copper wire) on a piece of brass. The solenoid coil is fed from a high-current operational amplifier

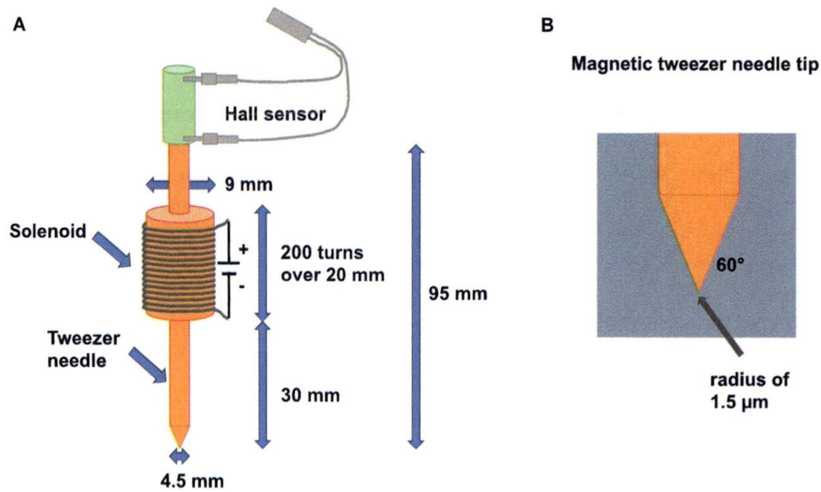

Fig. 1 Magnetic tweezer design. (A) The magnetic tweezer needle is usually magnetized by a solenoid with for example 200 windings. The produced magnetic field is captured through a Hall probe attached to the tail end of the magnetic forceps needle. (B) The needle features a pointed tip with an aperture angle of 60° and a radius of 1.5 µm to produce a strong magnetic field gradient.

(for instance, OPA549T; Texas Instruments, Dallas, TX). The tweezer needle is fixed on a micromanipulator (for example, InjectMan NI2; Eppendorf, Hamburg, Germany) to ensure accurate motion of the needle point. The location of the needle point with respect to a magnetic bead, such as a bead adhered to a cell or floating in oil, is determined using an inverted bright field microscope with a 40×, 0.6 NA objective with a large working distance. The pictures are taken with a CCD camera that is initiated via a data acquisition card, normally at a speed of 50 frames/s. The analog signal output of the data recording card delivers the nominal value of the magnetic induction for the following feedback loop. This allows the image recording to be synchronized with the force measurement protocol. Such synchronization is especially important when rapid motions of beads are captured after abrupt variations in force. Consequently, by offsetting the magnetic hysteresis of the core compound, it is possible to implement any force pattern with an amplitude of up to 100 nN on superparamagnetic beads of 5 µm diameter. An empirical equation involving a mere three free variables can be utilized to characterize the exerted force as a function of bead-to-needle spacing and magnetic inducement for a variety of superparamagnetic beads and for various core types. This equation can be integrated into the closed loop feedback to conduct force-controlled experiments. To visualize the diversity of this tweezer design for investigations in cell biology, steps of accelerating and decelerating force can be exploited to quantify the nonlinear (force-dependent) viscoelastic and plastic cell rheology of the NIH 3T3 mouse embryonic fibroblast cell line (Kah et al., 2020).

Due to the objective, different kinds of force-exertion procedures can be used: Firstly, a "force ramp" gauges the force-strain relationship with a progressively altering force; secondly, a "force clamp" tests the transition from one state to another with a constant force; and thirdly, a "force jump" causes a sudden alteration in the magnitude of the force to examine non-equilibrium reactions. The three tests deal in various manners with the elastic characteristics, the intermediate configurations and the dynamics of the transition and together result in a thermodynamic model.

Finally, a considerable advantage of magnetic forceps techniques lies in the fact that, among the different physical phenomena, magnetism offers significant advantages over light or electricity. Magnetic fields penetrate biological tissue unimpeded and are even used for medical treatments. An inimitable feature of this biophysical technique is that magnetic fields can be transformed into mechanical stimuli that can act as signals to regulate biological reactions.

2.2 Multipole magnetic tweezers

The implementation and characterization of multi-pole magnetic tweezers facilitates the positioning of small magnetic probes in the interior of living cells (De Vries et al., 2005). To date, magnetic tweezers can be classified into two groups. In the first group, the magnetic tweezers permit the application of high forces but are composed of just one or two poles and can thus only apply forces in a single direction. In the second group, the magnetic tweezers are composed of several poles and permit the application of forces in more than one direction but with extremely low forces. The magnetic tweezers illustrated here unite both elements in a single device: high forces in a direction that can be precisely controlled. For this purpose, magnetic structures in the micrometer range are produced employing clean room technologies. Magnetic flux gradients of $\nabla B = 8 \times 10^3\ \mathrm{T\,m^{-1}}$ can be obtained over the dimensions of an individual cell with these magnetic tweezers. This makes it possible to exert forces of up to 12 pN on para-magnetic spheres with a diameter of 350 nm, thereby maneuvering them through the cytoplasm of a living cell. It is anticipated that the present magnetic tweezers will be able to exert forces in the picoNewton range on beads with a diameter of just 100 nm.

The demands on techniques that enable nanomanipulation within cells are as listed in the following: Firstly, the force transducers employed should be tiny (<<1 μm); secondly, the forces which can be applied to the sample should be in the appropriate biological regime of at minimum a few pN; and thirdly, the forces need to be adjustable in terms of amplitude and direction so that the probe can be positioned where required. The utilization of magnetic forces could be a promising candidate to meet these demands. The contemporary implementation of magnetic tweezers on living cells is mainly based on microrheology (Bausch et al., 1999). Magnetic beads inserted inside cells have been demonstrated to be capable of exerting large forces (Hosu et al., 2003). Nevertheless, these magnetic tweezers have either one or two poles and thus they cannot simply be used to handle magnetic probes in various directions. In fact, the pole(s) could be maneuvered in relation to the bead so that the force on the bead can be altered in direction. A major disadvantage of this type of arrangement is that it is technically extremely challenging to obtain the necessary velocity and precision for relocating the pole(s) to enable precise manipulation. Other designs have been implemented that are easier to use and comprise several magnetic poles and thus enable two-dimensional (2D) or three-dimensional (3D) manipulation, but suffer from the inherent disadvantage of low forces (Amblard et al., 1996; Gosse & Croquette, 2002; Sacconi et al., 2005). A new kind of magnetic tweezers based on micrometer-sized magnetic pole structures that can be placed nearby a living cell is presented. It is demonstrated that it is possible in this manner to couple high forces with the ability for true 2D manipulation of the magnetic sphere inside living cells. These magnetic tweezers can potentially be utilized for the nanomanipulation of the molecular structures within living cells. The optical tweezers technique also offers the benefit of full 3D control and has also been applied to control beads within cells or on cell surfaces, for example for microrheology (Caspi et al., 2002; Laurent et al., 2002) or molecular manipulation (Peters et al., 1999). The benefits of magnetic tweezers are particularly apparent when studying intracellular characteristics. Optical tweezers exercise forces on microscopic targets that have a refractive index that is in opposition to their environments. As there are a large number of such items in a cell, optical tweezers cannot consistently operate selectively in the intracellular surroundings. In addition, the relatively high optical intensities needed can even hamper or destroy the specimen (Neuman et al., 1999). These complications of optical tweezers can be solved with magnetic fields affecting magnetic target particles or magnetic particles bound to non-magnetic objects. In these situations, no noticeable heating is to be anticipated, there are no limitations with regard to the refractive index of the particle, and triggering is even possible in completely closed and non-transparent spaces, for example, in closed containers.

3. Mechanosensitive proteins

Cellular mechanosensitivity originates from different but interconnected areas: the cytoskeleton and the plasma membrane (Fig. 2A). Elements of the cytoskeleton react to both bulk and local mechanical stresses, whereas receptors and ion channels on the cell surface respond to the tethering force of ligands and membrane tension. The resulting alterations in signal transmission and gene expression frequently result in long-lasting consequences for cell fate and performance (Ingber, 2006). In spite of the differing outcomes, shared issues have been identified in their molecular mechanisms (Hu et al., 2017). Mechanosensitive proteins alter

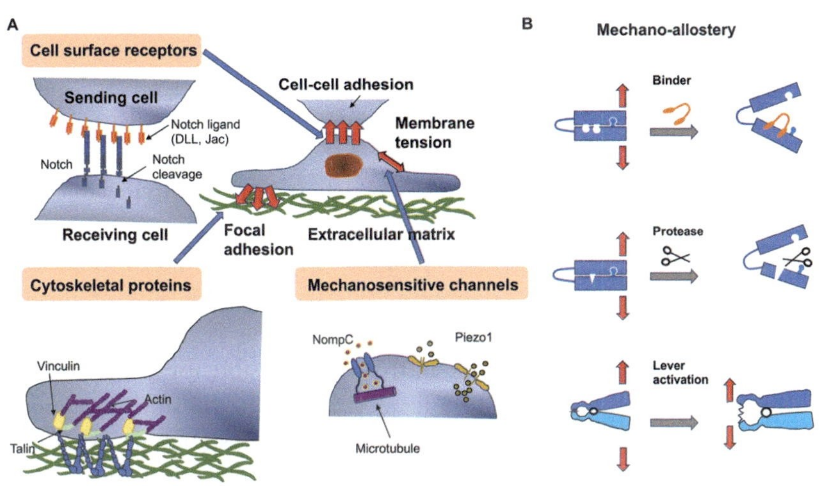

Fig. 2 Magnetic tweezers can be utilized to probe mechanosensitive proteins. (A) Cellular profile of mechanosensitivity. The arrows denote the typical direction of forces. Typical examples of mechanosensitive proteins that have been explored with magnetic tweezers are sketched. (B) Various modes of the mechanism of mechanical allostery impact the conformation of the mechanosensitive protein.

their conformations when exposed to forces, and the alteration in conformations can enlist other proteins or act as a target for proteolytic degradation. Certain proteins can use the force acting on one domain for advanced structural reorganizations in remote domains that are able to fulfill certain tasks such as opening of ion channel pores (Lin et al., 2019) (Fig. 2B). These mechanisms are analogous to allosteric regulation, with the difference that the mechanical force acts as a distant effector.

Magnetic tweezers are used in a growing number of applications to investigate intracellular transport and mechanosensing. A conventional strategy is to cover magnetic beads with a compound that guarantees their correct intracellular enrollment, use phagocytosis when the destination of the beads is intracellular, and then use a microprobe to exert magnetic forces on the beads. The beads can be employed in this situation to control the stiffness of the organelle (Collins et al., 2012; Mahowald et al., 2009), or just as a tool to pacify the cell and monitor downstream reactions (Hemphill et al., 2011). As a result, the magnetic tweezer method has been able to gain a foothold in this field and lead to notable results. The emphasis is on the investigation of stimulation mechanisms via specific receptor-ligand interactions or antibody-antigen interactions. Nevertheless, there is still great potential for improving the simple magnetic tweezer technique in terms of automation and combinability with other cell biological, biochemical or molecular methods.

3.1 Cytoskeletal proteins

The high accuracy of force exertion and spatial position measurement in single-molecule magnetic tweezers is beneficial for identifying fine variations in the force-dependent regions of mechanosensitive proteins. When the sites for force application are properly chosen, the consequent force-induced alteration can be interpreted as a simple indicator of mechanosensitivity. The first application examined in this context has been the investigation of titin, a giant protein that is responsible for muscle elasticity (Labeit & Kolmerer, 1995). This filamentous protein featuring microscale deformations has been well-suited to conduct protein stretching experiments using laser tweezers and atomic force microscopy (AFM) (Kellermayer et al., 1997; Marszalek et al., 1999; Rief et al., 1997). Advanced devices have uncovered more information about the mechanical nature of titin (Chen et al., 2015; Pang et al., 2018; Rico et al., 2013; Tapia-Rojo et al., 2019), which is assumed to be a complex mechanosensing system (Ibata & Terentjev, 2021; Puchner et al., 2008).

The force-dependent analysis has then been broadened to encompass smaller and increasingly dynamic proteins involved in mechanobiological processes, for instance the cytoskeletal focal adhesion protein talin. Talin and its adaptor proteins link actin filaments with the extracellular matrix through focal adhesion complexes (Kanchanawong et al., 2010; Liu et al., 2015), and this juncture then actively conveys mechanical forces (Goult et al., 2018). Single-molecule tweezers have provided an essential input to elucidate the mechanism of talin functionality (Dahal et al., 2022; Tapia-Rojo et al. 2020, 2023). In particular, it has been demonstrated that the elongation of talin reveals the usually concealed vinculin binding sites and thereby enlists vinculin proteins in reaction to mechanical forces (Fig. 2B) (del Rio et al., 2009). The subsequent research has fully identified the intermediate structures and the points of transition among them (Wang et al., 2021; Yao et al., 2016). The piconewton forces utilized in the *in vitro* tweezer tests have actually been found to be substantial for the *in vivo* capabilities of talin (Austen et al., 2015; Jiang et al., 2003; Margadant et al., 2011). Similar experiments have been also carried out with other actin-binding proteins, among them vinculin (Huang et al., 2017; Mierke et al., 2010), α-catenin (Buckley et al., 2014; Yao et al., 2014), filamin (Ferrer et al., 2008; Lee et al., 2009), and α-actinin (Grison et al., 2017; Le et al., 2017; Roca-Cusachs et al., 2013). All these investigations highlight the usefulness of molecular tweezers in studying the mechanosensitivity of the cytoskeleton.

Thereafter, magnetic beads coated with integrin ligands have been applied to mechanically investigate the connection between the extra-cellular matrix and the cytoskeleton. Using forces in the range of pN to 1 nN, an integral role of the focal adhesion protein vinculin for force transmission has been detected (Alenghat et al., 2000). Later, a high-force magnetic tweezers instrument has been built, whereby the distance between the magnet and the beads is reduced to only a few μm. This made it possible to exert forces up to 100 nN on 5 μm beads, a force that is well-suited to mechanically probe living cells (Kollmannsberger & Fabry, 2007), such as vinculin wildtype and vinculin deficient cells (Mierke et al., 2010) and FAK wildtype, FAK deficient and a kinase-dead FAK mutant, R454-FAK cells (Mierke et al., 2017). These forces lie in the similar regime as the forces applied by cells to their environment during matrix adhesion and migration and are expected to enable the mechanics of such cellular phenomena to be examined in depth. However, in this type of experiment it is crucial to place the magnet very near to the probe, which may be challenging for cells in 3D microenvironments, cell spheroids, organoids or tissues.

The talin-vinculin interface is a key mechanosensory system in cellular focal adhesions. How talin and vinculin react to forces and mutually control each other is still elusive. Through a combination of single-molecule magnetic tweezers experiments, molecular dynamics simulations, actin bundling experiments and adhesion assembly studies performed in living cells, a two-way allosteric network within vinculin as a regulator of talin-vinculin interaction can be characterized. A maturation mechanism of vinculin in the attachment of talin has been identified, which strengthens the attachment to talin at a rate of $0.03\,s^{-1}$ (Franz et al., 2023). This allosteric transition can effectively compete with the force-induced de-tethring of vinculin from talin solely at forces up to 10 pN. Mimicking allosteric activation through mutation leads to a vinculin molecule that bundles actin filaments and anchors to focal adhesions in a force-independent fashion. Thus, the allosteric switch restricts talin-vinculin cross-talk and focal adhesion development to intermediate force settings. The "allosteric vinculin mutant" is a precious molecular instrument to continue deciphering the mechanical and biochemical pathways at focal adhesions and beyond.

3.2 Cell surface membrane receptors

The strong attachment of Arg-Gly-Asp (RGD)-coated magnetic microparticles toward integrins linked to the cytoskeleton has been exploited multiple times to mechanically analyze the cytoskeleton by employing vastly divers cell types. The benefit of this experimental approach is that the magnetic beads do not have to be positioned inside living cells—it is sufficient for them to be on the surface. In an illustrative case, such particles have been attached to the surface of smooth muscle cells of the human respiratory tract (Fabry et al., 2001). However, when it comes to the interpretation of the findings of this type of cellular microrheological experiments, the specific nature of the receptor-ligand linkage needs to be considered, as it may change the cell adhesion and/or mechanotransduction of the cues (Hynes, 2002). At the same time, this kind of interaction between particle attachment and cell biochemistry points to additional intriguing opportunities for magnetomechanical regulation of various cell behaviors (Cho et al., 2012).

Beyond studying the characteristics of cells, an intriguing ability to drive cellular signaling circuits has recently been shown to be achieved by tweezing magnetic particles coated with specific biologically functional molecules onto cell surfaces (Cho et al., 2012) or within cells (Bonnemay et al., 2013; Hoffmann et al., 2013). The first approach could even be utilized to regulate cell apoptosis (Cho et al., 2012). In these kinds of experiments, magnetic nanoparticles have been covered with an antibody that targets the death receptor 4 (DR4) of DLD-1 colon cancer cells. The cells containing these particles have been exposed to a horizontal magnetic field, causing the particles to contract and cause the DR4 receptors to aggregate, effectively inducing apoptosis (Fig. 3). Nanoparticles are similar in their size to the receptors and can raise the local concentration of the receptor to the point where cellular signaling processes are activated.

Magnetic nanoparticles in cells have been tweaked with magnetic tweezers (Bonnemay et al., 2013; Cho et al., 2012) to direct cellular signal transduction. For this purpose, appropriately functionalized (or functionalized *in situ* in the cells) superparamagnetic nanoparticles have been introduced into or picked up by the cells, and their position has been targeted with microtip electromagnetic tweezers (Etoc et al. 2013, 2015). In a striking case, magnetic particles labeled with small Rho-GTPases, which are crucial for the control of cell migration and polarization, have been engineered and placed inside NIH 3T3 fibroblasts (Etoc et al., 2013). As the beads have been dragged to the cell borders, they triggered the generation of lamellipodia (Fig. 4). A somewhat similar procedure has been utilized for spatio-temporal guidance of microtubule nucleation in *Xenopus laevis* egg extracts using magnetic tweezers (Bonnemay et al., 2013; Hoffmann et al., 2013). In this instance, superparamagnetic nanoparticles have been conjoined with RanGTP proteins, which have a function in regulating the cytoskeleton (Soh et al., 2010). Magnetic tweaking of these nanoparticles resulted in local concentrations of RanGTP exceeding the nucleation threshold, thus permitting the regulation of microtubule growth at the targeted positions.

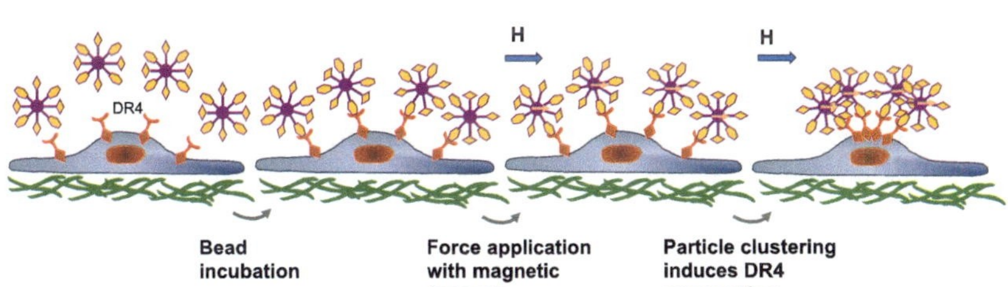

Fig. 3 Membrane-receptor clustering induced with magnetic tweezers. Schematic image of the magnetic tweezer-based force exertion that concentrates on cell surface receptors to induce signal transduction pathways inside the cell.

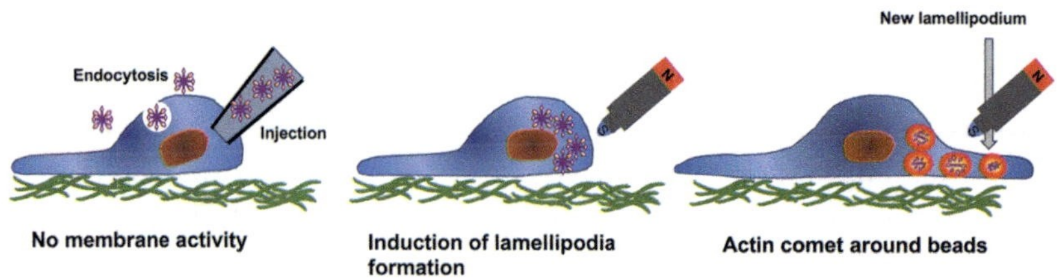

Fig. 4 Magnetic tweezers can be used to induce the formation of lamellipodia. Rho-GTPase-coated superparamagnetic beads are moved toward the cell borders to trigger the formation of a new lamellipodium.

Mechanical force has also been observed to activate receptor proteins at cell-cell contacts, one of which is the Notch family. Notch proteins represent transmembrane receptors that are crucially engaged during tissue formation and homeostasis (Siebel & Lendahl, 2017), and are triggered by attachment to ligands, such as Delta-like and Jagged families, on adjacent cell membranes. It has been hypothesized that ligand engagement and consequent endocytosis in the canonical route produces a pulling force that elongates Notch (Parks et al., 2000). This idea has been reinforced by the fact that the protease site in Notch for activation is typically concealed within an autoinhibitory configuration (Gordon et al., 2015). Subsequently, single-molecule tweezers have been used to investigate the forces involved in Notch-ligand interactions (Meloty-Kapella et al., 2012; Shergill et al., 2012; Stephenson & Avis, 2012), and ultimately a high-throughput assay has smartly proven that proteolytic sensitivity improves when Notch is tugged with low forces that uncover the cleavage site (Gordon et al., 2015). Noteworthy, the same piece of work demonstrated that piconewton forces are adequate to trigger Notch activation in living cells. However, some other research using force-measuring DNA nanostructures and plasma membrane force probes has provided systematic quantitative evidence of the force requirements for Notch activation (Luca et al., 2017; Wang & Ha, 2013), which additionally points to the mechanosensitivity of Notch signal transduction.

A further typical classic example of receptor-ligand linkage investigated with tweezers concerns the crosstalk between von Willebrand factor (VWF) and the platelet receptor glycoprotein Ibα (GPIbα), which facilitates the adhesion of platelets to blood vessel walls (Ruggeri & Mendolicchio, 2007). It has long been recognized that vascular adhesion is especially efficient in the presence of high shear forces such as those in the circulation, and many issues arise about the nature of receptor-ligand attachment that triggers the mechanism. These issues have been tacked effectively with the help of single-molecule force techniques. Previous studies have established the force-dependent engagement between the A1 domain of VWF and GPIbα reliably (Arya et al., 2002; Yago et al., 2008), whereas the subsequent optical tweezers experiments have analyzed the single-molecule kinetics of the identical pair at a higher force resolution (Kim et al., 2010, 2015). Most interestingly, VWF shows also force-induced unfolding and proteolytic scission (Zhang et al., 2009), in a similar manner to the Notch receptors discussed above. Collectively, these findings have deepened the understanding of how VWF activity is regulated by shear flow in the initial phase of hemostasis. Among the other surface receptors that have been investigated with single-molecule tweezers are cadherins, integrins, selectins and T-cell receptors (Chen et al., 2017).

In addition, the stiffness of aortic tissue, which is associated with cardiovascular disease, has been investigated using magnetic tweezers. To assess the stiffness of the cell cortex of A7r5 rat vascular smooth muscle cells, a GRGDNP (H-Gly-Arg-Gly-Asp-Asn-Pro-OH) peptide (RGD), which is necessary for tethering to transmembrane integrin receptors within A7r5 cells, has been covalently attached to magnetic beads. The beads have been subsequently incubated with the cells at 37 °C and 5% CO_2 to encourage attachment. A magnetic microneedle consisting of a highly permeable nickel alloy exhibiting low hysteresis and enclosed by a copper solenoid has been utilized as a magnetic tweezer tip probe. This probe has been positioned 150–200 μm distant from the attached beads at the starting point of the measurement. Cortical stiffness has been computed as the stiffness of bead-integrin adhesion, a ratio between the exerted force and the movement of the magnetic beads that adhere to the cell (Saphirstein et al., 2013). The results show that lysophosphatidic acid (LPA), which is an activator of myosin, enhances cell contractility and increases cortical stiffness, and PP2, which is a small molecule inhibitor of Src-dependent FA recycling, prevents the LPA-induced rise in cortical stiffness.

In addition, bovine aortic endothelial cells have been analyzed by targeting PECAM-1 with anti-PECAM-1 coated superparamagnetic beads (Collins et al., 2012).

3.3 Membrane sensitive channels

What are the criteria to define a mechanosensitive channel? In 2010, the definition has been made by Árnadóttir and Chalfie, who recommended a number of criteria (Árnadóttir & Chalfie, 2010). The first criterium is that mechanosensitive channel has to be located within a mechanosensory organ. The second requirement involves the elimination of the mechanosensitive channels of the sensory cell, which rules out the mechanical reaction as a direct and not an indirect effect, for example, of developmental defects. The third criterium is that modification of the physical characteristics of these channels via mutagenesis have to be mirrored through changes in the mechanical reaction. Finally, the fourth prerequisite is that the channel which is heterologous expressed can be force activated. A key question is how mechanical cues from the surroundings are transmitted to convert the mechanosensitive channel from a closed to an open state. Two models for force transmission and the gating have been developed: In the first model, which is referred to as membrane tension model, a force exerted on the lipid bilayer produces a membrane tension that opens the channel. In the second model, which is referred to as the tether model, a force exerted on the cell or extracellular matrix is passed through a connector linking the channel to ectomembrane constituents to open the channel. The decisive difference between the two model mechanisms is not the involvement of the membrane, but rather whether any additional elements are needed for force transmission.

Mechanosensitive channels can be defined as transmembrane ion channels that are triggered by mechanical stimuli (Jin et al., 2020). They are recognized to fulfill key functions in heat and touch perception, pain and proprioception. Because of the structural intricacy of mechanosensitive channels, the technique for gauging stretch under bidirectional tension has not yet been extended to full-length proteins. Nevertheless, a number of investigations have used the ability of nanotweezers to examine how small forces influence the key areas of mechanosensitive channels and how they behave in a physiological manner. The piezo channel, which is a large, homotrimeric protein forming a propeller-shaped structure, serves as an example of mechanosensitive channels (Saotome et al., 2018). The three paddles of a piezo channel are strongly curved and give the receiving membrane the shape of a nanodome (Guo & MacKinnon, 2017). Most interestingly, the dome structure has been observed to reversibly become flattened when mechanical disturbances are introduced using AFM (Lin et al., 2019), which could account for channel opening through membrane tension (Cox et al., 2016). In another investigation, the various areas of Piezo1 have been precisely cultivated with magnetic nanoparticles and identified as the areas causing mechanosensitivity (Wu et al., 2016). The schematic of the alteration of the open state of Piezo through inactivation can be described as follows by a step-based mechanism. The first step denotes that the Piezo is in a closed-state highlighting dome-like C-terminal extracellular domain (CED) and inner pore helices. The second step is the Piezo in tension-induced open condition, whereby the CED is subject to a pulling force through a magnetic bead. There is a force exerted through pore helixes toward inactivation of the channel. When the CED moves off the channel, it alters the inactivation through crosstalk with the pore helices. The third step depicts the piezo in a tension-induced open configuration at a negative maintaining potential. Larger cross-talk with the pore helixes is provided by a stronger force toward the inactivation of the channel. In this step, there are positive charges present within the pore helices. In addition, the Coulombic force keeps the pore helices and the neighboring CED in a rapidly inactivating condition. The fourth step represents the piezo when it is in a voltage-induced open condition and is maintained at a positive sustaining potential. In this phase, a weaker force is applied for inactivation. Coulombic force triggers dislocated pore helices and CED similar to magnetic attraction to trigger a gradual inactivating condition (Wu et al., 2017). Consequently, nano-tweezers become also valuable to control the response of mechanosensitive channels via inducing membrane deformations in advanced systems. Cell membranes have been impacted with optical tweezers to generate cavities. It was demonstrated that tapping triggers Ca^{2+} transients in cells containing Piezo1 (Falleroni et al., 2018). Recently, it has been revealed that tension in cell membranes is not spread across large physical distances and induces piezo-channels in the near vicinity alone, indicating a limited mechanosensitivity to subcellular areas (Shi et al., 2018). Lastly, tugging Piezo1 with magnetic nanoparticles can even elicit behavioral alterations in the brains of freely moving mice (Lee et al., 2021), highlighting the inherent specificity and effectiveness of force application concepts.

The ankyrin repeat, which is a member of the transient receptor (TRP) family, is hypothesized to act as a "gate spring" in mechanotransduction within hair cells (Sotomayor et al., 2005). Single-molecule stretching with AFM has been used to determine the elasticity of ankyrin repeats and demonstrate that their behavior is reminiscent of a spring (Lee et al., 2006). Later work revealed that ankyrin repeats in NompC, which is a member of the TRP family, attach to microtubules and are integral for mechanosensitivity *in vivo* (Zhang et al., 2015), pointing to a cytoskeleton-dependent machinery for gating.

3.4 Nuclear mechanical analysis

As it is common knowledge that nuclei are exposed to forces and respond to them, the direct impact of forces on the nucleus has been examined (Guilluy et al., 2014). Insulated nuclei can respond to force by modifying their stiffness to resist the external tension (Guilluy et al., 2014). Using magnetic tweezers, it has been determined that the application of force to the Link complex participant nesprin-1 results in a reinforcement of the nucleus and thus a nuclear stiffening. This nuclear stiffening requires an undamaged nuclear lamina and emerin, which is a protein present on the inner nuclear membrane (Guilluy et al., 2014). However, chromatin or nuclear actin do not appear to be involved in nuclear stiffening. Specifically, emerin undergoes tyrosine phosphorylation when force is exerted and controls the mechanical response of the nucleus when tension is applied. Finally, it has been found that mechanotransduction is not restricted to cell surface receptors and cell adhesions, but is also present within the cell nucleus (Guilluy et al., 2014).

Generally, it is common knowledge that both the nucleoskeleton and the chromatin are involved in the mechanical characteristics of the cell nucleus (Dahl & Kalinowski, 2011; Fischer et al., 2020; Miroshnikova et al., 2017; Pajerowski et al., 2007). To find out whether modifications of the mechanical characteristics of the DNA influence the stiffening of the cell nucleus through force excitation, isolated cell nuclei that have been exposed to trichostatin A, which works as a histone deacetylase inhibitor and facilitates DNA decondensation, have been investigated. Actually, treatment with trichostatin A failed to suppress the force-triggered nuclear stiffening, although trichostatin A increased the average nuclear size 2.3-fold (Guilluy et al., 2014). The exposure of nuclei to DNAse I produced analogous findings (Guilluy et al., 2014), which confirmed the assumption that chromatin and DNA do not favor the modification of the nucleus due to external forces. Even though the chromatin and DNA did not cope with the stiffening of the cell nucleus due to the mechanical stress exerted on the LINC complex, it is possible that the force directly changes the structure of the chromatin. The impact of heating on the mechanical reaction of the nuclei to mechanical stress has not yet been clarified, but appears to be very intriguing, as soft matter adjusts to temperature changes by modifying its mechanical phenotype (Aermes et al., 2021).

3.5 Magnetic guidance of intracellular signal transmission with magnetic tweezers

In comparison to optical tweezers, magnetic tweezers have led to several investigations on spatial intracellular disturbances due to the mechanical manipulation of magnetic nanoparticles. While there exist (at least ordinarily) no organelles that act as permanent or inducible magnets within a living cell, superparamagnetic nanoparticles can be either microinjected or endocytosed into cells and utilized as force sensors. A decisive benefit of magnetic tweezers over optical tweezers for investigations on living cells lies in the fact that the magnetic fields are non-invasive and that the force generated can be extensively parallelized (Tseng et al., 2012). Magnetic nanoparticles exist in a multitude of sizes and coverings where conjugation of ligands and antibodies may be fairly simple as many commercially available custom assays are abundant. Thus, force-transmitting grips directed to various molecules to be of specific interest can be easily produced to implement magnetically driven, long-distance mechanical actuation of a wide array of signal transduction pathways and cellular architectures (Bonnemay et al., 2015).

Spatial manipulation by magnetic-mechanical actuation is achieved through the translocation of nanoparticles that have been made functional with proteins, which can proceed in two ways: firstly, ion channels or receptors located at the cell membrane or secondly, intracellular spreading of signaling proteins. Attaching magnetic nanoparticles to the mechanosensitive ion channel TREK-1 enabled the use of a highly targeted force and consequently channel activation (Hughes et al., 2008). In a similar way, functionalized magnetic nanoparticles placed on the surface of a single hair cell bundle of a North American bullfrog provided ultrafast mechanical excitation of hair cells to investigate mechanotransduction in sensory systems (Lee et al., 2014). In other instances, the mechanical forces are exerted on the force grips to organize tethered proteins in a spatial manner, where the mechanics of the proteins have no direct involvement. Similar to multivalent ligands for receptor agglomeration and activation, the oligomerization of individual IgE/FcεRI receptor complexes within mast cells can be used to mimic calcium signaling by a magnetic switch (Mannix et al., 2008). Since intracellular signaling usually starts from the plasma membrane of the cell, local targeting of ion channels and receptors is a convenient mechanism for achieving effective remote control of cellular signal transduction. Since magnetic nanoparticles can be placed within a cell, direct distant control of intracellular signal transduction is feasible without activation of cell surface channels or receptors. This approach provided the inspiration for a number of pioneering investigations into the magnetic guidance of intracellular signal transduction.

Whereas endocytosis is usually assumed to downregulate inbound cues, signaling endosomes offer physical gateways for switching between various signaling routes (Pálfy et al., 2012). Thus, endocytosed magnetic nanoparticles may be arranged as endosomes for signaling. In the course of axonal recovery after trauma, neurotrophin-activated tropomyosin-related kinase B receptors are being endocytosed and serve as signaling endosomes. In another study, magnetic nanoparticles functionalized using agonist antibodies of tropomyosin-related kinase B have undergone endocytosis inside primary neurons. In the complete lack of a magnetic field, the signaling endosomes activated via the nanoparticles have been trafficked into the developing neurites and stimulated their growth. However, when the transport of signaling endosomes has been delayed by a strongly localized magnetic field gradient, the motility of the growth cones has been substantially impaired (Steketee et al., 2011).

Rho-GTPases have a central function in the assembly and structure of the actin cytoskeleton, and their precise spatio-temporal regulation governs dynamic morphological alterations occurring during cell migration. Magnetic particles of 500 nanometers in size coupled to the Rac-GTP exchange factor TIAM-1 have been introduced into fibroblasts and utilized as nanoscale probes to investigate the activity of molecular complexes within living cells. Mechanical actuation to move a magnetic particle to the anterior margin of the cell triggered protrusion activity and an actin comet tail from the particle (Fig. 4 right image) (Etoc et al., 2013). Five-hundred-nanometer particles can be large enough to obstruct micropipettes, so low particle concentrations have to be used and their movement is often restricted due to intracellular obstacles. The use of smaller functionalized magnetic nanoparticles as small as 15 nm enables the intracellular diffusion movement of nanoparticles to be exploited to produce an intracellular protein gradient driven by magnetic forces with a time resolution of a few tens of seconds (Etoc et al., 2015). The insertion of magnetic nanoparticles via microinjection and endocytosis each have their advantages and disadvantages. Thus specific biological settings exists where they are most appropriately implemented. Another scenario in which distant magnetic guidance of cell signaling is utilized is a membraneless system where direct input of magnetic nanoparticles to a cell-free extract eliminates the requirement for endocytosis and microinjection. A cell extract droplet has been established as a model system for the study of microtubule nucleation and composition through magnetic guidance. It has been demonstrated that the gradient of the GTPase RanGTP, that is powered by the GTP exchange factor RCC1, can stabilize and facilitate microtubule assembly within extracts (Kaláb et al., 2006). Ran-conjugated magnetic nanoparticles can be incorporated into the extract and embodied in individual emulsion droplets. Clustering of Ran nanoparticles causes microtubule nucleation when exposed to a magnetic field (Hoffmann et al., 2013). Dynein motors are able to orchestrate the polarized microtubule arrays, controlling their orientation through the direction of magnetic forces. Overall, these investigations underline the diversity and efficiency of controlling signal transmission at a distance through magnetic mechanical actuation.

4. 3D applications of magnetic tweezers in cells

There has been the development of a powerful hexapole magnetic tweezer system that can be employed for closed-loop 3D swarm control measurements (Zhang et al., 2019). The magnetic tweezers system enables more versatile applications at the microscale, including microparticle swarm guidance, cell penetration and cell therapeutics. This experimental analysis illustrates the closed-loop navigation of a microparticle swarm that is free to travel in both 2D and 3D landscapes. In addition, the applicability of the magnetic nanotweezers can be expanded to *in vivo* situations (Cho et al. 2012, 2016). Magnetic nanoparticles have been inserted into the yolk of a zebrafish embryo to activate ovarian TNF receptors (OTRs). A uniform magnetic field has then been created to encourage the accumulation of OTRs by exploiting the dipolar attraction and create apoptotic incentives (Cho et al., 2012). A characteristic phenotypic deficiency of the tail section of the zebrafish has been detected in the magnetically stimulated group compared to the negative controls. Such an approach can be translated to a higher animal model by implementing the idea of the magnetic tandem apoptosis trigger (mTAT) in a system for the clinically relevant treatment of multidrug-resistant cancer (MDR) (Cho et al., 2016). The mTAT system can induce both extrinsic, for example, DR-4 signaling, and intrinsic, for example, DNA injury, apoptotic responses through magnetically guided receptor assembly and intracellular delivery of anticancer agents. This ability led to the complete eradication of MDR cancer in a mouse xenograft model, which could not be accomplished with chemical pharmacotherapy alone.

In contrast to dipolar attraction, the utilization of the tensile mode *in vivo* is commonly limited by the exponential decay of the magnetic field strength. Nevertheless, when the cells are loaded with aggregates of magnetic nanoparticles ($\sim 10^5$), they can undergo an attraction in the piconewton regime even when they are just a few centimeters distant from the magnet. Such cells bearing magnetic nanoparticles, referred to as magnetically labeled cells (mag-cells), can be drawn to an external magnet and thereby delivered to a target location in an *in vivo* tissue like the brain of a rat (Yun et al., 2018).

Using suitable magnetic nanoparticles and improved magnetic arrays, magnetic nanotweezers can be utilized for controlling individual cells and subcellular transduction events. More precisely, magnetic nanoparticles can impart a force on targeted Notch receptors in a gradient magnetic field with a magnet-to-particle distance of below one micrometer, resulting in mechanical triggering of the Notch receptor signal transduction route. In a Gal4/UAS reporter system, the transcriptional profiling of candidate cells can be set up consistently at each target location of interest with single cell resolution. Apoptosis of individual cells has also been verified with the mTAT system (Cho et al., 2016). A few micromagnets applied a uniform magnetic field exclusively to the focused cell area, which induced DR 4 assemblage and initiated the apoptosis signaling response cascade. Magnetic nanotweezers also capable to investigate the dynamics of subcellular receptor signals when exposed to a magnetic gradient field (Kim et al., 2017; Seo et al., 2016). Magnetic nanoparticles can induce the local accumulation of cadherins to promote the formation of stable adherens junctions in a particular subcellular area.

While strong magnetic fields are impossible to "focus" in the same way as light, modern microfabrication techniques permit the creation of microstructures that enable the field gradients—and the consequent magnetic forces—to be limited to tiny dimensions. This capability lays the cornerstone for magnetic tweezers, the traditional version of which is capable of manipulating magnetic objects. More specifically, the so-called negative magnetophoretic tweezers have been designed to capture and manipulate entirely non-magnetic particles, when suspended in a highly magnetic solution. These two types of magnetic tweezers are two complementary techniques suitable for various types of uses. This progress lays the theoretical foundation for both modes and highlights their specific fields of application, ranging from 2D and 3D colloid handling to live cell capture, monitoring cell functionality and single molecule testing. While the manipulation of magnetic objects, such as objects with positive magnetic susceptibility, in non-magnetic fluids with magnetic fields is conceptually easy, a somewhat less intuitive option is to magnetically manipulate non-magnetic particles, such as those with magnetic susceptibility near zero or even negative, while not applying magnetic particles to them. This kind of tugging is feasible when the magnetic susceptibility of the target bead is lower than that of the enclosing medium - in such circumstances, the force applied to the bead is conducted from high-field to low-field areas, for instance, "away" from the origin of the magnetic field. This combination is referred to as "negative magnetophoresis." How can negative magnetophoresis enable effective "focusing" of magnetic field minima to very small dimensions and manipulation of particles in both 2D and 3D by creating substrates or "pens" that include magnetic and non-magnetic regions? When the magnetic fluids are paramagnetic salt solutions, colloids as small as submicron can be captured and transported in completely enclosed magnetofluidic enclosures. Moreover, these colloids can be crystallized and even kept stably suspended in 3D. When the highly sensitive fluid is composed of biocompatible nanoparticles, it is possible to achieve analogous performance in the stimulation of living cells, involving cell sorting and the establishment of organized cell clusters. Finally, negative magnetophoretic tweezing is a nascent technique, but may be particularly useful for studying the 3D organization and dynamics of cells due to the capability of levitation capture.

5. Conclusions and future directions

The magnetic tweezer device has developed very-well in the past 20 years and become a traditional biophysical device that is easy to handle and less costly. There is still a question regarding the future of nano-precision magnetic tweezer. Definitely the hallmark contributions of magnetic tweezers seem to be the analysis of mechanosensitive assemblies to uncover their molecular regulatory mechanisms. Moreover, it will also be possible to focus on how these systems perform *in vivo*. Proteins with unconventional motifs and intrinsically rearranged regions could especially benefit from tweezer measurements, which provide a plethora of insights into their ever-changing transformations. Another possibility are correlative measurements in a tweezer setup with fluorescence optics or an electrophysiology manipulation rig, which facilitate experiments with primary cells, cell spheroids or organoids to obtain information on the mechanobiology of native cell tissues. Notwithstanding the solid advancements in live-cell tweezer approaches (Arbore et al., 2019; Mohammed et al., 2019), efforts to collect high-resolution data with live cells will face another level of challenges due to various background noise sources. Nonetheless, this appears to be an intriguing path that will address essential open questions in mechanobiology on the level of single molecules. Finally, magnetic tweezers will enable the transfer from traditional 2D measurements toward 3D measurements. Moreover, the combination with other biophysical cell probing techniques, such as optical tweezers will be highly beneficial. Subsequently, the precise intracellular navigation of magnetic particles within the cytoplasm, organelles or nucleus of cells will provide further insights in mechanical properties of these compartments.

References

Aermes, C., Hayn, A., Fischer, T., & Mierke, C. T. (2020). Environmentally controlled magnetic nano-tweezer for living cells and extracellular matrices. *Scientific Reports, 10*, 13453. https://doi.org/10.1038/s41598-020-70428-w.

Aermes, C., Hayn, A., Fischer, T., & Mierke, C. T. (2021). Cell mechanical properties of human breast carcinoma cells depend on temperature. *Scientific Reports, 11*, 10771. https://doi.org/10.1038/s41598-021-90173-y.

Alenghat, F. J., Fabry, B., Tsai, K. Y., Goldmann, W. H., & Ingber, D. E. (2000). Analysis of cell mechanics in single vinculin-deficient cells using a magnetic tweezer. *Biochemical and Biophysical Research Communications, 277*, 93–99. https://doi.org/10.1006/bbrc.2000.3636.

Amblard, F., Maggs, A. C., Yurke, B., Pargellis, A. N., & Leibler, S. (1996). Subdiffusion and anomalous local viscoelasticity in actin networks. *Physical Review Letters, 77*, 4470–4473. https://doi.org/10.1103/PhysRevLett.77.4470.

Arbore, C., Perego, L., Sergides, M., & Capitanio, M. (2019). Probing force in living cells with optical tweezers: From single-molecule mechanics to cell mechanotransduction. *Biophysical Reviews, 11*, 765–782. https://doi.org/10.1007/s12551-019-00599-y.

Árnadóttir, J., & Chalfie, M. (2010). Eukaryotic mechanosensitive channels. *Annual Review of Biophysics, 39*, 111–137. https://doi.org/10.1146/annurev.biophys.37.032807.125836.

Arya, M., Anvari, B., Romo, G. M., Cruz, M. A., Dong, J.-F., McIntire, L. V., et al. (2002). Ultralarge multimers of von Willebrand factor form spontaneous high-strength bonds with the platelet glycoprotein Ib-IX complex: Studies using optical tweezers. *Blood, 99*, 3971–3977. https://doi.org/10.1182/blood-2001-11-0060.

Ashkin, A., Dziedzic, J. M., Bjorkholm, J. E., & Chu, S. (1986). Observation of a single-beam gradient force optical trap for dielectric particles. *Optics Letters, 11*, 288. https://doi.org/10.1364/OL.11.000288.

Austen, K., Ringer, P., Mehlich, A., Chrostek-Grashoff, A., Kluger, C., Klingner, C., et al. (2015). Extracellular rigidity sensing by talin isoform-specific mechanical linkages. *Nature Cell Biology, 17*, 1597–1606. https://doi.org/10.1038/ncb3268.

Basoli, F., Giannitelli, S. M., Gori, M., Mozetic, P., Bonfanti, A., Trombetta, M., et al. (2018). Biomechanical characterization at the cell scale: Present and prospects. *Frontiers in Physiology, 9*, 1449. https://doi.org/10.3389/fphys.2018.01449.

Bausch, A. R., Möller, W., & Sackmann, E. (1999). Measurement of local viscoelasticity and forces in living cells by magnetic tweezers. *Biophysical Journal, 76*, 573–579. https://doi.org/10.1016/S0006-3495(99)77225-5.

Bessalova, V., Perov, N., & Rodionova, V. (2016). New approaches in the design of magnetic tweezers–current magnetic tweezers. *Journal of Magnetism and Magnetic Materials, 415*, 66–71. https://doi.org/10.1016/j.jmmm.2016.03.038.

Bonnemay, L., Hoffmann, C., & Gueroui, Z. (2015). Remote control of signaling pathways using magnetic nanoparticles. *WIREs Nanomedicine and Nanobiotechnology, 7*, 342–354. https://doi.org/10.1002/wnan.1313.

Bonnemay, L., Hostachy, S., Hoffmann, C., Gautier, J., & Gueroui, Z. (2013). Engineering spatial gradients of signaling proteins using magnetic nanoparticles. *Nano Letters, 13*, 5147–5152. https://doi.org/10.1021/nl402356b.

Buckley, C. D., Tan, J., Anderson, K. L., Hanein, D., Volkmann, N., Weis, W. I., et al. (2014). The minimal cadherin-catenin complex binds to actin filaments under force. *Science (New York, N. Y.), 346*, 1254211. https://doi.org/10.1126/science.1254211.

Caspi, A., Granek, R., & Elbaum, M. (2002). Diffusion and directed motion in cellular transport. *Physical Review E, 66*, 011916. https://doi.org/10.1103/PhysRevE.66.011916.

Chen, H., Yuan, G., Winardhi, R. S., Yao, M., Popa, I., Fernandez, J. M., et al. (2015). Dynamics of equilibrium folding and unfolding transitions of titin immunoglobulin domain under constant forces. *Journal of the American Chemical Society, 137*, 3540–3546. https://doi.org/10.1021/ja5119368.

Chen, Y., Ju, L., Rushdi, M., Ge, C., & Zhu, C. (2017). Receptor-mediated cell mechanosensing. *MBoC, 28*, 3134–3155. https://doi.org/10.1091/mbc.e17-04-0228.

Cho, M. H., Kim, S., Lee, J.-H., Shin, T.-H., Yoo, D., & Cheon, J. (2016). Magnetic tandem apoptosis for overcoming multidrug-resistant cancer. *Nano Letters, 16*, 7455–7460. https://doi.org/10.1021/acs.nanolett.6b03122.

Cho, M. H., Lee, E. J., Son, M., Lee, J.-H., Yoo, D., Kim, J., et al. (2012). A magnetic switch for the control of cell death signalling in in vitro and in vivo systems. *Nature Materials, 11*, 1038–1043. https://doi.org/10.1038/nmat3430.

Collins, C., Guilluy, C., Welch, C., O'Brien, E. T., Hahn, K., Superfine, R., et al. (2012). Localized tensional forces on PECAM-1 elicit a global mechanotransduction response via the integrin-RhoA pathway. *Current Biology, 22*, 2087–2094. https://doi.org/10.1016/j.cub.2012.08.051.

Cox, C. D., Bae, C., Ziegler, L., Hartley, S., Nikolova-Krstevski, V., Rohde, P. R., et al. (2016). Removal of the mechanoprotective influence of the cytoskeleton reveals PIEZO1 is gated by bilayer tension. *Nature Communications, 7*, 10366. https://doi.org/10.1038/ncomms10366.

Dahal, N., Sharma, S., Phan, B., Eis, A., & Popa, I. (2022). Mechanical regulation of talin through binding and history-dependent unfolding. *Science Advances, 8*, eabl7719. https://doi.org/10.1126/sciadv.abl7719.

Dahl, K. N., & Kalinowski, A. (2011). Nucleoskeleton mechanics at a glance. *Journal of Cell Science, 124*, 675–678. https://doi.org/10.1242/jcs.069096.

De Vlaminck, I., Henighan, T., van Loenhout, M. T. J., Burnham, D. R., & Dekker, C. (2012). Magnetic forces and DNA mechanics in multiplexed magnetic tweezers. *PLoS One, 7*, e41432. https://doi.org/10.1371/journal.pone.0041432.

De Vries, A. H. B., Krenn, B. E., Van Driel, R., & Kanger, J. S. (2005). Micro magnetic tweezers for nanomanipulation inside live cells. *Biophysical Journal, 88*, 2137–2144. https://doi.org/10.1529/biophysj.104.052035.

del Rio, A., Perez-Jimenez, R., Liu, R., Roca-Cusachs, P., Fernandez, J. M., & Sheetz, M. P. (2009). Stretching single talin rod molecules activates vinculin binding. *Science (New York, N. Y.), 323*, 638–641. https://doi.org/10.1126/science.1162912.

Dulin, D., Cui, T. J., Cnossen, J., Docter, M. W., Lipfert, J., & Dekker, N. H. (2015). High spatiotemporal-resolution magnetic tweezers: Calibration and applications for DNA dynamics. *Biophysical Journal, 109*, 2113–2125. https://doi.org/10.1016/j.bpj.2015.10.018.

Dulin, D., Lipfert, J., Moolman, M. C., & Dekker, N. H. (2013). Studying genomic processes at the single-molecule level: Introducing the tools and applications. *Nature Reviews. Genetics, 14*, 9–22. https://doi.org/10.1038/nrg3316.

Etoc, F., Lisse, D., Bellaiche, Y., Piehler, J., Coppey, M., & Dahan, M. (2013). Subcellular control of Rac-GTPase signalling by magnetogenetic manipulation inside living cells. *Nature Nanotechnology, 8*, 193–198. https://doi.org/10.1038/nnano.2013.23.

Etoc, F., Vicario, C., Lisse, D., Siaugue, J.-M., Piehler, J., Coppey, M., et al. (2015). Magnetogenetic control of protein gradients inside living cells with high spatial and temporal resolution. *Nano Letters, 15*, 3487–3494. https://doi.org/10.1021/acs.nanolett.5b00851.

Fabry, B., Maksym, G. N., Butler, J. P., Glogauer, M., Navajas, D., & Fredberg, J. J. (2001). Scaling the microrheology of living cells. *Physical Review Letters, 87*, 148102. https://doi.org/10.1103/PhysRevLett.87.148102.

Falleroni, F., Torre, V., & Cojoc, D. (2018). Cell mechanotransduction with piconewton forces applied by optical tweezers. *Front. Cell. Neurosci. 12*, 130. https://doi.org/10.3389/fncel.2018.00130.

Ferrer, J. M., Lee, H., Chen, J., Pelz, B., Nakamura, F., Kamm, R. D., et al. (2008). Measuring molecular rupture forces between single actin filaments and actin-binding proteins. *Proceedings of the National Academy of Sciences of the United States of America, 105*, 9221–9226. https://doi.org/10.1073/pnas.0706124105.

Fischer, T., Hayn, A., & Mierke, C. T. (2020). Effect of nuclear stiffness on cell mechanics and migration of human breast cancer cells. *Frontiers in Cell and Developmental Biology, 8*, 393. https://doi.org/10.3389/fcell.2020.00393.

Fisher, J. K., Vicci, L., Cribb, J., O'Brien, E. T., Taylor, R. M., & Superfine, R. (2006). Magnetic force micromanipulation systems for the biological sciences. *Nano, 01*, 191–205. https://doi.org/10.1142/S1793292006000276.

Franz, F., Tapia-Rojo, R., Winograd-Katz, S., Boujemaa-Paterski, R., Li, W., Unger, T., et al. (2023). Allosteric activation of vinculin by talin. *Nature Communications, 14*, 4311. https://doi.org/10.1038/s41467-023-39646-4.

Gordon, W. R., Zimmerman, B., He, L., Miles, L. J., Huang, J., Tiyanont, K., et al. (2015). Mechanical allostery: Evidence for a force requirement in the proteolytic activation of notch. *Developmental Cell, 33*, 729–736. https://doi.org/10.1016/j.devcel.2015.05.004.

Gosse, C., & Croquette, V. (2002). Magnetic tweezers: Micromanipulation and force measurement at the molecular level. *Biophysical Journal, 82*, 3314–3329. https://doi.org/10.1016/S0006-3495(02)75672-5.

Goult, B. T., Yan, J., & Schwartz, M. A. (2018). Talin as a mechanosensitive signaling hub. *Journal of Cell Biology, 217*, 3776–3784. https://doi.org/10.1083/jcb.201808061.

Grison, M., Merkel, U., Kostan, J., Djinović-Carugo, K., & Rief, M. (2017). α-Actinin/titin interaction: A dynamic and mechanically stable cluster of bonds in the muscle Z-disk. *Proceedings of the National Academy of Sciences of the United States of America, 114*, 1015–1020. https://doi.org/10.1073/pnas.1612681114.

Guck, J., Ananthakrishnan, R., Mahmood, H., Moon, T. J., Cunningham, C. C., & Käs, J. (2001). The optical stretcher: A novel laser tool to micromanipulate cells. *Biophysical Journal, 81*, 767–784. https://doi.org/10.1016/S0006-3495(01)75740-2.

Guilluy, C., Osborne, L. D., Van Landeghem, L., Sharek, L., Superfine, R., Garcia-Mata, R., et al. (2014). Isolated nuclei adapt to force and reveal a mechanotransduction pathway in the nucleus. *Nature Cell Biology, 16*, 376–381. https://doi.org/10.1038/ncb2927.

Guo, Y. R., & MacKinnon, R. (2017). Structure-based membrane dome mechanism for Piezo mechanosensitivity. *eLife, 6*, e33660. https://doi.org/10.7554/eLife.33660.

Hemphill, M. A., Dabiri, B. E., Gabriele, S., Kerscher, L., Franck, C., Goss, J. A., et al. (2011). A possible role for integrin signaling in diffuse axonal injury. *PLoS One, 6*, e22899. https://doi.org/10.1371/journal.pone.0022899.

Hoffmann, C., Mazari, E., Lallet, S., Le Borgne, R., Marchi, V., Gosse, C., et al. (2013). Spatiotemporal control of microtubule nucleation and assembly using magnetic nanoparticles. *Nature Nanotechnology, 8*, 199–205. https://doi.org/10.1038/nnano.2012.246.

Hosu, B. G., Jakab, K., Bánki, P., Tóth, F. I., & Forgacs, G. (2003). Magnetic tweezers for intracellular applications. *Review of Scientific Instruments, 74*, 4158–4163. https://doi.org/10.1063/1.1599066.

Hu, X., Margadant, F. M., Yao, M., & Sheetz, M. P. (2017). Molecular stretching modulates mechanosensing pathways. *Protein Science, 26*, 1337–1351. https://doi.org/10.1002/pro.3188.

Huang, D. L., Bax, N. A., Buckley, C. D., Weis, W. I., & Dunn, A. R. (2017). Vinculin forms a directionally asymmetric catch bond with F-actin. *Science (New York, N. Y.), 357*, 703–706. https://doi.org/10.1126/science.aan2556.

Hughes, S., McBain, S., Dobson, J., & El Haj, A. J. (2008). Selective activation of mechanosensitive ion channels using magnetic particles. *Journal of the Royal Society, Interface / the Royal Society, 5*, 855–863. https://doi.org/10.1098/rsif.2007.1274.

Huhle, A., Klaue, D., Brutzer, H., Daldrop, P., Joo, S., Otto, O., et al. (2015). Camera-based three-dimensional real-time particle tracking at kHz rates and Ångström accuracy. *Nature Communications, 6*, 5885. https://doi.org/10.1038/ncomms6885.

Hynes, R. O. (2002). Integrins. *Cell, 110*, 673–687. https://doi.org/10.1016/S0092-8674(02)00971-6.

Ibata, N., & Terentjev, E. M. (2021). Why exercise builds muscles: Titin mechanosensing controls skeletal muscle growth under load. *Biophysical Journal, 120*, 3649–3663. https://doi.org/10.1016/j.bpj.2021.07.023.

Ingber, D. E. (2006). Cellular mechanotransduction: Putting all the pieces together again. *The FASEB Journal, 20*, 811–827. https://doi.org/10.1096/fj.05-5424rev.

Iskratsch, T., Wolfenson, H., & Sheetz, M. P. (2014). Appreciating force and shape—The rise of mechanotransduction in cell biology. *Nature Reviews. Molecular Cell Biology, 15*, 825–833. https://doi.org/10.1038/nrm3903.

Jiang, G., Giannone, G., Critchley, D. R., Fukumoto, E., & Sheetz, M. P. (2003). Two-piconewton slip bond between fibronectin and the cytoskeleton depends on talin. *Nature, 424*, 334–337. https://doi.org/10.1038/nature01805.

Jin, P., Jan, L. Y., & Jan, Y.-N. (2020). Mechanosensitive ion channels: Structural features relevant to mechanotransduction mechanisms. *Annual Review of Neuroscience, 43*, 207–229. https://doi.org/10.1146/annurev-neuro-070918-050509.

Kah, D., Dürrbeck, C., Schneider, W., Fabry, B., & Gerum, R. C. (2020). High-force magnetic tweezers with hysteresis-free force feedback. *Biophysical Journal, 119*, 15–23. https://doi.org/10.1016/j.bpj.2020.05.018.

Kaláb, P., Pralle, A., Isacoff, E. Y., Heald, R., & Weis, K. (2006). Analysis of a RanGTP-regulated gradient in mitotic somatic cells. *Nature, 440*, 697–701. https://doi.org/10.1038/nature04589.

Kanchanawong, P., Shtengel, G., Pasapera, A. M., Ramko, E. B., Davidson, M. W., Hess, H. F., et al. (2010). Nanoscale architecture of integrin-based cell adhesions. *Nature, 468*, 580–584. https://doi.org/10.1038/nature09621.

Kellermayer, M. S. Z., Smith, S. B., Granzier, H. L., & Bustamante, C. (1997). Folding-unfolding transitions in single titin molecules characterized with laser tweezers. *Science (New York, N. Y.), 276*, 1112–1116. https://doi.org/10.1126/science.276.5315.1112.

Kim, J., Seo, D., Lee, J., Southard, K. M., Lim, Y., Kim, D., et al. (2017). Single-cell mechanogenetics using monovalent magnetoplasmonic nanoparticles. *Nature Protocols, 12*, 1871–1889. https://doi.org/10.1038/nprot.2017.071.

Kim, J., Zhang, C.-Z., Zhang, X., & Springer, T. A. (2010). A mechanically stabilized receptor–ligand flex-bond important in the vasculature. *Nature, 466*, 992–995. https://doi.org/10.1038/nature09295.

Kim, T.-J., Joo, C., Seong, J., Vafabakhsh, R., Botvinick, E. L., Berns, M. W., et al. (2015). Distinct mechanisms regulating mechanical force-induced Ca2+ signals at the plasma membrane and the ER in human MSCs. *eLife, 4*, e04876. https://doi.org/10.7554/eLife.04876.

Kodama, T., Osaki, T., Kawano, R., Kamiya, K., Miki, N., & Takeuchi, S. (2013). Round-tip dielectrophoresis-based tweezers for single micro-object manipulation. *Biosensors and Bioelectronics, 47*, 206–212. https://doi.org/10.1016/j.bios.2013.03.022.

Kollmannsberger, P., & Fabry, B. (2007). BaHigh-force magnetic tweezers with force feedback for biological applications. *Review of Scientific Instruments, 78*, 114301. https://doi.org/10.1063/1.2804771.

Kollmannsberger, P., Mierke, C. T., & Fabry, B. (2011). Nonlinear viscoelasticity of adherent cells is controlled by cytoskeletal tension. *Soft Matter, 7*, 3127–3132. https://doi.org/10.1039/C0SM00833H.

Labeit, S., & Kolmerer, B. (1995). Titins: Giant proteins in charge of muscle ultrastructure and elasticity. *Science (New York, N. Y.), 270*, 293–296. https://doi.org/10.1126/science.270.5234.293.

Landgraf, F. J. G., Emura, M., & de Campos, M. F. (2008). On the Steinmetz hysteresis law. *Journal of Magnetism and Magnetic Materials, 320*, e531–e534. https://doi.org/10.1016/j.jmmm.2008.04.011.

Lansdorp, B. M., Tabrizi, S. J., Dittmore, A., & Saleh, O. A. (2013). A high-speed magnetic tweezer beyond 10,000 frames per second. *Review of Scientific Instruments, 84*, 044301. https://doi.org/10.1063/1.4802678.

Laurent, V. M., Hénon, S., Planus, E., Fodil, R., Balland, M., Isabey, D., et al. (2002). Assessment of mechanical properties of adherent living cells by bead micromanipulation: Comparison of magnetic twisting cytometry vs optical tweezers. *Journal of Biomechanical Engineering, 124*, 408–421. https://doi.org/10.1115/1.1485285.

Le, S., Hu, X., Yao, M., Chen, H., Yu, M., Xu, X., et al. (2017). Mechanotransmission and mechanosensing of human alpha-actinin 1. *Cell Reports, 21*, 2714–2723. https://doi.org/10.1016/j.celrep.2017.11.040.

Lee, G., Abdi, K., Jiang, Y., Michaely, P., Bennett, V., & Marszalek, P. E. (2006). Nanospring behaviour of ankyrin repeats. *Nature, 440*, 246–249. https://doi.org/10.1038/nature04437.

Lee, H., Pelz, B., Ferrer, J. M., Kim, T., Lang, M. J., & Kamm, R. D. (2009). Cytoskeletal deformation at high strains and the role of cross-link unfolding or unbinding. *Cellular and Molecular Bioengineering, 2*, 28–38. https://doi.org/10.1007/s12195-009-0048-8.

Lee, J., Shin, W., Lim, Y., Kim, J., Kim, W. R., Kim, H., et al. (2021). Non-contact long-range magnetic stimulation of mechanosensitive ion channels in freely moving animals. *Nature Materials, 20*, 1029–1036. https://doi.org/10.1038/s41563-020-00896-y.

Lee, J.-H., Kim, J., Levy, M., Kao, A., Noh, S., Bozovic, D., et al. (2014). Magnetic nanoparticles for ultrafast mechanical control of inner ear hair cells. *ACS Nano, 8*, 6590–6598. https://doi.org/10.1021/nn5020616.

Lin, Y.-C., Guo, Y. R., Miyagi, A., Levring, J., MacKinnon, R., & Scheuring, S. (2019). Force-induced conformational changes in PIEZO1. *Nature, 573*, 230–234. https://doi.org/10.1038/s41586-019-1499-2.

Liu, J., Wang, Y., Goh, W. I., Goh, H., Baird, M. A., Ruehland, S., et al. (2015). Talin determines the nanoscale architecture of focal adhesions. *Proceedings of the National Academy of Sciences of the United States of America, 112*, E4864–E4873. https://doi.org/10.1073/pnas.1512025112.

Luca, V. C., Kim, B. C., Ge, C., Kakuda, S., Wu, D., Roein-Peikar, M., et al. (2017). Notch-Jagged complex structure implicates a catch bond in tuning ligand sensitivity. *Science (New York, N. Y.), 355*, 1320–1324. https://doi.org/10.1126/science.aaf9739.

Mahowald, J., Arcizet, D., & Heinrich, D. (2009). Impact of external stimuli and cell micro-architecture on intracellular transport states. *Chemphyschem: a European Journal of Chemical Physics and Physical Chemistry, 10*, 1559–1566. https://doi.org/10.1002/cphc.200900226.

Mannix, R. J., Kumar, S., Cassiola, F., Montoya-Zavala, M., Feinstein, E., Prentiss, M., et al. (2008). Nanomagnetic actuation of receptor-mediated signal transduction. *Nature Nanotechnology, 3*, 36–40. https://doi.org/10.1038/nnano.2007.418.

Margadant, F., Chew, L. L., Hu, X., Yu, H., Bate, N., Zhang, X., et al. (2011). Mechanotransduction in vivo by repeated talin stretch-relaxation events depends upon vinculin. *PLoS Biology, 9*, e1001223. https://doi.org/10.1371/journal.pbio.1001223.

Marszalek, P. E., Lu, H., Li, H., Carrion-Vazquez, M., Oberhauser, A. F., Schulten, K., et al. (1999). Mechanical unfolding intermediates in titin modules. *Nature, 402*, 100–103. https://doi.org/10.1038/47083.

Meloty-Kapella, L., Shergill, B., Kuon, J., Botvinick, E., & Weinmaster, G. (2012). Notch ligand endocytosis generates mechanical pulling force dependent on dynamin, epsins, and actin. *Developmental Cell, 22*, 1299–1312. https://doi.org/10.1016/j.devcel.2012.04.005.

Mierke, C. T. (2021). Bidirectional mechanical response between cells and their microenvironment. *Frontiers in Physics, 9*, 749830. https://doi.org/10.3389/fphy.2021.749830.

Mierke, C. T., Fischer, T., Puder, S., Kunschmann, T., Soetje, B., & Ziegler, W. H. (2017). Focal adhesion kinase activity is required for actomyosin contractility-based invasion of cells into dense 3D matrices. *Scientific Reports*, 7, 42780. https://doi.org/10.1038/srep42780.

Mierke, C. T., Kollmannsberger, P., Zitterbart, D. P., Diez, G., Koch, T. M., Marg, S., et al. (2010). Vinculin facilitates cell invasion into three-dimensional collagen matrices. *Journal of Biological Chemistry, 285*, 13121–13130. https://doi.org/10.1074/jbc.M109.087171.

Mierke, C. T., Puder, S., Aermes, C., Fischer, T., & Kunschmann, T. (2020). Effect of PAK inhibition on cell mechanics depends on Rac1. *Frontiers in Cell and Developmental Biology, 8*, 13. https://doi.org/10.3389/fcell.2020.00013.

Mierke, C. T., Zitterbart, D. P., Kollmannsberger, P., Raupach, C., Schlötzer-Schrehardt, U., Goecke, T. W., et al. (2008). Breakdown of the endothelial barrier function in tumor cell transmigration. *Biophysical Journal, 94*, 2832–2846. https://doi.org/10.1529/biophysj.107.113613.

Miroshnikova, Y. A., Nava, M. M., & Wickström, S. A. (2017). Emerging roles of mechanical forces in chromatin regulation. *Journal of Cell Science*jcs. 202192. https://doi.org/10.1242/jcs.202192.

Mohammed, D., Versaevel, M., Bruyère, C., Alaimo, L., Luciano, M., Vercruysse, E., et al. (2019). Innovative tools for mechanobiology: Unraveling outside-in and inside-out mechanotransduction. *Frontiers in Bioengineering and Biotechnology*, 7, 162. https://doi.org/10.3389/fbioe.2019.00162.

Neuman, K. C., Chadd, E. H., Liou, G. F., Bergman, K., & Block, S. M. (1999). Characterization of photodamage to *Escherichia coli* in optical traps. *Biophysical Journal, 77*, 2856–2863. https://doi.org/10.1016/S0006-3495(99)77117-1.

Neuman, K. C., Lionnet, T., & Allemand, J.-F. (2007). Single-Molecule micromanipulation techniques. *Annual Review of Materials Research, 37*, 33–67. https://doi.org/10.1146/annurev.matsci.37.052506.084336.

Neuman, K. C., & Nagy, A. (2008). Single-molecule force spectroscopy: Optical tweezers, magnetic tweezers and atomic force microscopy. *Nature Methods, 5*, 491–505. https://doi.org/10.1038/nmeth.1218.

Pajerowski, J. D., Dahl, K. N., Zhong, F. L., Sammak, P. J., & Discher, D. E. (2007). Physical plasticity of the nucleus in stem cell differentiation. *Proceedings of the National Academy of Sciences, 104*, 15619–15624. https://doi.org/10.1073/pnas.0702576104.

Pálfy, M., Reményi, A., & Korcsmáros, T. (2012). Endosomal crosstalk: Meeting points for signaling pathways. *Trends in Cell Biology, 22*, 447–456. https://doi.org/10.1016/j.tcb.2012.06.004.

Pang, S. M., Le, S., & Yan, J. (2018). Mechanical responses of the mechanosensitive unstructured domains in cardiac titin. *Biology of the Cell, 110*, 65–76. https://doi.org/10.1111/boc.201700061.

Parks, A. L., Klueg, K. M., Stout, J. R., & Muskavitch, M. A. (2000). Ligand endocytosis drives receptor dissociation and activation in the Notch pathway. *Development (Cambridge, England), 127*, 1373–1385.

Peters, I. M., Van Kooyk, Y., Van Vliet, S. J., De Grooth, B. G., Figdor, C. G., & Greve, J. (1999). 3D single-particle tracking and optical trap measurements on adhesion proteins. *Cytometry, 36*, 189–194. https://doi.org/10.1002/(SICI)1097-0320(19990701)36:3<189::AID-CYTO7>3.0.CO;2-3.

Puchner, E. M., Alexandrovich, A., Kho, A. L., Hensen, U., Schäfer, L. V., Brandmeier, B., et al. (2008). Mechanoenzymatics of titin kinase. *Proceedings of the National Academy of Sciences of the United States of America, 105*, 13385–13390. https://doi.org/10.1073/pnas.0805034105.

Ravi, V. M., Will, P., Kueckelhaus, J., Sun, N., Joseph, K., Salié, H., et al. (2022). Spatially resolved multi-omics deciphers bidirectional tumor-host interdependence in glioblastoma. *Cancer Cell, 40*, 639–655.e13. https://doi.org/10.1016/j.ccell.2022.05.009.

Rico, F., Gonzalez, L., Casuso, I., Puig-Vidal, M., & Scheuring, S. (2013). High-speed force spectroscopy unfolds titin at the velocity of molecular dynamics simulations. *Science (New York, N. Y.), 342*, 741–743. https://doi.org/10.1126/science.1239764.

Rief, M., Gautel, M., Oesterhelt, F., Fernandez, J. M., & Gaub, H. E. (1997). Reversible unfolding of individual titin immunoglobulin domains by AFM. *Science (New York, N. Y.), 276*, 1109–1112. https://doi.org/10.1126/science.276.5315.1109.

Roca-Cusachs, P., del Rio, A., Puklin-Faucher, E., Gauthier, N. C., Biais, N., & Sheetz, M. P. (2013). Integrin-dependent force transmission to the extracellular matrix by -actinin triggers adhesion maturation. *Proceedings of the National Academy of Sciences, 110*, E1361–E1370. https://doi.org/10.1073/pnas.1220723110.

Ruggeri, Z. M., & Mendolicchio, G. L. (2007). Adhesion mechanisms in platelet function. *Circulation Research, 100*, 1673–1685. https://doi.org/10.1161/01.RES.0000267878.97021.ab.

Sacconi, L., Tolić-Nørrelykke, I. M., Antolini, R., & Pavone, F. S. (2005). Combined intracellular three-dimensional imaging and selective nanosurgery by a nonlinear microscope. *Journal of Biomedical Optics, 10*, 014002. https://doi.org/10.1117/1.1854675.

Saotome, K., Murthy, S. E., Kefauver, J. M., Whitwam, T., Patapoutian, A., & Ward, A. B. (2018). Structure of the mechanically activated ion channel Piezo1. *Nature, 554*, 481–486. https://doi.org/10.1038/nature25453.

Saphirstein, R. J., Gao, Y. Z., Jensen, M. H., Gallant, C. M., Vetterkind, S., Moore, J. R., et al. (2013). The focal adhesion: A regulated component of aortic stiffness. *PLoS One, 8*, e62461. https://doi.org/10.1371/journal.pone.0062461.

Sarkar, R., & Rybenkov, V. V. (2016). A guide to magnetic tweezers and their applications. *Frontiers in Physics, 4*. https://doi.org/10.3389/fphy.2016.00048.

Seifert, M., van Nies, P., Papini, F. S., Arnold, J. J., Poranen, M. M., Cameron, C. E., et al. (2020). Temperature controlled high-throughput magnetic tweezers show striking difference in activation energies of replicating viral RNA-dependent RNA polymerases. *Nucleic Acids Research, 48*, 5591–5602. https://doi.org/10.1093/nar/gkaa233.

Seo, D., Southard, K. M., Kim, J., Lee, H. J., Farlow, J., Lee, J., et al. (2016). A mechanogenetic toolkit for interrogating cell signaling in space and time. *Cell, 165*, 1507–1518. https://doi.org/10.1016/j.cell.2016.04.045.

Shergill, B., Meloty-Kapella, L., Musse, A. A., Weinmaster, G., & Botvinick, E. (2012). Optical tweezers studies on notch: Single-molecule interaction strength is independent of ligand endocytosis. *Developmental Cell, 22*, 1313–1320. https://doi.org/10.1016/j.devcel.2012.04.007.

Shi, Z., Graber, Z. T., Baumgart, T., Stone, H. A., & Cohen, A. E. (2018). Cell membranes resist flow. *Cell, 175*, 1769–1779.e13. https://doi.org/10.1016/j.cell.2018.09.054.

Shon, M. J., Rah, S.-H., & Yoon, T.-Y. (2019). Submicrometer elasticity of double-stranded DNA revealed by precision force-extension measurements with magnetic tweezers. *Science Advances, 5*, eaav1697. https://doi.org/10.1126/sciadv.aav1697.

Siebel, C., & Lendahl, U. (2017). Notch signaling in development, tissue homeostasis, and disease. *Physiological Reviews, 97*, 1235–1294. https://doi.org/10.1152/physrev.00005.2017.

Smith, S. B., Finzi, L., & Bustamante, C. (1992). Direct mechanical measurements of the elasticity of single DNA molecules by using magnetic beads. *Science (New York, N. Y.), 258*, 1122–1126. https://doi.org/10.1126/science.1439819.

Soh, S., Byrska, M., Kandere-Grzybowska, K., & Grzybowski, B. A. (2010). Reaction-diffusion systems in intracellular molecular transport and control. *Angewandte Chemie International Edition, 49*, 4170–4198. https://doi.org/10.1002/anie.200905513.

Sotomayor, M., Corey, D. P., & Schulten, K. (2005). In search of the hair-cell gating spring. *Structure (London, England: 1993), 13*, 669–682. https://doi.org/10.1016/j.str.2005.03.001.

Steketee, M. B., Moysidis, S. N., Jin, X.-L., Weinstein, J. E., Pita-Thomas, W., Raju, H. B., et al. (2011). Nanoparticle-mediated signaling endosome localization regulates growth cone motility and neurite growth. *Proceedings of the National Academy of Sciences of the United States of America, 108*, 19042–19047. https://doi.org/10.1073/pnas.1019624108.

Stephenson, N. L., & Avis, J. M. (2012). Direct observation of proteolytic cleavage at the S2 site upon forced unfolding of the Notch negative regulatory region. *Proceedings of the National Academy of Sciences of the United States of America, 109*. https://doi.org/10.1073/pnas.1205788109.

Strick, T. R., Allemand, J.-F., Bensimon, D., Bensimon, A., & Croquette, V. (1996). The elasticity of a single supercoiled DNA molecule. *Science (New York, N. Y.), 271*, 1835–1837. https://doi.org/10.1126/science.271.5257.1835.

Strick, T. R., Allemand, J.-F., Bensimon, D., & Croquette, V. (1998). Behavior of supercoiled DNA. *Biophysical Journal, 74*, 2016–2028. https://doi.org/10.1016/S0006-3495(98)77908-1.

Tanase, M., Biais, N., & Sheetz, M. (2007). Magnetic tweezers in cell biology. *Methods in Cell Biology* (pp. 473–493). Elsevier. ⟨https://doi.org/10.1016/S0091-679X(07)83020-2⟩.

Tapia-Rojo, R., Alonso-Caballero, Á., & Fernández, J. M. (2020). Talin folding as the tuning fork of cellular mechanotransduction. *Proceedings of the National Academy of Sciences of the United States of America, 117*, 21346–21353. https://doi.org/10.1073/pnas.2004091117.

Tapia-Rojo, R., Eckels, E. C., & Fernández, J. M. (2019). Ephemeral states in protein folding under force captured with a magnetic tweezers design. *Proceedings of the National Academy of Sciences of the United States of America, 116*, 7873–7878. https://doi.org/10.1073/pnas.1821284116.

Tapia-Rojo, R., Mora, M., Board, S., Walker, J., Boujemaa-Paterski, R., Medalia, O., et al. (2023). Enhanced statistical sampling reveals microscopic complexity in the talin mechanosensor folding energy landscape. *Nature Physics, 19*, 52–60. https://doi.org/10.1038/s41567-022-01808-4.

Timonen, J. V. I., & Grzybowski, B. A. (2017). Tweezing of magnetic and non-magnetic objects with magnetic fields. *Advanced Materials, 29*, 1603516. https://doi.org/10.1002/adma.201603516.

Tseng, P., Judy, J. W., & Di Carlo, D. (2012). Magnetic nanoparticle–mediated massively parallel mechanical modulation of single-cell behavior. *Nature Methods, 9*, 1113–1119. https://doi.org/10.1038/nmeth.2210.

Wang, H., & Wang, J. (2017). How cryo-electron microscopy and X-ray crystallography complement each other. *Protein Science, 26*, 32–39. https://doi.org/10.1002/pro.3022.

Wang, X., & Ha, T. (2013). Defining single molecular forces required to activate integrin and notch signaling. *Science (New York, N. Y.), 340*, 991–994. https://doi.org/10.1126/science.1231041.

Wang, Y., Yao, M., Baker, K. B., Gough, R. E., Le, S., Goult, B. T., et al. (2021). Force-dependent interactions between talin and full-length vinculin. *Journal of the American Chemical Society, 143*, 14726–14737. https://doi.org/10.1021/jacs.1c06223.

Wu, J., Goyal, R., & Grandl, J. (2016). Localized force application reveals mechanically sensitive domains of Piezo1. *Nature Communications, 7*, 12939. https://doi.org/10.1038/ncomms12939.

Wu, J., Young, M., Lewis, A. H., Martfeld, A. N., Kalmeta, B., & Grandl, J. (2017). Inactivation of mechanically activated Piezo1 ion channels is determined by the C-terminal extracellular domain and the inner pore helix. *Cell Reports, 21*, 2357–2366. https://doi.org/10.1016/j.celrep.2017.10.120.

Yago, T., Lou, J., Wu, T., Yang, J., Miner, J. J., Coburn, L., et al. (2008). Platelet glycoprotein Ibα forms catch bonds with human WT vWF but not with type 2B von Willebrand disease vWF. *The Journal of Clinical Investigation*, JCI35754. https://doi.org/10.1172/JCI35754.

Yang, T., Park, C., Rah, S.-H., & Shon, M. J. (2022). Nano-precision tweezers for mechanosensitive proteins and beyond. *Molecules and Cells, 45*, 16–25. https://doi.org/10.14348/molcells.2022.2026.

Yao, M., Goult, B. T., Klapholz, B., Hu, X., Toseland, C. P., Guo, Y., et al. (2016). The mechanical response of talin. *Nature Communications, 7*, 11966. https://doi.org/10.1038/ncomms11966.

Yao, M., Qiu, W., Liu, R., Efremov, A. K., Cong, P., Seddiki, R., et al. (2014). Force-dependent conformational switch of α-catenin controls vinculin binding. *Nature Communications, 5*, 4525. https://doi.org/10.1038/ncomms5525.

Yun, S., Shin, T.-H., Lee, J.-H., Cho, M. H., Kim, I.-S., Kim, J., et al. (2018). Design of magnetically labeled cells (Mag-Cells) for in vivo control of stem cell migration and differentiation. *Nano Letters, 18*, 838–845. https://doi.org/10.1021/acs.nanolett.7b04089.

Zhang, W., Cheng, L. E., Kittelmann, M., Li, J., Petkovic, M., Cheng, T., et al. (2015). Ankyrin repeats convey force to gate the NOMPC mechanotransduction channel. *Cell, 162*, 1391–1403. https://doi.org/10.1016/j.cell.2015.08.024.

Zhang, X., Halvorsen, K., Zhang, C.-Z., Wong, W. P., & Springer, T. A. (2009). Mechanoenzymatic cleavage of the ultralarge vascular protein von Willebrand factor. *Science (New York, N. Y.), 324*, 1330–1334. https://doi.org/10.1126/science.1170905.

Zhang, X., Rogowski, L. W., & Kim, M. J. (2019). 3D micromanipulation of particle swarm using a hexapole magnetic tweezer. *2019 IEEE/RSJ International Conference on Intelligent Robots and Systems (IROS)* (pp. 1581–1586) Macau, China: Presented at the 2019 IEEE/RSJ International Conference on Intelligent Robots and Systems (IROS), IEEE. ⟨https://doi.org/10.1109/IROS40897.2019.8968300⟩.

Zhou, Y., Tang, Q., Zhao, X., Zeng, X., Chong, C., & Yan, J. (2023). A novel design for magnetic tweezers with wide-range temperature control. *Biophysical Journal, 122*, 3860–3868. https://doi.org/10.1016/j.bpj.2023.08.004.

9780443222290